BIOCHEMISTRY AND MOLECULAR BIOLOGY IN THE POST GENOMIC ERA

SULFATED POLYSACCHARIDES

BIOCHEMISTRY AND MOLECULAR BIOLOGY IN THE POST GENOMIC ERA

Additional books in this series can be found on Nova's website under the Series tab.

Additional e-books in this series can be found on Nova's website under the e-book tab.

BIOCHEMISTRY AND MOLECULAR BIOLOGY IN THE POST GENOMIC ERA

SULFATED POLYSACCHARIDES

MIGUEL GAMA
HELENA BONCIANI NADER
AND
HUGO ALEXANDRE DE OLIVEIRA ROCHA
EDITORS

nova publishers
New York

Library of Congress Cataloging-in-Publication Data

ISBN: 978-1-63482-975-5
Library of Congress Control Number: 2015940237

Published by Nova Science Publishers, Inc. † New York

CONTENTS

PREFACE

In the post-genomic era, science is still challenged to explain the biosynthesis of complex polysaccharides and glycoconjugates. Unlike nucleic acids and proteins, the information needed for their biosynthesis is not clearly contained to this day within the genome of the various organisms. This means that no biosynthetic code has been revealed yet. As a result, there will be millions of structurally distinct, functional chemical species at the end of their biosynthesis.

This book offers an up-to-date view on sulfated polysaccharide structure and function state of the art in different life kingdoms: *bacteria, protista, plantae, fungi and animalia*. The structure, activities and current thinking on the interplay between these two vital features (as well as ways to study them) are reviewed in the present book. The growing economical interest in sulfated polysaccharides due to their potential biotechnological use in different areas, such as pharmaceutical and food industries are also commented.

The information within the chapters adds to other prior available literature;our intention is to extend and further develop the discussions related to Glycobiology.

ABBREVIATIONS AND ACRONYMS

ACE	Dynamic affinity Capillary Electrophoresis
AEP	Anionic Emulsion Polymerization
AFM	Atomic Force Microscopy
AM	Acrylamide
AmB	Amphotericin B
Ara	Arabinose
ARH	Arginine-Rich Hexapeptide
APP	Amyloid Precursor Protein
aPTT	activated Partial Thromboplastin Time
APTT	Activated Partial Thrombin Time
AT	Antithrombin
bFGF	basic Fibroblast Growth Factor
BGL	Blood Glucose Levels
BMT	Bone Marrow Transplantation
BSA	Bovine Serum Albumin
BTSA	N,O-Bis-(Trimethylsilyl)-Acetamide
C4ST	Chondroitin 4-O-sulfotransferases
C5-EP	Chondroitin-glucuronate C5 epimerase
C6ST	Chondroitin 6-O-sulfotransferases
CAC	Critical Aggregation Concentrations
Ca-Sp	Calcium-Spirulan
CD	Circular Dichroism
CDSNS	Desulfated N-sulfated heparin
CE	Capillary Electrophoresis
c-GlcNAc6ST	N-acetylglucosamine-6-O-sulfotransferase
CHST6	Carbohydrate sulfotransferase 6 gene
CLSM	Confocal Laser Scanning Microscopy
COPD	Chronic Obstructive Pulmonary Disease
COSY	COrrelation SpectrocopY
CRP	Complement Regulatory Protein
CS	Chondroitin Sulfate
CS	Chitosan
CSPGs	Chondroitin Sulfate Proteoglycans
CTAB	Cetyltrimethyl ammonium bromide
CVB3	Coxsackie virus

1D	Unidimensional
2D	Bidimensional
3D	Three-dimensional
D4ST-1	Dermatan 4-O-sulfotransferase-1
DAPHP	Diaminopyridine group
DBS	Dried Blood Spot
DCC	Dicyclohexylcarbodiimide
DEPT	Distortionless Enhancement by Polarization Transfer
DF	Dietary Fibre
Df	deferoxamine
DHB	2,5-Dihydroxybenzoic Acid
DLS	Dynamic Light Scattering
DMB	Dimethylmethylene blue
DMEM	Dulbecco's Modified Eagle's Medium
DMF	Dimethylformamide
DmS	Dermatan Sulfate
DNase	Deoxyribonuclease
DOX	Doxorubicin
DPPH	2,2-Diphenyl-1-picrylhydrazyl
DS	Dermatan Sulfate
Ds	Degree of Sulfation
DSC	Differential Scanning Calorimetry
ECM	Extracellular Matrix
EDC	1-Ethyl-3-[3-dimethylamino]propyl]carbodiimide
EDXA	Energy-Dispersive X-ray Analysis
EMCV	Encephalomyocarditis virus
EPR	Enhanced Permeability and Retention
ER	Endoplasmic Reticulum
ERT	Enzyme Replacement Therapy
ESI	ElectroSpray Ionization
ESI IT MS	ElectroSpray Ion Trap Mass Spectrometer
ES-MS	ElectroSpray-Mass Spectrometry
EXT	Exostosin
FAB-MS	Fast Atom Bombardament-Mass Spectrometry
FACS	Fluorescence Activated Cell Sorter
FBS	Fetal Bovine Serum
FDA	US Food and Drug Administration
FGF	Fibroblast Growth Factor
FITC-BSA	Fluorescein Isothiocyanate conjugate Bovine Serum Albumin
FPHP	Folate-poly(ethylene glycol)-conjugated Heparin
Fru	Fructose
FTIR	Fourier Transform Infrared Spectroscopy
Fuc	Fucose
G4S	N-acetylgalactosamine-4-sulfatase
GAG	Glycosaminoglycan

Gal	Galactose
GalA	Galacturoni acid
GalNAc	N-Acetylgalactosamine
GalNAcT-I	GalNAc transferase I
GalNAcT-II	GalNAc transferase II
GALNS	N-acetylgalactosamine-6-sulfate sulfatase
GAS	Glucuronate 2-sulfatase
GC	Gas Chromatography
GC-MS	Gas Chromatography Mass Spectroscopy
Glc	Glucose
GLC	Gas-Liquid Chromatography
GlcA	Glucuronic acid
GlcAT-II	GlcA transferase II
GlcAU	Glucuronic acid
GlcN	Glucosamine
GlcNAc	N-Acetylglucosamine
GlcNAc6ST	GlcNAc-6-sulfotransferase
GlcNS	Glucosamine-N-sulfate
GMA	Glycidyl methacrylate
GnT	N-acetylglucosaminyltransferase
GPC	Gel Permeation Chromatography
GPI	Glycosylphosphatidylinositol
GTases	Glycosyltransferases
Gy	Gray
HA	Hyaluronan
HBV	Hepatitis B Virus
HCMV	Human Cytomegalovirus
HE	Heparin
HeLa	Human cervix adenocarcinoma cells
Hep	Heparin
HFDs	High Fat Diets
HILIC	Amide hydrophilic interaction chromatography
HIV	Human immunodeficiency virus
HLC	Heparin-like compounds
HMBC	Heteronuclear Multiple Bond Correlation
HMPV	Human metapneumovirus
HOBT	1-Hydroxy-benzotriazole hydrate
HPLC	High Performance Liquid Chromatography
HPSEC	High Performance Size Exclusion Chromatography
HPV	Human Papilloma Virus
HR	High Resolution
HS	Heparan Sulfate
HSLC	Heparan Sulfate-Like Compounds
HSQC	Bidimensional heteronuclear ^{13}C-^{1}H spectroscopy
HSV	Herpes Simplex Virus
HVEM	Herpes Virus Entry Mediators

IAV	Influenza A Virus
IdoA	Iduronic Acid
IdoAU	Iduronic Acid
IFV-A	Influenza virus type A
IL	Interleukin
IMTA	Integrated multitrophic aquaculture
IR	Infrared
JEV	Japanese Encephalitis Virus
K_D	Dissociation Constant
KS	Keratan Sulfate
KS-Gal6ST	Keratan sulfate Gal-6-sulfotransferase
LALS	Low Angle Light Scattering
LC	Liquid Chromatography
LC-MS/MS	Liquid Chromatography Tandem Mass Spectrometry
LMWHP	Low Molecular Weight Heparin
LRRs	Leucine-Rich Repeats
LPS	Lipopolysaccharides
LS	Light Scattering
MAA	Methacrylic anhydride
MALDI	Matrix-Assisted Laser Desorption and Ionization
MALS	Multi Angle Light Scattering
Man	Manose
MB	Methylene Blue
MCD	Macular Corneal Dystrophy
MCDK	Kidney epithelial cells extracted from an African green monkey
MES	2-Morpholinoethane sulphonic acid
MPS	Mucopolysaccharidoses
MRI	Magnetic Resonance Imaging
MS	Mass Spectrometry
MSC	Mesenchymal Stem Cells
MT-4	Human T cell leukaemia
MTSTFA	N-Methyl-N-(Trimethylsilyl)-Trifluoroacetamide
MW	Molecular Weight
m/z	Mass over Charge
NASH	Nonalcoholic steatohepatitis
Na-Sp	Sodium-Spirulan
Nd	Not determinate
NDST	N-deacetylase/N-sulfotransferase
NeuAc	N-acetylneuraminic acid
NGT	N-acetylglucosamine transporter
NHS	N-hydroxysuccinimide
NMR	Nuclear Magnetic Resonance
NO	Nitric oxide
nOe	nuclear Overhauser effect
NPs	Nanoparticles

NRE	Non-Reducing End
OA	Osteoarthritis
OSCS	Oversulfated chondroitin sulfate
OST	O-Sulfotransferase
PAGE	PolyAcrylamide Gel Electrophoresis
PAP	3′-Phosphoadenosine 5′-phosphate
PBS	Phosphate Buffered Saline
PC	Paper Chromatography
PCA	Principal Component Analysis
PCL	Poly-ε-caprolactone
PDLLA	Poly(D,L-lactic acid)
PEG	Poly(ethylenglycol)
PEI	Polyethylenimine
PG	Proteoglycan
γ-PGA	Poly(γ-glutamic acid)
PGs	Prostaglandins
PIBCA	Poly(isobutylcyanoacrylate)
pK	Logarithm of the Dissociation Constant
PLGA	Poly(lactic-*co*-glycolic acid)
PLLA	Poly(L-lactide)
PMMA-b-PMAETMA	Poly (methyl methacrylate-b-trimethylaminoethyl methacrylate)
PNPS	p-Nitrophenol sulfate
ppm	Parts per Million
PT	Prothrombin Time
PTR	Proteoglycan Tandem Repeat
PVA	Poly(vinyl alcohol)
Q-TOF	Quadrupole Time-Of-Flight
RAFT	Reversible Addition-Fragmentation chain-Transfer
RALS	Right Angle Light Scattering
Rha	Rhamnose
Rib	Ribose
RNase A	Ribonuclease A
ROS	Oxygen species
RPIP-UHPLC	Reverse Phase Ion-pairing ultra-high pressure liquid chromatography
RREP	Redox Radical Emulsion Polymerization
RSM	Response-Surface Methodologies
RT	Room temperature
SAX	Strong Anion Exchange
SDF-1	Stromal cell-derived factor 1
SEC	Size Exclusion Chromatography
SED	Spondyloepiphyseal Dysplasia
SEM	Scanning Electron Microscopy
SLRPs	Small Leucine-Rich Proteoglycans
SNA-MS	Noncovalent Affinity Mass Spectroscopy

SPR	Surface Plasmon Resonance
SPs	Sulfated Polysaccharides
STases	Sulfotransferases
STD	Saturation Transfer Difference
TBA	Tetrabutyl ammonium
TEM	Transmission Electron Microscopy
TLC	Thin-Layer Chromatography
Tre	Trehalose
TOCSY	TOtal Correlation SpectroscopY
TOF	Time-of-Flight
TPP	Tripolyphosphate
UroA	Uronic acid
USP	United State Pharmacopeia
UST	Uronyl 2-O-sulfotransferase
VEGF	Vascular Endothelial Growth Factor
Vt	Tidal volume
XRD	X-ray diffraction
Xyl	Xylose

In: Sulfated Polysaccharides ISBN: 978-1-63482-975-5
Editors: M. Gama, H.B. Nader, H.A. de Oliveira Rocha © 2015 Nova Science Publishers, Inc.

Chapter 1

HEPARAN SULFATE AND HEPARIN: STRUCTURE AND FUNCTION

Marcelo A. Lima*[1,2*]*, Renan P. Cavalheiro*[1]*,
Maria C. Z. Meneghetti*[1]*, Leonardo T. D. B. Nobre*[1]*,
***Gustavo M. Viana*[1]*, Edwin A. Yates*[2,1] *and Helena B. Nader*[1]

[1]Departamento de Bioquímica, Escola Paulista de Medicina,
Universidade Federal de São Paulo (UNIFESP), São Paulo, Brazil
[2]Department of Biochemistry, Institute of Integrative Biology,
University of Liverpool, Liverpool, Great Britain

ABSTRACT

Heparan sulfate and heparin are members of a broad family of sulfated, linear polysaccharides, the glycosaminoglycans. Heparan sulfate is ubiquitously found on the cell surface and in the extracellular matrix of all multicellular animals, whereas the presence of heparin is scattered in the evolutionary tree, being found primarily in the cytoplasmic granules of mast cells of some vertebrate and invertebrate species. Heparin and heparan sulfate share a common underlying structure, where the relative amounts of alpha-L-iduronic acid residues and the degree of sulfation, including the N-sulfation of the alpha-D-glucosamine residue are more abundant in heparin. Both glycosaminoglycans are active in a range of developmental, regulatory and pathophysiological situations and are widely studied. Due to its localization at the cell surface and the extracellular matrix, heparan sulfate seem to be involved in a wide variety of biological process including cell growth, differentiation, adhesion, endocytosis, among others. The true biological role of heparin is yet to be demonstrated, but it is known that pharmacologically this glycosaminoglycan plays an important role in hemostasis, inflammation and angiogenesis. Those different activities are a result of interactions with specific proteins that lead to the biological effects. The molecular interactions involved in the binding have been characterized to various degrees, including identification of

* Corresponding Author: Marcelo A. Lima, Departamento de Bioquímica, Universidade Federal de São Paulo (UNIFESP), Rua Três de Maio, São Paulo, SP 04044-020 Brazil, Email: mlima@unifesp.br

glycosaminoglycan-binding motifs in some proteins as well as the minimum saccharide sequence in the heparin/heparan sulfate polysaccharide chain. Heparan sulfate and heparin have considerable potential structural heterogeneity and it is apparent that the relationship between structure and activity involves redundancy, even though their biosynthesis is strongly regulated. The structure, activities and current thinking on the structure activity relationship of these molecules will be reviewed.

1. INTRODUCTION

Life as we know it - in conventional biological terms - is made possible by the flow of information within biological systems, where genetic information carried by DNA may be transcript and then translated into more diverse molecules, which eventually form cells and tissues. Such a view is termed *The Central Dogma of Life* and this process occurs for any known life form. However, the term only encapsulates molecules that go through template-driven biosynthetic pathways and ignores, for example, glycosylation - the addition of carbohydrates to proteins - which is the most abundant protein post-translational modification in complex organisms and is responsible for many of the fine-control mechanisms that coordinate events that allow complex life to exist.

Multicellular and complex life begun with the Extracellular Matrix (ECM) appearance as it is seems as the representative structure of all multicellular organisms. The ECM is composed of proteins such as collagens, laminin, fibronectin and various glycoconjugates that are produced by cells, exported and arranged in a heterogeneous macromolecular network. This provides structural support, tissue organization and orientation (tissue biomechanics), but also the necessary substrate for cell growth, migration, proliferation, adhesion, differentiation and invasion [1], besides the role of acting in various sensory crossroads of cell-cell and cell-microenvironment relationships.

One of the ECM's foundations is the glycoconjugates (carbohydrates covalently bound to proteins and/or lipids) [2], which are synthesized in the secretory biosynthetic pathway of cells (Endoplasmic Reticulum/Golgi apparatus) by a process called glycosylation. These glycoconjugates are organized into macromolecular networks that will, in fact, act in diverse cellular dynamics mentioned above. Such glycoconjugates structures are the most abundant class of structurally diverse and heterogeneous molecules present in the ECM and cell surface, forming the so-called glycocalyx.

Unlike nucleic acids and proteins, the information needed for the glycoconjugates biosynthesis is not clearly contained – up to now – within the genome of multicellular organisms [3] meaning that glycoconjugates do not have an identified biosynthetic code. Their diverse and heterogeneous structures are the result of the action of several glycosyltransferases, which are capable of polymerizing carbohydrate chains as well other classes of enzymes such as sulfotransferases and epimerases, which alter their substitution pattern and stereochemistry at specific sites along a given carbohydrate polymer. As a result, at the end of saccharide biosynthesis, there will be a large number of structurally distinct functional chemical entities. It's noteworthy that all of this synthesis process is quite elaborate and regulated in the various organisms and their tissues [3], but the driving force behind this event is yet to be uncovered.

Amongst the glycoconjugates, two mainstream complex polysaccharides are Heparan sulfate (HS) and Heparin (Hep), members of a broad family of sulfated, linear polysaccharides, the glycosaminoglycans (GAGs). Heparan sulfate is ubiquitously found on the cell surface and in the extracellular matrix of all multicellular animals, whereas the presence of heparin is scattered along the evolutionary tree, being found primarily in cytoplasmic granules of mast cells of some vertebrate and invertebrate species [4, 5].

2. HEPARAN SULFATE AND HEPARIN STRUCTURE

HS and heparin are members of the GAGs family and with the exception of hyaluronic acid, are present in tissues as proteoglycans, where the polysaccharide chains are covalently bound to a protein backbone. GAGs are linear polysaccharides consisting of repetitive disaccharide units, composed by a hexosamine (D-glucosamine or D-galactosamine) and an uronic acid sugar (D-glucuronic acid, L-iduronic) or D-galactose, linked by glycosidic linkages. Different GAGs are discriminated according to the type of hexosamine, type of uronic acid and glycosidic bond (α or β) and its position (1→3 and 1→4), presence and position of the sulfate substitution, being classified as heparin (Hep), HS, dermatan sulfate (DS), chondroitin sulfate (CS), keratan sulfate (KS), and hyaluronic acid (HA) [6, 7].

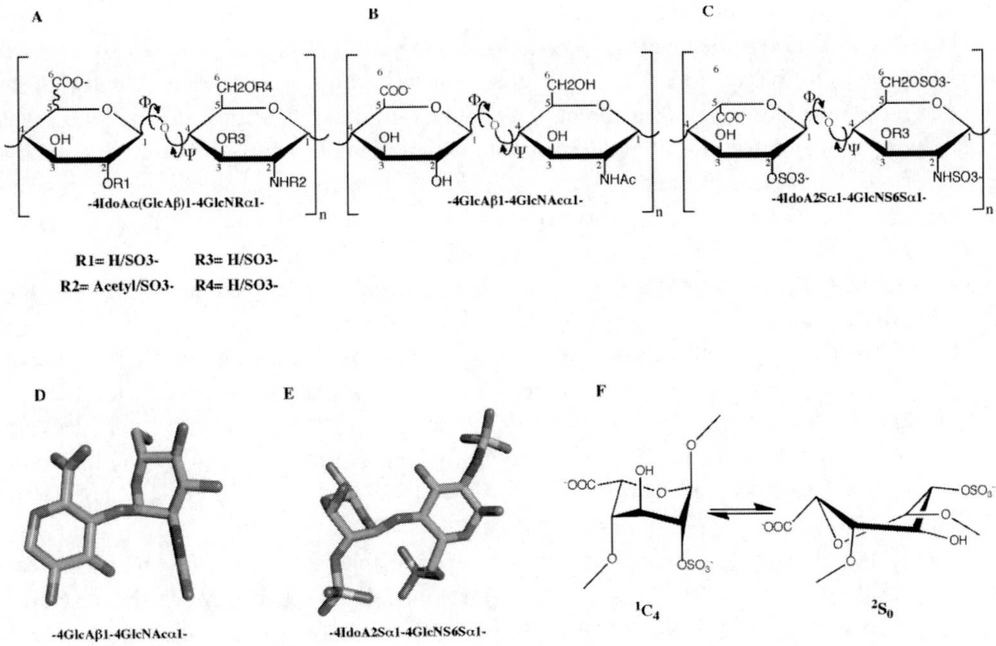

Figure 1. HS and Hep Structural Features. (A) Possible substitution pattern of HS and Hep, (B) HS major disaccharide, (C) Hep major disaccharide, (D) 3D carton for the HS major constituent disaccharide, (E) 3D carton for the Hep major constituent disaccharide and (F) α-L-iduronate-2-O-sulfate conformation equilibrium. phi (Φ) and psi (ψ) denote the glycosidic linkage torsion angles.

HS and Hep chains are polymerized in a non-template driven process with basic backbone consisting of β1–4-linked D-glucuronic acid (GlcA) and α1–4-linked N-acetyl-D-glucosamine (GlcNAc) where a series of enzymatic processing reactions may occur, in which N-acetylglucosamine residues become N-deacetylated and N-sulfated; glucuronic acids undergo epimerization at C5 to L-iduronic acid (IdoA); and sulfate groups can be added at C2 of the uronic acids, at C6 of the glucosamine residues, and in fewer cases, at C3 of the glucosamine residues [8, 9]. However, these substitutions are not complete, that is, do not modify the entire polymer, giving rise to the term substitution pattern, where for a simple disaccharide unit 48 substituent combinations can occur (Figure 1A). Despite the possible structural diversity, there are underlying structures for both HS and Hep where HS is usually composed of 40-60% of GlcA-GlcNAc (Figure 1B and D) while heparin, at least 60-70% of its constituents disaccharide is IduA2S-GlcNS,6S (Figure 1C and 1E) [10-12], consequently the most obvious differences between HS and Hep, besides their tissue distribution, is the GlcA to IdoA ratio and sulfation and N-acetylation degree. Hence, Hep may be seen as a more sulfated, HS analog present exclusively in mast cells.

All together, such features give rise to 4 structural peculiarities: 1) the abundance of different disaccharides (composition); 2) the order of these disaccharides (sequence); 3) gathering of similar disaccharides along the chain (domains); 4. stereochemical variation in monosaccharides (Figure 1F) and their glycosidic linkages as well as interaction with its surrounding chemical environment (conformation). To tackle these peculiarities, orthogonal analytical methodologies are required (Figure 2). For instance, HS and Hep can be directly analyzed by agarose gel electrophoresis (Figure 2A) where the relative abundance of such saccharides can be directly assessed. Low abundance samples – cell cultures for example – can have their GAGs radiolabeled via ^{35}Sulfate incorporation prior analysis. Moreover, these very samples can be degraded using specific heparin lyases and further analyzed by chromatography (Figure 2B) with direct UV and radioactivity detection or, in same cases, with fluorescence detection after derivativation with fluorescent probes via reductive amination of their reducing-end (Figure 2C), being the abundance of their composing disaccharides measured. As for conformational studies, Nuclear Magnetic Resonance can be used to detect subtle changes in iduronic acid equilibrium (Figure 2D). Ultimately, via such extensive structural diversity, HS and Hep are able to bind and regulate a vast inventory of proteins involved in numerous biological processes.

HS and Hep differ significantly in terms of their composing disaccharides. Likewise, the domain organization of both is quite different where HS has short *S* domains (IdoA2S and GlcNS residues on its majority), lengthy *NA* domains (GlcA and GlcNAc residues) and variable stretches of mixed disaccharides in between *S* and *NA* domains [14, 15]. On the other hand, Hep is composed, primarily, of IdoA2S-GlcNS,6S, *S-like* domain, hence their underlying 3D organization differ substantially.

Owing the abundance and easy access to Hep and the fact that it also binds proteins whose natural ligands are HS, several Hep structural models have been proposed, yet, models for HS are less abundant and until recently it was only available for a protein:HS crystal in which the HS saccharide corresponded only to a tetrasaccharide [16]. Figure 3 shows four conformation models for two HS dp24 and two Hep d24 and, as easily noticed, the differences in primary composition, substitution pattern, domain structure, distinct glycosidic linkage geometry and its torsion angles phi (Φ) and psi (ψ) as well as the iduronate conformational equilibrium do, according to the conventional definitions, make HS fundamentally different to heparin. Furthermore, HS has been considered to have higher bending and flexibility when compared to Hep [14].

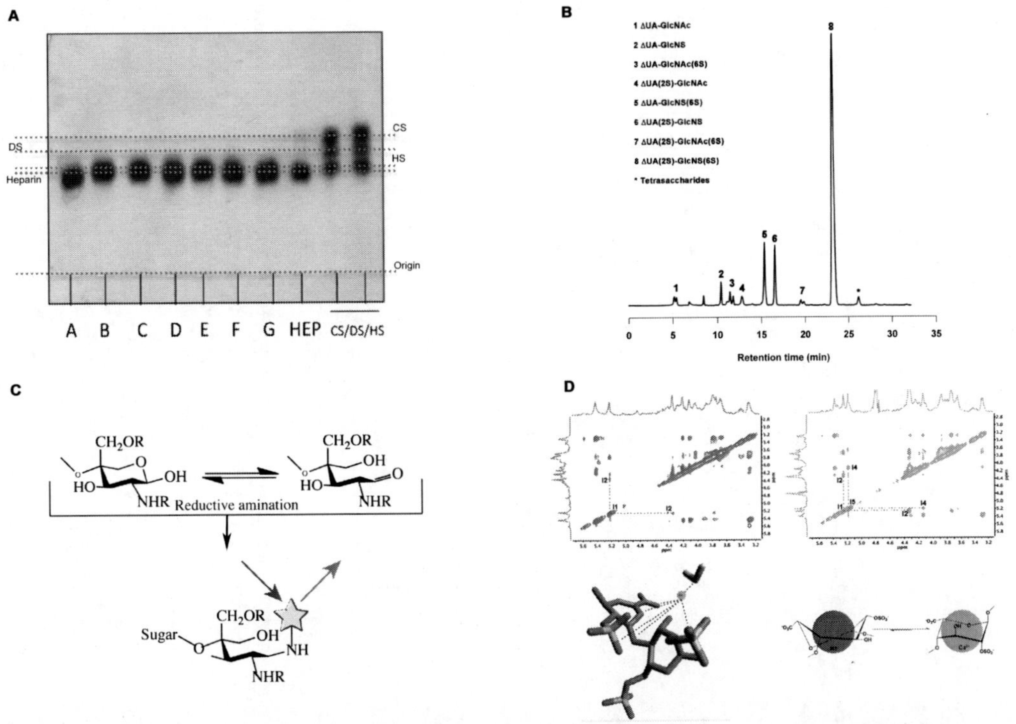

Figure 2. Workflow for the systematic analysis of HS and Hep structural peculiarities. (A) Agarose gel electrophoresis in 1,3-diaminopropane acetate buffer 0.05 M, pH 9.0 of HS and Hep. The different GAGs are separated based on their interaction with diamines. (B) Strong Anion Exchange chromatographic separation of HS and Hep disaccharides generated by lyase action, (C) Schematic representation of a HS and Hep disaccharide derivatization with fluorescent probes and (D) 2D NMR studies for the determination of iduronic acid conformation equilibrium, adapted from [13].

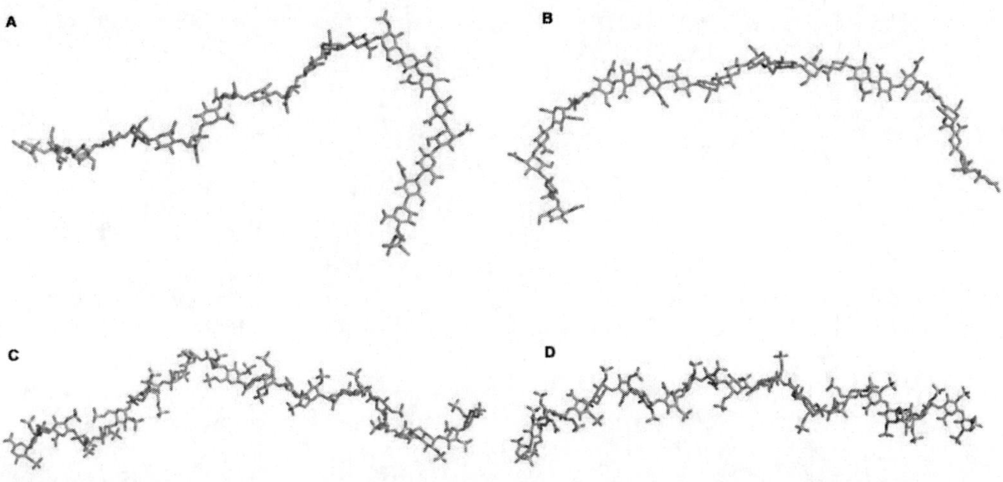

Figure 3. Potential HS and Hep conformational states. (A) Bent conformation of a HS dp24, (B) Extended conformation of a HS dp24, (C and D) Two possible conformations for a Hep dp24. Cartoon representations were generated in Pymol [17] using PDB files available in [14] and PDB file 3IRJ, respectively.

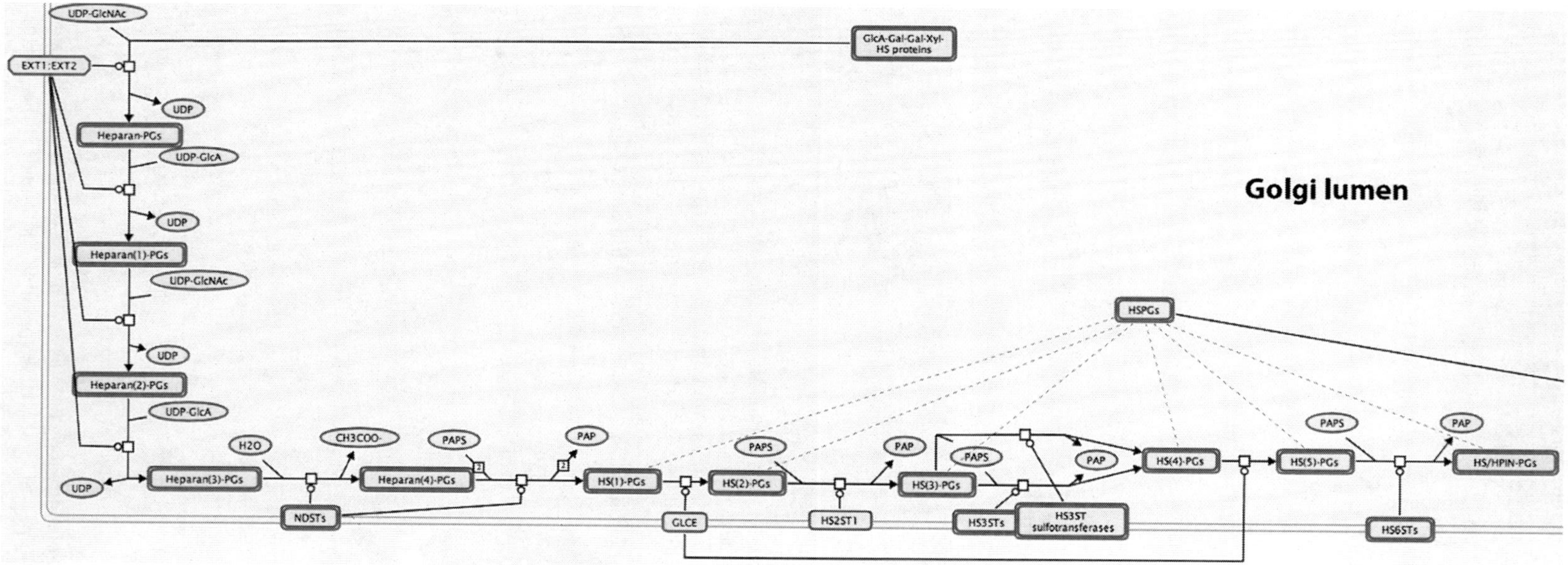

Figure 4. GAG biosynthetic machinery. HS and Hep biosynthesis process is highlighted in blue. Key enzymes for their synthesis are xilosyl, galactosyl and glucuronosyltransferase, which polymerase the tetrasaccharide linkage region; EXT1 and 2 – bifunctional glycosyltransferases that catalyze the transfer of both GlcNAc and GlcA from their respective sugar-nucleotides – that polymerase the HS and Hep chains which are then modified by sulfotransferases and epimerases. Figure was generated using the Reactome FI Cytoscape app [21].

In addition to intrinsic structure variation, chemical environment variability has to be taken into account when dealing with conformation states of both HS and Hep. One simple example is when different cations are present within a given system.

Several authors have shown that different cations can bind and influence HS/Hep polymeric structure where local and global conformation parameters are deeply affected. In Ca(2+)-heparin, for instance, the α-L-iduronate-2-O-sulfate residues preferentially adopt $^{1}C_4$ (Figure 1F) conformation [13, 18] and in the case of Cu(2+)-heparin, specific coordination to four points in the structure happen throughout the heparin chain; the carboxylic acid group and the ring oxygen of α-L-iduronate-2-O-sulfate, the glycosidic oxygen between this residue and the adjacent glucosamine and its 6-O-sulfate group which leads to a more rigid polysaccharide polymer [19].

Furthermore, using systematically modified analogues of heparan sulfate – all made from pharmaceutical heparin by selective chemical modification – the relationship between substitution pattern and conformational states was explored. Using these model compounds, it was shown that only a fraction of possible sequences were sufficient to provide diverse backbone conformations and that the unique information content potentially available in HS sequences, defined ultimately by conformation, is vastly lower than the potential sequence diversity [20]. Nonetheless, HS and Hep biosynthetic machinery (Figure 4) is extremely large but not fully comprehended, although highly regulated [3].

3. Heparan Sulfate and Heparin Structure-Fuction

HS and Hep are involved with various biological functions that are related to their tissue distribution and the result of their structural diversity. While HS is a omnipresent structure of the cell surface and extracellular matrix of multicellular organisms, having wide influence on biological processes, the function of biologically available heparin, in contrast to that supplied pharmaceutically, is not clear, and is most likely linked to inflammatory responses when secreted by mast cells during inflammatory events. Dynamic temporal and spatial variations in their structure underpin the concept of the 'heparanome' [22] and facilitate a complex interactome [23, 24] where, upon HS and Hep binding, an entire protein-protein interaction network can be affected (Figure 5A and B) and, furthermore, a given network may be involved in quite distinct biological processes (Figure 5C). Likewise, heparin – when present – will be available for protein binding to an interactome of analogous complexity.

In fact, HS and Hep have been shown to participate in a wide range of biological activities such as cell-cell communication [25], cell growth and proliferation [26], morphogenesis [27], angiogenesis [28], tumor progression [29], viral infection [30, 31] and many others. Additionally, reviving data from several decades ago [32, 33], functions for nuclear HS are being further investigated and evidence points to possible functions in hampering cell proliferation, inhibition of DNA topoisomerase I activity and inhibition of gene transcription [34].

As mentioned, the activities of both HS and Hep are believed to be possible via their extensive structural diversity where, as shown in Figure 3, HS and Hep are able to bind and regulate considerable Protein-Protein Networks. Yet, the nature of Hep and HS-protein interactions are at constant debate and theories that try to explain them are as follow: 1)

Interactions are driven merely by net charge [35, 36]; 2) Selectivity results from clusters of exact disaccharides [37]; 3) Requirement for chains of specific length and sequence [38, 39]; 4) Selectivity is determined by the three dimensional conformation of a given saccharide structure [24, 40] and 5) The presence of "key sulfate groups".

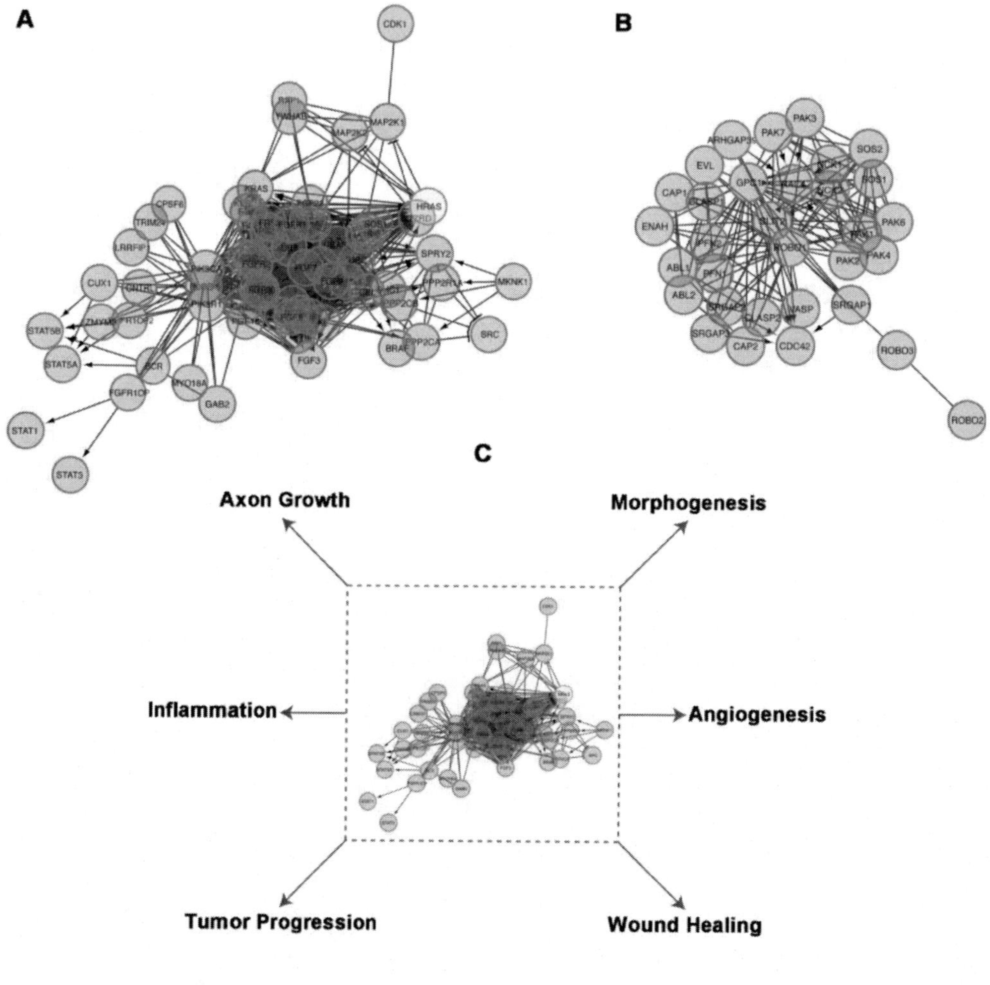

Figure 5. Protein-Protein Network. (A) FGF-FGFR Signaling, (B) Signaling by Robo Receptor and (C) FGF-FGFR Signaling in different biological processes. Protein Networks were generated using the Reactome FI Cytoscape app [21].

Several studies have been done to try to extract definite structure-activity relationship of HS and Hep [41-52]. Nonetheless, in most cases, Hep or HS samples exhibiting limited structural diversity were utilized. So far, therefore, only a limited collection of structural diversity for both saccharide and protein has been assessed. Furthermore, the use of defined oligosaccharides end-up masking natural sequence complexity in a systematic way that, in some cases, makes extrapolation to the full-length polysaccharides risky. From the protein perspective, several attempts have been made to define specific protein sequences that are the target for HS and Hep binding. Consensus sequences such as X-B-B-X-B-X and X-B-B-B-X-

X-B-X where B is a basic residue and X is a hydropathic residue have been identified [53]. Nonetheless, it is important to emphasize, once again, that these sequences have been identified using a limited collection of proteins and, furthermore, using techniques in which certain characteristics – binding mode – can be favored. Additionally, these sequences are not necessarily linear, that is, within a certain protein stretch, they may be found in tridimensional space [54, 55].

Most of this debate came about with the discovery of the heparin pentasaccharide active for Antithrombin (AT) in which a "key sulfate" at C3 of the central glucosamine residue (Figure 6A) was disseminated as pivotal for AT binding and activation (Figure 6B), therefore, anticoagulant activity. In fact, a closer look at the original papers that first described this specific sequence suggests such structure as the one possessing highest affinity under the experimental conditions employed, high salt elution, therefore, selective for highly charged species [56, 57].

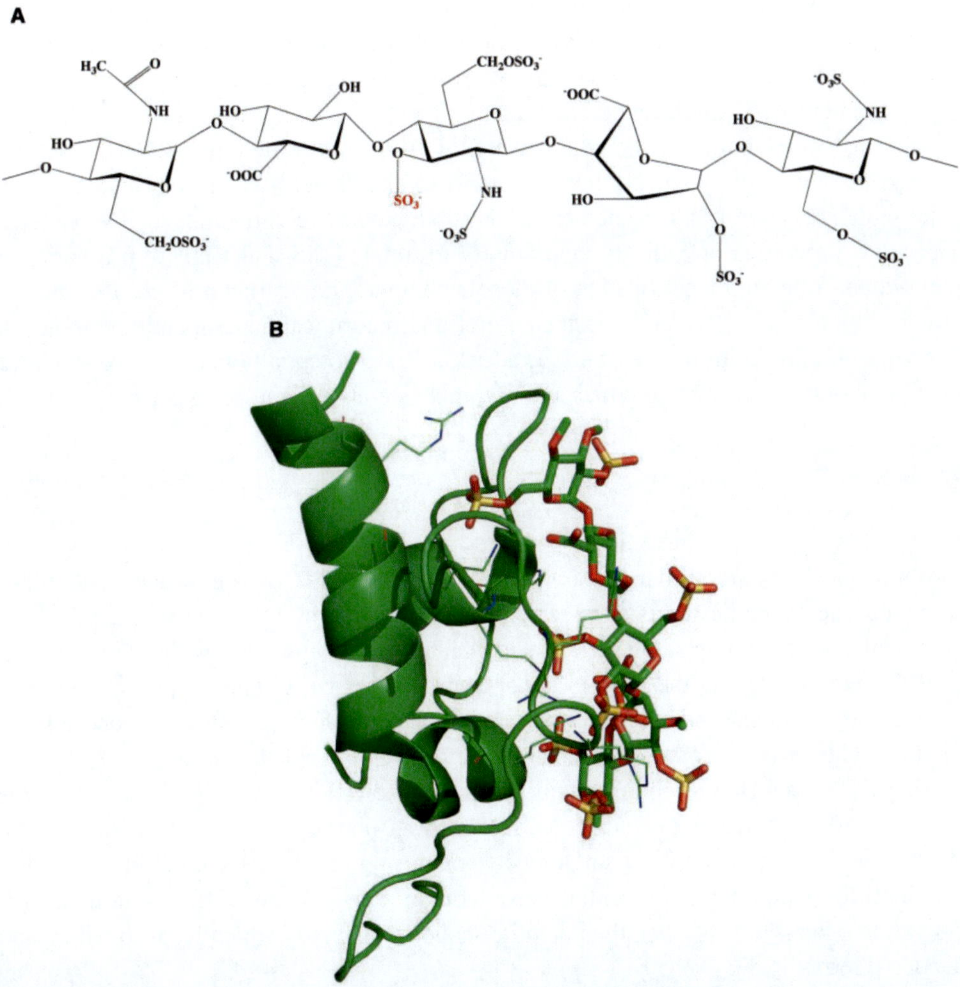

Figure 6. Heparin-AT interaction. (A) Heparin pentasaccharide active for AT. "Key sulfate" is in red and (B) Schematic close-up representation of the AT-pentasaccharide complex (pdb 1E03). Important amino acids (K11, R13, R46, R47, K114, L125, R129) are shown in lines. Image was generated in Pymol [17].

Recently, accumulated data suggests that, however, that net charge plays a significant role in the affinity of heparin for AT and that the 3-O-sulfated group in the central glucosamine unit of the heparin binding pentasaccharide is not essential and, sometimes, not even related to AT activation [58, 59]. Also, residues adjacent to the pentasaccharide influence its activity [60, 61] and, even non-carbohydrate structures can bind both the pentasaccharide binding site and the extended heparin binding sites of AT, leading to enhance anticoagulant activity [62]. Furthermore, other saccharide structures achieve the structural requirements of AT binding and, it has proposed that the AT stabilization, rather than specific secondary structural changes, may be a key factor of its anticoagulant activity [58]. Biologically, though, HS and Hep function can be seen as one of the processes ignored by the terms *The Central Dogma of Life* and the "*magic bullet approach*" once it biosynthesis is a non-template driven process where conventional approaches of event inhibition and gene knock-out do not allow the full understanding of its functionally as whole. Furthermore, unlike some carbohydrate-protein interactions, hence activity, HS and Hep exhibit redundancy between sulfation pattern, conformation and bio-activity, providing several unique properties: 1) a particular activity can be elicited by several distinct structures; 2) distinct compounds can have the same type of activity but, be active to different extents; and 3) single structures can exhibit several activities.

These properties should not be confused with GAGs possessing universal activity (or no selectivity) or, their activity being related primarily to charge density: Rather, the relationship between sequence, conformation and activity is subtle and there is redundancy. Nonetheless, such peculiar properties provide the opportunity to employ HS and Hep from a variety of sources which, following identification of those with appropriate structures, can then be used to influence biological activities, as in the case of pharmaceutical heparin and low molecular weight heparin of different sources, and, because there are potentially many active structures, those with favorable on-target activities and minimal side-effects can be selected.

Conclusion

Several questions are still to be answered regarding HS and Hep structure, therefore, function and one of the key facts necessary to tackle them is the full understating of HS and Hep biosynthesis. As pointed out, such process cannot be encapsulated within the terms *The Central Dogma of Life* and the "*magic bullet approach*" once it happens in a non-template driven fashion and transcends conventional approaches of single event inhibition and gene knock-out studies which, for this case, do not lead to much information.

Perhaps HS and Hep structure-function can be viewed as an example of a complex biological system where redundancy is certain and its plasticity is brought to life via the several possibilities available for a single of its composing disaccharide unit yet, as a whole, the unique information content potentially available in their polymers, defined ultimately by conformation, is vastly lower than the potential sequence diversity which leads to redundancy in functional terms.

REFERENCES

[1] Pickup, M.W., Mouw, J.K., Weaver, V.M., 2014. The extracellular matrix modulates the hallmarks of cancer. *EMBO reports*.15, 12, 1243 – 1253.

[2] Vallen, M.J., van der Steen, S.C., van Tilborg, A.A., Massuger, L.F., van Kuppevelt, T.H., 2014. Sulfated sugars in the extracellular matrix orchestrate ovarian cancer development: 'When sweet turns sour'. *Gynecologic oncology*., 135, 2, 371-381.

[3] Kreuger, J., Kjellen, L., 2012. Heparan sulfate biosynthesis: regulation and variability. *The journal of histochemistry and cytochemistry : official journal of the Histochemistry Society*. 60, 898-907.

[4] Nader, H.B., Ferreira, T.M., Toma, L., Chavante, S.F., Dietrich, C.P., Casu, B., Torri, G., 1988. Maintenance of heparan sulfate structure throughout evolution: chemical and enzymic degradation, and 13C-n.m.r.-spectral evidence. *Carbohydrate research*. 184, 292-300.

[5] Dreyfuss, J.L., Regatieri, C.V., Jarrouge, T.R., Cavalheiro, R.P., Sampaio, L.O., Nader, H.B., 2009. Heparan sulfate proteoglycans: structure, protein interactions and cell signaling. *Anais da Academia Brasileira de Ciencias*. 81, 409-29.

[6] Dietrich, C.P., Tersariol, I.L., Toma, L., Moraes, C.T., Porcionatto, M.A., Oliveira, F.W., Nader, H.B., 1998. Structure of heparan sulfate: identification of variable and constant oligosaccharide domains in eight heparan sulfates of different origins. *Cell Mol Biol (Noisy-le-grand)*. 44, 417-29.

[7] Sampaio, L.O., Nader, H.B., 2006. Emergence and structural characteristics of chondroitin sulfates in the animal kingdom. *Adv Pharmacol*. 53, 233-51.

[8] Nader, H.B., Lopes, C.C., Rocha, H.A., Santos, E.A., Dietrich, C.P., 2004. Heparins and heparinoids: occurrence, structure and mechanism of antithrombotic and hemorrhagic activities. *Current pharmaceutical design*. 10, 951-66.

[9] Casu, B., Naggi, A., Torri, G., 2015. Re-visiting the structure of heparin. *Carbohydrate research*. 403, 60-68.

[10] Andrade, G.P., Lima, M.A., de Souza Junior, A.A., Fareed, J., Hoppensteadt, D.A., Santos, E.A., Chavante, S.F., Oliveira, F.W., Rocha, H.A., Nader, H.B., 2013. A heparin-like compound isolated from a marine crab rich in glucuronic acid 2-O-sulfate presents low anticoagulant activity. *Carbohydrate polymers*. 94, 647-54.

[11] Nader, H.B., Dietrich, C.P., Buonassisi, V., Colburn, P., 1987. Heparin sequences in the heparan sulfate chains of an endothelial cell proteoglycan. *Proceedings of the National Academy of Sciences of the United States of America*. 84, 3565-9.

[12] Perlin, A.S., Mackie, D.M., Dietrich, C.P., 1971. Evidence for a (1 leads to 4)-linked 4-O-(-L-idopyranosyluronic acid 2-sulfate)-(2-deoxy-2-sulfoamino-D-glucopyranosyl 6-sulfate) sequence in heparin. Long-range H-H coupling in 4-deoxy-hex-4-enopyranosides. *Carbohydrate research*. 18, 185-94.

[13] Cordula, C.R., Lima, M.A., Shinjo, S.K., Gesteira, T.F., Pol-Fachin, L., Coulson-Thomas, V.J., Verli, H., Yates, E.A., Rudd, T.R., Pinhal, M.A., Toma, L., Dietrich, C.P., Nader, H.B., Tersariol, I.L., 2014. On the catalytic mechanism of polysaccharide lyases: evidence of His and Tyr involvement in heparin lysis by heparinase I and the role of Ca^{2+}. *Molecular bioSystems*. 10, 54-64.

[14] Khan, S., Fung, K.W., Rodriguez, E., Patel, R., Gor, J., Mulloy, B., Perkins, S.J., 2013. The solution structure of heparan sulfate differs from that of heparin: implications for function. *The Journal of biological chemistry*. 288, 27737-51.

[15] Tersariol, I.L., Ferreira, T.M., Medeiros, M.G., Porcionatto, M.A., Moraes, C.T., Abreu, L.R., Nader, H.B., Dietrich, C.P., 1994. Sequencing of heparan sulfate proteoglycans: identification of variable and constant oligosaccharide regions in eight heparan sulfate proteoglycans of different origins. *Brazilian journal of medical and biological research = Revista brasileira de pesquisas medicas e biologicas / Sociedade Brasileira de Biofisica ... [*et al.*]*. 27, 2097-102.

[16] Shaya, D., Zhao, W., Garron, M.L., Xiao, Z., Cui, Q., Zhang, Z., Sulea, T., Linhardt, R.J., Cygler, M., 2010. Catalytic mechanism of heparinase II investigated by site-directed mutagenesis and the crystal structure with its substrate. *The Journal of biological chemistry*. 285, 20051-61.

[17] Schrodinger, LLC, *The PyMOL Molecular Graphics System, Version 1.3r1*, 2010.

[18] Rabenstein, D.L., Robert, J.M., Peng, J., 1995. Multinuclear magnetic resonance studies of the interaction of inorganic cations with heparin. *Carbohydrate research*. 278, 239-56.

[19] Rudd, T.R., Skidmore, M.A., Guimond, S.E., Guerrini, M., Cosentino, C., Edge, R., Brown, A., Clarke, D.T., Torri, G., Turnbull, J.E., Nichols, R.J., Fernig, D.G., Yates, E.A., 2008. Site-specific interactions of copper(II) ions with heparin revealed with complementary (SRCD, NMR, FTIR and EPR) spectroscopic techniques. *Carbohydrate research*. 343, 2184-93.

[20] Rudd, T.R., Yates, E.A., 2010. Conformational degeneracy restricts the effective information content of heparan sulfate. *Molecular bioSystems*. 6, 902-8.

[21] Wu, G., Dawson, E., Duong, A., Haw, R., Stein, L., 2014. ReactomeFIViz: a Cytoscape app for pathway and network-based data analysis. *F1000Res*. 3, 146.

[22] Turnbull, J., Powell, A., Guimond, S., 2001. Heparan sulfate: decoding a dynamic multifunctional cell regulator. *Trends Cell Biol*. 11, 75-82.

[23] Ori, A., Wilkinson, M.C., Fernig, D.G., 2011. A systems biology approach for the investigation of the heparin/heparan sulfate interactome. *The Journal of biological chemistry*. 286, 19892-904.

[24] Ori, A., Wilkinson, M.C., Fernig, D.G., 2008. The heparanome and regulation of cell function: structures, functions and challenges. *Front Biosci*. 13, 4309-38.

[25] Shriver, Z., Capila, I., Venkataraman, G., Sasisekharan, R., 2012. Heparin and heparan sulfate: analyzing structure and microheterogeneity. *Handbook of experimental pharmacology*. 159-76.

[26] Matsuo, I., Kimura-Yoshida, C., 2013. Extracellular modulation of Fibroblast Growth Factor signaling through heparan sulfate proteoglycans in mammalian development. *Current Opinion in Genetics & Development*. 23, 399-407.

[27] Shimokawa, K., Kimura-Yoshida, C., Nagai, N., Mukai, K., Matsubara, K., Watanabe, H., Matsuda, Y., Mochida, K., Matsuo, I., 2011. Cell surface heparan sulfate chains regulate local reception of FGF signaling in the mouse embryo. *Developmental cell*. 21, 257-72.

[28] Dreyfuss, J.L., Regatieri, C.V., Lima, M.A., Paredes-Gamero, E.J., Brito, A.S., Chavante, S.F., Belfort, R., Jr., Farah, M.E., Nader, H.B., 2010. A heparin mimetic

isolated from a marine shrimp suppresses neovascularization. *Journal of thrombosis and haemostasis : JTH*. 8, 1828-37.

[29] Raman, K., Kuberan, B., 2010. Chemical Tumor Biology of Heparan Sulfate Proteoglycans. *Current chemical biology*. 4, 20-31.

[30] Baleux, F., Arien, K., Desjardins, D., Michiels, J., Coic, Y.M., Connell, B.J., Bonnaffe, D., Bouchemal, K., Le Grand, R., Vanham, G., Dereuddre-Bosquet, N., Lortat-Jacob, H., 2014. Mini CD4-heparan Sulfate Mimetic Conjugates Display Sub Nanomolar Anti-HIV-1 Activity and Protect Macaques against a SHIV162P3 Vaginal Route Challenge. *AIDS research and human retroviruses*. 30 Suppl 1, A262.

[31] Choudhary, S., Marquez, M., Alencastro, F., Spors, F., Zhao, Y., Tiwari, V., 2011. Herpes simplex virus type-1 (HSV-1) entry into human mesenchymal stem cells is heavily dependent on heparan sulfate. *Journal of biomedicine & biotechnology*. 2011, 264350.

[32] Fedarko, N.S., Conrad, H.E., 1986. A unique heparan sulfate in the nuclei of hepatocytes: structural changes with the growth state of the cells. *The Journal of cell biology*. 102, 587-99.

[33] Ishihara, M., Fedarko, N.S., Conrad, H.E., 1986. Transport of heparan sulfate into the nuclei of hepatocytes. *The Journal of biological chemistry*. 261, 13575-80.

[34] Kovalszky, I., Hjerpe, A., Dobra, K., 2014. Nuclear translocation of heparan sulfate proteoglycans and their functional significance. *Biochimica et biophysica acta*. 1840, 2491-7.

[35] Mulloy, B., 2005. The specificity of interactions between proteins and sulfated polysaccharides. *An Acad Bras Cienc*. 77, 651-64.

[36] Rudd, T.R., Skidmore, M.A., Guerrini, M., Hricovini, M., Powell, A.K., Siligardi, G., Yates, E.A., 2010. The conformation and structure of GAGs: recent progress and perspectives. *Curr Opin Struct Biol*. 20, 567-74.

[37] Jastrebova, N., Vanwildemeersch, M., Rapraeger, A.C., Gimenez-Gallego, G., Lindahl, U., Spillmann, D., 2006. Heparan sulfate-related oligosaccharides in ternary complex formation with fibroblast growth factors 1 and 2 and their receptors. *J Biol Chem*. 281, 26884-92.

[38] Jastrebova, N., Vanwildemeersch, M., Lindahl, U., Spillmann, D., 2010. Heparan sulfate domain organization and sulfation modulate FGF-induced cell signaling. *J Biol Chem*. 285, 26842-51.

[39] Lindahl, U., Backstrom, G., Hook, M., Thunberg, L., Fransson, L.A., Linker, A., 1979. Structure of the antithrombin-binding site in heparin. *Proceedings of the National Academy of Sciences of the United States of America*. 76, 3198-202.

[40] Lindahl, U., Li, J.P., 2009. Interactions between heparan sulfate and proteins-design and functional implications. *Int Rev Cell Mol Biol*. 276, 105-59.

[41] Walker, A., Turnbull, J.E., Gallagher, J.T., 1994. Specific heparan sulfate saccharides mediate the activity of basic fibroblast growth factor. *The Journal of biological chemistry*. 269, 931-5.

[42] Turnbull, J.E., Fernig, D.G., Ke, Y., Wilkinson, M.C., Gallagher, J.T., 1992. Identification of the basic fibroblast growth factor binding sequence in fibroblast heparan sulfate. *The Journal of biological chemistry*. 267, 10337-41.

[43] Guimond, S.E., Turnbull, J.E., 1999. Fibroblast growth factor receptor signalling is dictated by specific heparan sulphate saccharides. *Current biology : CB*. 9, 1343-6.

[44] Gambarini, A.G., Miyamoto, C.A., Lima, G.A., Nader, H.B., Dietrich, C.P., 1993. Mitogenic activity of acidic fibroblast growth factor is enhanced by highly sulfated oligosaccharides derived from heparin and heparan sulfate. *Molecular and cellular biochemistry*. 124, 121-9.

[45] Kreuger, J., Salmivirta, M., Sturiale, L., Gimenez-Gallego, G., Lindahl, U., 2001. Sequence analysis of heparan sulfate epitopes with graded affinities for fibroblast growth factors 1 and 2. *The Journal of biological chemistry*. 276, 30744-52.

[46] Pye, D.A., Vives, R.R., Hyde, P., Gallagher, J.T., 2000. Regulation of FGF-1 mitogenic activity by heparan sulfate oligosaccharides is dependent on specific structural features: differential requirements for the modulation of FGF-1 and FGF-2. *Glycobiology*. 10, 1183-92.

[47] Bianchini, P., Osima, B., Parma, B., Nader, H.B., Dietrich, C.P., 1985. Lack of correlation between "in vitro" and "in vivo" antithrombotic activity of heparin fractions and related compounds. Heparan sulfate as an antithrombotic agent "in vivo". *Thrombosis research*. 40, 597-607.

[48] Brito, A.S., Cavalcante, R.S., Palhares, L.C., Hughes, A.J., Andrade, G.P., Yates, E.A., Nader, H.B., Lima, M.A., Chavante, S.F., 2014. A non-hemorrhagic hybrid heparin/heparan sulfate with anticoagulant potential. *Carbohydrate polymers*. 99, 372-8.

[49] Chavante, S.F., Santos, E.A., Oliveira, F.W., Guerrini, M., Torri, G., Casu, B., Dietrich, C.P., Nader, H.B., 2000. A novel heparan sulphate with high degree of N-sulphation and high heparin cofactor-II activity from the brine shrimp Artemia franciscana. *International journal of biological macromolecules*. 27, 49-57.

[50] Cosmi, B., Fredenburgh, J.C., Rischke, J., Hirsh, J., Young, E., Weitz, J.I., 1997. Effect of nonspecific binding to plasma proteins on the antithrombin activities of unfractionated heparin, low-molecular-weight heparin, and dermatan sulfate. *Circulation*. 95, 118-24.

[51] Dietrich, C.P., Paiva, J.F., Castro, R.A., Chavante, S.F., Jeske, W., Fareed, J., Gorin, P.A., Mendes, A., Nader, H.B., 1999. Structural features and anticoagulant activities of a novel natural low molecular weight heparin from the shrimp Penaeus brasiliensis. *Biochimica et biophysica acta*. 1428, 273-83.

[52] Lima, M.A., Viskov, C., Herman, F., Gray, A.L., de Farias, E.H., Cavalheiro, R.P., Sassaki, G.L., Hoppensteadt, D., Fareed, J., Nader, H.B., 2013. Ultra-low-molecular-weight heparins: precise structural features impacting specific anticoagulant activities. *Thromb Haemost*. 109, 471-8.

[53] Cardin, A.D., Weintraub, H.J., 1989. Molecular modeling of protein-glycosaminoglycan interactions. *Arteriosclerosis*. 9, 21-32.

[54] Esko, J.D., Linhardt, R.J., *Proteins that Bind Sulfated Glycosaminoglycans*, in *Essentials of Glycobiology*, Varki, A., Cummings, R.D., Esko, J.D., Freeze, H.H., Stanley, P., Bertozzi, C.R., Hart, G.W., Etzler, M.E., Editors. 2009, Cold Spring Harbor Laboratory Press. The Consortium of Glycobiology Editors, La Jolla, California: Cold Spring Harbor (NY).

[55] Torrent, M., Nogues, M.V., Andreu, D., Boix, E., 2012. The "CPC clip motif": a conserved structural signature for heparin-binding proteins. *PloS one*. 7, e42692.

[56] Atha, D.H., Lormeau, J.C., Petitou, M., Rosenberg, R.D., Choay, J., 1985. Contribution of monosaccharide residues in heparin binding to antithrombin III. *Biochemistry*. 24, 6723-9.

[57] Atha, D.H., Lormeau, J.C., Petitou, M., Rosenberg, R.D., Choay, J., 1987. Contribution of 3-O- and 6-O-sulfated glucosamine residues in the heparin-induced conformational change in antithrombin III. *Biochemistry*. 26, 6454-61.

[58] Lima, M.A., Hughes, A.J., Veraldi, N., Rudd, T.R., Hussain, R., Brito, A.S., Chavante, S.F., Tersariol, I.I., Siligardi, G., Nader, H.B., Yates, E.A., 2013. Antithrombin stabilisation by sulfated carbohydrates correlates with anticoagulant activity. *MedChemComm*. 4, 870-873.

[59] Chavante, S.F., Brito, A.S., Lima, M., Yates, E., Nader, H., Guerrini, M., Torri, G., Bisio, A., 2014. A heparin-like glycosaminoglycan from shrimp containing high levels of 3-O-sulfated D-glucosamine groups in an unusual trisaccharide sequence. *Carbohydrate research*. 390, 59-66.

[60] Guerrini, M., Guglieri, S., Beccati, D., Torri, G., Viskov, C., Mourier, P., 2006. Conformational transitions induced in heparin octasaccharides by binding with antithrombin III. *The Biochemical journal*. 399, 191-8.

[61] Guerrini, M., Guglieri, S., Casu, B., Torri, G., Mourier, P., Boudier, C., Viskov, C., 2008. Antithrombin-binding octasaccharides and role of extensions of the active pentasaccharide sequence in the specificity and strength of interaction. Evidence for very high affinity induced by an unusual glucuronic acid residue. *The Journal of biological chemistry*. 283, 26662-75.

[62] Henry, B.L., Connell, J., Liang, A., Krishnasamy, C., Desai, U.R., 2009. Interaction of antithrombin with sulfated, low molecular weight lignins: opportunities for potent, selective modulation of antithrombin function. *The Journal of biological chemistry*. 284, 20897-908.

In: Sulfated Polysaccharides
Editors: M. Gama, H. Nader, H. de Oliveira Rocha
ISBN: 978-1-63482-975-5

Chapter 2

CHONDROITIN AND KERATAN SULFATE

Shunji Tomatsu[1,*], Tsutomu Shimada[1], Pravin Patel[1], Robert W. Mason[1], Tadahisa Mikami[2], Hiroshi Kitagawa[2], Adriana M. Montaño[3] and Tadao Orii[4]

[1]Nemours/Alfred I. duPont Hospital for Children, Wilmington, DE, US
[2]Department of Biochemistry, Kobe Pharmaceutical University, Japan
[3]Department of Pediatrics and Department of Biochemistry and Molecular Biology, Saint Louis University, St. Louis, MO, US
[4]Department of Pediatrics, Gifu University, Gifu, Japan

ABSTRACT

Two of the major *glycosaminoglycans* (GAGs) found in cartilage and cornea are chondroitin sulfate (CS) and keratan sulfate (KS). Both GAGs are covalently attached to the core protein of *proteoglycans* (PGs). Their concentration and distribution within the matrix are considerably variable depending upon age, species, tissue, morphological location within the tissue, and physiological or pathological condition. These GAGs, which are distributed throughout the whole body, play important physiological roles associated with inflammation, growth, coagulation, fibrinolysis, lipolysis, and cell-matrix biology.

CS consists of *N*-acetylgalactosamine and glucuronic acid which form the repeating disaccharide unit of the polymer, and comprises a chain of about 40 repeating units. CS is the most prevalent GAG in cartilage and cornea. Concentrations of chondroitin-4 sulfate gradually decreases with age, while chondroitin-6 sulfate concentration increases with age.

KS consists of disaccharide repeating units of *N*-acetylglucosamine and galactose, and has a variable chain length and a variable degree of sulfation. KS comprises of mono-sulfated or di-sulfated disaccharide, and generally the level of sulfation increases with age. KS is present at low levels in fetal and newborn cartilage, and its concentration increases with maturation, to become over 50% of the total GAG content of this tissue.

* Corresponding Author: Tel.: 302-298-7336, Fax: 302-651-6888, E-mail: stomatsu@nemours.org. The first three authors should be regarded as joint first authors.

Some types of mucopolysaccharidoses (MPS IV, VI, and VII) are inherited metabolic disorders characterized by excessive accumulation of CS and/or KS in tissues, including cartilage and cornea, leading to characteristic and systemic skeletal dysplasia. In inherited macular corneal dystrophy the characteristic abnormal corneal transparency is caused by altered KS structure that is not sulfated, while a sub-type of spondyloepiphyseal dysplasia is caused by altered sulfation of CS. Moreover, some types of cancer, osteoarthritis, and spinal cord injury lead to alterations in CS and KS, suggesting that both of these GAGs play a wide range of pathophysiological roles. To quantify CS and KS, highly sensitive and accurate tandem mass spectrometry methods have been established. This chapter describes an historical view of synthesis and degradation pathways, structure, tissue distribution, physiological roles, and the pathological conditions associated with CS and KS. An LC-MS/MS assay system for CS and KS is also described.

1. Introduction

Chondroitin sulfate (CS) and keratan sulfate (KS), repeating units of D-glucuronic acid (GlcA) and *N*-acetyl-D-acetylgalactosamine (GalNAc) or D-galactose (Gal) and *N*-acetyl-D-glucosamine (GlcNAc), are identified as the major glycosaminoglycans (GAGs) in cartilage and cornea. The existence of CS and KS has been recognized since the 1880s and 1930s, respectively. These GAGs have been known to attach to over a dozen different proteins to form proteoglycans (PGs) in a variety of tissues, playing important physiological roles (Table 1). In the last three decades, tremendous research related to the structure and the biological functions of CS and KS has allowed researchers to identify the specific enzymes/genes involved in the biosynthesis and degradation of CS and KS as well as the molecular mechanisms that regulate the expression and sulfation pattern. Marked progress has also been achieved in development of experimental conditions to study the secretion and regulation of CS and KS, both *in vitro* and *in vivo*. It is also of great interest that a relationship between altered levels of these two GAGs and several pathological conditions have been elucidated over the last two decades. Accumulation of KS and/or CS is seen in several types of mucopolysaccharidoses (MPS), a group of lysosomal storage disorder, which manifest as progressive systemic skeletal dysplasia, joint abnormality, and corneal clouding (Table 2). Lack of sulfation of keratan to yield KS leads to genetic macular corneal dystrophy (MCD) with serious corneal opacity. CS and/or KS are also associated with the common diseases such as neural plasticity, spinal cord injury, osteoarthritis, cancers, obesity, and diabetic nephropathy, proving that these GAGs are pathological factors as well as signal induced factors and are critical as biomarkers. CS- and KS-specific endoglycosidases and anti-CS or KS monoclonal antibodies have enabled development of sensitive detection methods for CS and KS.

Further, development of rapid, accurate, and reproducible tandem mass spectrometry assay systems have enabled measurement of CS and KS in a variety of biological specimens from range of different species.

This technology is now being applied to diagnosis, prognosis, monitoring therapeutic efficacy, and pathogenesis of genetic and acquired disorders.

Table 1. Correlation between CS or KS and pathophysiological condition

	Chondroitin sulfate	Keratan sulfate
Main location	Cartilage, cornea, brain	Cartilage,cornea, brain
Physiological role	growth development, cell proliferation, tissue hydration, mechnical strenght of the skin	growth, tissue hydration, cell migration and proliferation role in embryonic development, tissue repair, inflammatory response
Related genetic disorders	MPS IVA, MPS VI, MPS VII, spondyloepiphyseal dysplasia, congenital joint dislocations	MPS IVA, MPS IVB, MCD type 1 and type 2
Other related disorders	Cancer, osteoarthritis, spinal cord injury, neuronal plasticity	Cancer: prostate, pancreatic, breast, colorectal, and cervical cancers, Osteoarthritis, spinal cord injury, unilateral nigrostriatal axotomy, amyotrophic lateral sclerosis, obesity, heart failure, hepatic fibrosis
Use for therapy	opthamologic diseases, degenerative joint disease	chronic obstructive pulmonary disease, epithelial defects of the cornea, rheumatoid arthritis

Studies of CS and KS have led to the publication of around 15,000 and 2,500 research articles, respectively (using CS or KS as keywords in a PubMed search in July, 2014).This chapter summarizes the most recent advances in our understanding of the structure, tissue distribution, biosynthesis and degradation pathways, physiological and pathological conditions, therapeutic use, and novel assay system for CS and KS.

2. Chondroitin Sulfate (CS)

2.1. Historical Review

Chondroitin sulfate (CS) was first isolated from cartilage in 1884 [1], although the structure of CS remained unknown until described by Levene in 1915. CS is composed of repeating units of D-glucuronic acid (GlcA) and *N*-acetyl-D-galactosamine (GalNAc) [2, 3]. CS is synthesized as a long chain polymer, linked at its reducing end to a core protein. CS is widely distributed on cell surfaces and within extra/ pericellular matrices in the form of proteoglycans (CSPGs). CS chains are covalently attached to the corc protein at a serine residue within an amino acid consensus sequence and the number of chains bound to single core protein will depend on the number of these consensus sequences in the protein. CSPGs are involved not only in various physiological roles such as cytokinesis, morphogenesis, embryonic development, osteogenesis, and neuronal plasticity, but also in pathological roles including skeletal disorders, mucopolysaccharidoses, glial scar formation after brain injury, and infections with viruses and bacteria [4].

Table 2. Clinical features and GAGs in Mucopolysaccharidoses (MPS)*

Disorder	Deficient Enzyme	Trait	Chromosome	Primary storage GAG(s)	Coarse facies	Mental retardation	Incomplete ossfication	Joint restriction
MPS I (Hurler)	α-L-Iduronidase (IDUA)	AR	4p16.3	DS, HS	+++++	+++++	+++	++++
MPS II (Hunter)	Iduronate-2-sulfatase (IDS)	XR	Xq28	DS, HS	++++	+++++	+++	+++
MPS IIIA (Sanfillipo A)	Heparan-N-sulfatase (SGSH)	AR	17q25.3	HS	++	+++++	+	+
MPS IIIB (Sanfillipo B)	α-N-Acetylglucoaminidase (NAGLU)	AR	17q21	HS	++	+++++	+	+
MPS IIIC (Sanfillipo C)	α-Glucosaminidase acetyltransferase (HGSNAT)	AR	8p11-q13	HS	++	+++++	+	+
MPS IIID (Sanfillipo D)	N-Acetylglucosamine 6-sulfatase (GNS)	AR	12q14	HS	++	+++++	+	+
MPS IVA (Morquio A)	Galactose 6-sulfatase, N-acetylgalactosamine-6-sulfate sufatase (GALNS)	AR	16q24.3	C6S, KS	+	-	+++++	+++++
MPS IVB (Morquio B)	β-Galactosidase (GLB1)	AR	3p21.33	KS	+	-	++	++
MPS VI (Maroteaux-Lamy)	N-Acetylgalactosamine-4-sulfatase (G4S)	AR	5q13.3	C4S, DS	++	-	++++	++++
MPS VII (Sly)	β-D-Glucuronidase (GUSB)	AR	7q21-q22	C4, 6S, DS, HS	++	+++	++	+++

* Clinical features represent a classical MPS. AR and XR indicate autosomal recessive and X-linked. CS, DS, HS, and KS denote chondroitin sulfate, dermatan sulfate, heparan sulfate, and keratan sulfate, respectively.

Accumulating evidence indicates that most functions of CSPGs are conducted through the CS moieties, with core proteins seeming to act merely as a scaffold [4]. CS moieties *in vivo* vary considerably in the size and the number of CS chains per core protein and in the position and degree of sulfation changes with age. Such structural complexity is able to support multifaceted functions of CSPGs.

2.2. Classification and Structure

CS is a linear polymer, without branches, and its sulfate groups make CS highly anionic. The disaccharide units are linked by two β ester bonds, (GlcAβ1→3GalNAc) and (GalNAcβ1→4GlcA), in the polymer chain.

The common repeating unit of CS (Figure 1) can be modified by sulfotransferases to produce different subtypes of CS. There are six known subtypes of CS, CS-(A, C-E, H, K), each of which is defined by its sulfation state with respect to the *O*-linked R groups at C4 and C6 of GalNAc and C2 of GlcA (Figure1).

Uniquely, CS-H contains iduronic acid (IdoA) instead of GlcA in some disaccharide units. CS-K is sulfated on C3 of GlcA, instead of the more common C2, and is only found in marine invertebrates (discussed in a further chapter) [5]. The differences between CS subtypes are listed in Table 3. Physiologically, CS-A (C4S) and CS-C (C6S) are the most abundant form of CS in the human body [4].

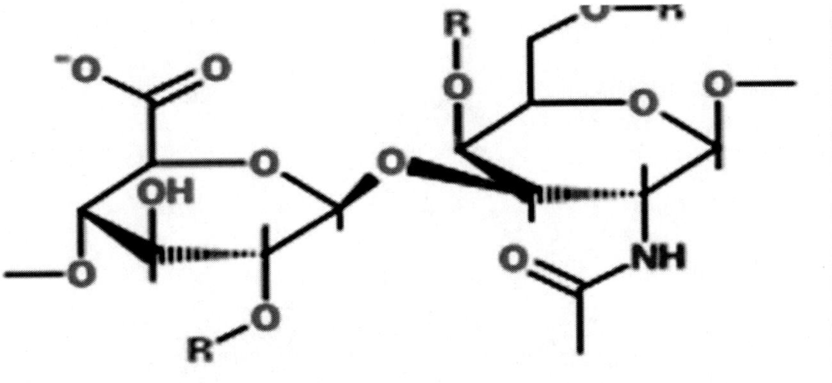

Figure 1. Structure of the repeating disaccharide units of CS. On the left is D-glucuronic acid (GlcA) and right is *N*-acetylgalactosamine (GalNAc). R denotes the potential sites for sulfation. R_2 is on C2 of GlcA, R_4 is on C4 of GalNAc, and R_6 is on C6 of GalNAc.

Table 3. Classification of CS

Subtype	Name	Other Name	Sulaftion State R2	R4	R6
CS-O	GlcA-GalNAc	chondroitin	H	H	H
CS-A	GlcA-GalNAc(4S)	chondroitin-4-sulfate (C4S)	H	SO_3	H
CS-C	GlcA-GalNAc(6S)	chondroitin-6-sulfate (C6S)	H	H	SO_3
CS-D	GlcA(2S) – GalNAc(6S)	chondroitin-2,6-sulfate (C2,6S)	SO_3	H	SO_3
CS-E	GlcA– GalNAc(4S,6S)	chondroitin-4,6-sulfate (C4,6S)	H	SO_3	SO_3
CS-H	IdoA–GalNAc(4S,6S)	chondroitin-4, 6-sulfate (CS-H)	H	SO_3	SO_3
CS-K	GlcA(3S) – GalNAc(4S)	chondroitin-3,4-sulfate (C3,4S)	SO_3*	SO_3	H

* The sulfate group on GlcA in CS-K is found on C3 instead of C2.

2.2.1. Biosynthesis

The synthesis of CS is conducted in three major phases within the Golgi apparatus [4, 6]. All glycosyltransferases responsible for CS biosynthesis utilize distinct UDP-sugars as a donor substrate. Since UDP-sugars, including UDP-GalNAc and UDP-GlcA, are synthesized in the cytosol, the first phase involves uptake of UDP-sugars into the lumen of the endoplasmic reticulum (ER)/Golgi via corresponding UDP-sugar transporters (Figure 2) [6]. In phase two, sequential addition of individual monosaccharide units occurs to form unsulfated CS polymers. In the last phase, CS is sulfated. Phases two and three take place during the greater synthesis of CSPGs (Figure 2) (see '2.4.3.1. Synthesis of Linker Region and CS Polymer Chain').

2.3. Degradation

Catabolism of CSPGs involves two overarching features, degradation of the core protein and catabolism of the polysaccharide.

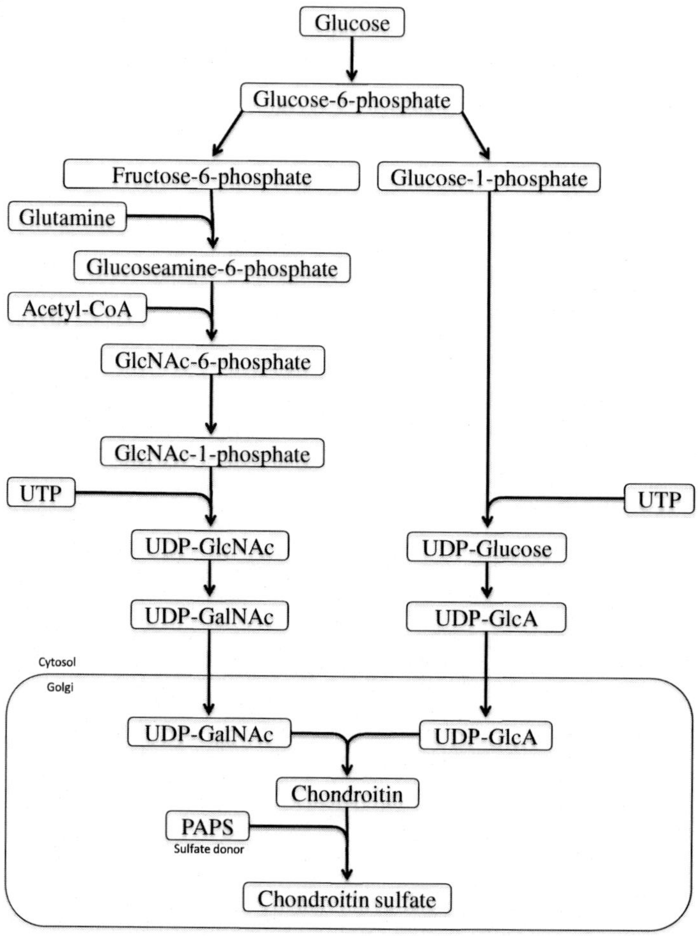

Figure 2. Abbreviated synthetic pathway for CS; from glucose to CS.

The exact catabolic pathway also depends on how the CSPG is expressed in the cell (secretion, cell surface, or intracellular) and where CS is located in the body. Degradation of the polymer chain takes place in a directional manner from the terminal saccharide, the non-reducing end (NRE), to the core protein [7]. The complete degradation of the CS polysaccharides takes place in a stepwise manner, starting at the cell surface and ending within the lysosomes.

The catabolism of polymeric CS starts with the hydrolysis of the large CS polymer into smaller CS oligosaccharides. The cleavage of CS into smaller units can be catalyzed by three endo-hyaluronidase enzymes; hyaluronidase-1, hyaluronidase-4, and hyaluronidase PH-20 [8]. While the primary substrate of hyaluronidase-1 and hyaluronidase PH-20 is hyaluronan, these two enzymes can also hydrolyze CS [8-10]. Uniquely, hyaluronidase-4 is specific to CS

[11]. The hyaluronidases cleave the GalNAcβ1→4GlcA glycosidic bond randomly along the CS polymer, generating CS molecules with a variable length, but usually produce molecules of two or three CS units [8, 12-14]. While hyaluronidase-4 and hyaluronidase PH-20 are linked to the cell surface by a glycosylphosphatidylinositol (GPI), hyaluronidase-1 is located within the lysosomes [8, 15]. After initial degradation, the CS oligosaccharides are delivered into the lysosomes. Once in the lysosome, two enzymes, *N*-acetylgalactosamine-4-sulfatase (G4S) and *N*-acetylgalactosamine-6-sulfatase (GALNS), hydrolyze the sulfate ester bonds at C4 or C6 of GalNAc, respectively [8, 16-18]. Glucuronate 2-sulfatase (GAS) hydrolyzes the sulfate ester bonds on C2 of GlcA [19, 20] (Table 4, Figure 3).

After the sulfate groups have been removed, the oligosaccharide needs to be broken down into saccharide monomers. Monomers are cleaved from the oligosaccharide from the non-reducing end (NRE) by different enzymes, depending on the species of the terminal saccharide.

If GlcA is present, β-glucuronidase hydrolyzes the GlcAβ1-3GalNAc glycosidic bond [21]. If GalNAc is present, β-hexosaminidases, of which there are three types (A, B, and S), hydrolyze the GalNAcβ1→4GlcA glycosidic bond [8, 22, 23]. The name of each enzyme, the catabolic reaction, and the position in the catabolic pathway is listed in Table 4, and an abbreviated pathway is shown in Figure 3.

Table 4. The stepwise catabolism of CS

Step	Location	Enzyme(s)	Activity	Reaction Catalyzed
1	Cell Surface	Hyaluronidase-4 Hyaluronidase PH-20	Hydrolysis of the glycosidic bond	GalNAcβ1-4GlcA→ GalNAc + GlcA
	Lysosome	Hyaluronidase-1		
2	Lysosome	N-acetylgalactosamine-4-sulfatase (G4S)	Hydrolysis of the sulfate ester bonds of GalNAc	$C4\text{-}OSO_3^{-}\rightarrow C4\text{-}OH$
		N-acetylgalactosamine-6-sulfate sulfatase (GALNS)	Hydrolysis of the sulfate ester bonds of GalNAc	$C6\text{-}OSO_3^{-}\rightarrow C6\text{-}OH$
		Glucuronate 2-sulphatase (GAS)	Hydrolysis of the sulfate ester bonds of GlcA	$C2\text{-}OSO_3^{-}\rightarrow C4\text{-}OH$
3	Lysosome	β-glucuronidase	Hydrolysis of the glycosidic bond, GlcA at the NRE	GlcAβ1-3GalNAc → GlcA + GalNAc
		β-hexosaminidase A β-hexosaminidase B β-hexosaminidase S	Hydrolysis of the glycosidic bond, GalNAc at the NRE	GalNAcβ1-4GlcA→ GalNAc + GlcA

* The reactions are catalyzed by the enzymes in the order shown.

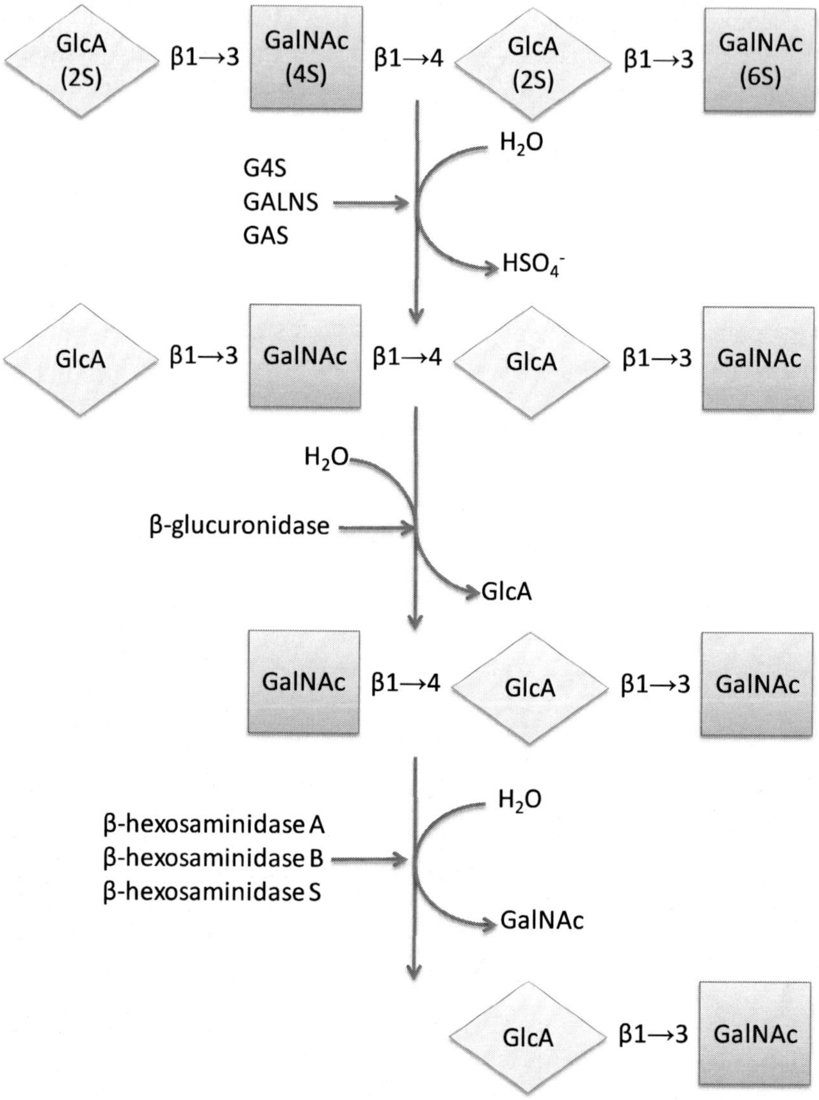

Figure 3. Catabolic pathway for a CS tetrasaccharide. Initial breakdown of the CS polymer is not shown.

2.4. Proteoglycans

2.4.1. Types

Chondroitin sulfate proteoglycans (CSPGs) include multiple CS polymers linked at one end to a central core protein.

There are 14 types of CSPGs, each of which is unique in its protein core. Each CSPG can be linked to any of the sub-types of CS, along with other GAGs (Table 5). Each CSPG is found in specific tissue types and has its own physiological function(s). Further descriptions of the physiological functions of CSPGs are described below.

2.4.2. Core Protein

Lecticans and SLRPs constitute the majority of CSPGs found in the body, and their structures are described below. The protein structures of two small leucine-rich proteoglycans are also described. Eight other CSPGs are known; however, these CSPGs do not belong to any common CSPG protein family. Discussion of the structure of these CSPGs is omitted.

2.4.2.1. Lecticans

Lecticans have an N- and C- terminal globular regions connected by a flexible central domain. The N-terminal globular region (G1) is made of a larger immunoglobulin (Ig)-like loop and two smaller proteoglycan tandem repeat (PTR) domains [25]. Aggrecan uniquely has a second N-terminal globular region (G2) composed of the two smaller PTR domains without an Ig-like loop [25].

Table 5. Distribution of CSPG

CSPGs	Number of Chondroitin Sulfate Chains	Main Tissue Distribution	References
Lecticans (Aggrecan			
Aggrecan	~100, ~120 GAGs	Cartilage; secreted	[24, 25]
Brevican	0-4, ~3 GAGs	Brain; secreted	[24, 25]
Neurocan	1-2, ~7 GAGs	Brain; secreted	[24, 25]
Versican	12-15, ~20 GAGs	Connective tissue; Aorta; Brain; secreted	[24, 25] [24, 25]
SLRPs (Small Leucine-Rich Proteoglycans)			
Biglycan	1-2	Connective tissue; secreted	[24, 26]
Decorin	1	Connective tissue; secreted	[24, 26]
Other			
CD44	1-4	Lymphocytes; membrane bound	[24]
Invariant chain	1	Antigen-processing cells; membrane bound	[24]
Leprecan	1-2	Basement-membranes; secreted	[24]
NG2	2-3	Neural cells; membrane bound	[24]
Phosphacan	2-5	Brain; membrane bound	[24]
Serglycin	10-15	Intracellular granules; myeloid cells	[24]
Thrombomodulin	1	Endothelial cells; membrane bound	[24]
Type IX collagen, α2 chain	1	Cartilage; vitreous humor; secreted	[24]

The central domains are very diverse with respect to their size and sequence and contain multiple sites for GAG attachment. The C-terminal region (G3) is composed of a C-type lectin domain bordered by complement regulatory protein (CRP)- and EGF-like domains [25].

Lecticans lack cysteine residues, which likely enables the central domain to adopt a mainly linear structure. The number of potential GAG attachment sites varies within the lectican family. Each lectican can also vary in the types of GAGs attached (Table 3).

2.4.2.2. SLRPs (Small Leucine-Rich PGs)

SLRPs are characterized by a domain of tandem leucine-rich repeats (LRRs), bordered by clusters of conserved cysteine residues [26, 27]. The only SLRPs known to bind CS are decorin and biglycan. The N-terminal domain of these proteins contain site(s) for GAG attachment, one in decorin and two in biglycan [28, 29]. They contain 10 LRRs within a central domain of the protein [28].

2.4.3. Linker Region

Each CS chain is attached to a serine residue of the core protein by a linker region, which has the same structure for all CSPGs. The structure of this linker region is [GlcAβ1→3Galβ1→3Galβ1→4Xylβ1-*O*-Ser] [30, 31]. All CS polymers are linked glycosidically, by an *O*-glycosyl bond, to serine in the protein core via a tetrasaccharide linker region. The linker region is made of [glucuronic acid→ galactose→ galactose→ xylose→ serine] and is common to all PGs [32].

2.4.3.1. Synthesis of Linker Region and CS Polymer Chain

The synthesis of CSPG starts when the newly translated protein core is delivered into the ER/Golgi [4, 6, 32]. In the ER/Golgi, distinct monosaccharides are individually used to build the linker region, and the CS polymer is synthesized by sequential addition of single sugars to the linker [4, 32] (Figure 4).

First, UDP-xylose transferase (XylT, UDP-Xyl transferase) catalyzes the addition of xylose to a serine residue of the protein core. Then, galactose transferases catalyze the addition of two galactose units; GalT-I (UDP-Gal transferase I) adds one unit to xylose and then GalT-II (UDP-Gal transferase II) adds the second galactose unit [4, 6]. Finally, GlcAT-I (UDP-GlcA transferase I) catalyzes the addition of GlcA to Gal [4, 6]. The addition of GalNAc to the GlcA of the linker region directs the start of CS polymer synthesis [4, 6, 32].

Chondroitin polymer synthesis starts once the synthesis of the linker region is completed. Polymerization takes place by alternate addition of unsulfated GlcA and GalNAc [4, 6, 32, 33]. Polymer synthesis is highly structured such that only alternate saccharides are added.

The first saccharide linked to the terminal GlcA of the linker region largely controls which GAG is going to be produced [4, 6]. To begin polymer synthesis of CS, GalNAc transferase I (GalNAcT-I) catalyzes the transfer of GalNAc to the linker region [4, 6, 32]. Then, alternating activity of GlcA transferase II (GlcAT-II), which catalyzes the transfer of GlcA to the growing polymer, and GalNAc transferase II (GalNAcT-II), which catalyzes the transfer of GalNAc to the growing polymer, catalyzes CS polymer synthesis [4, 6, 32, 34, 35]. There are six homologous glycosyltransferases responsible for chondroitin polymer synthesis. Four of these enzymes, ChSy-1, ChSy-2, ChSy-3, and ChPF, constitute chondroitin polymerase complexes that possess dual enzymatic activities, GlcAT-II and GalNAcT-II (Figure 4). Note that while we state here that the addition of GalNAc to the linker region and the addition of GalNAc during CS polymer synthesis are catalyzed by different enzymes, some evidence suggests that one enzyme can catalyze both transferase reactions [36].

Indeed, the other two enzymes, ChGn-1 and ChGn-2, possess both GalNAcT-I and GalNAcT-II activities, and therefore, are involved in chain initiation and elongation of chondroitin polymer (Figure 4).

In the next phase of CSPG synthesis, the unsulfated CS polymer is sulfated by sulfotransferases [4, 37]. Chondroitin 4-*O*-sulfotransferases (C4ST) and 6-*O*-sulfotransferases (C6ST) generate C4S and C6S, respectively [4, 37].

C4S is additionally sulfated by GalNAc 4-sulfate 6-*O*-sulfotransferase (4,6-ST) to form CS-4,6 (CS-E). C6S is further sulfated by uronyl 2-*O*-sulfotransferase (UST) to form CS-2,6 (CS-D) (Figure 5).

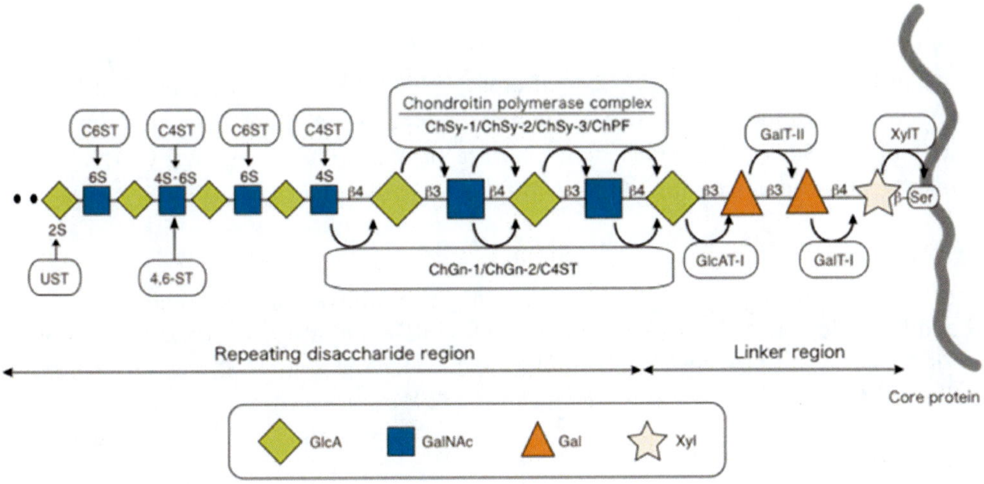

Figure 4. CS biosynthetic machineries.

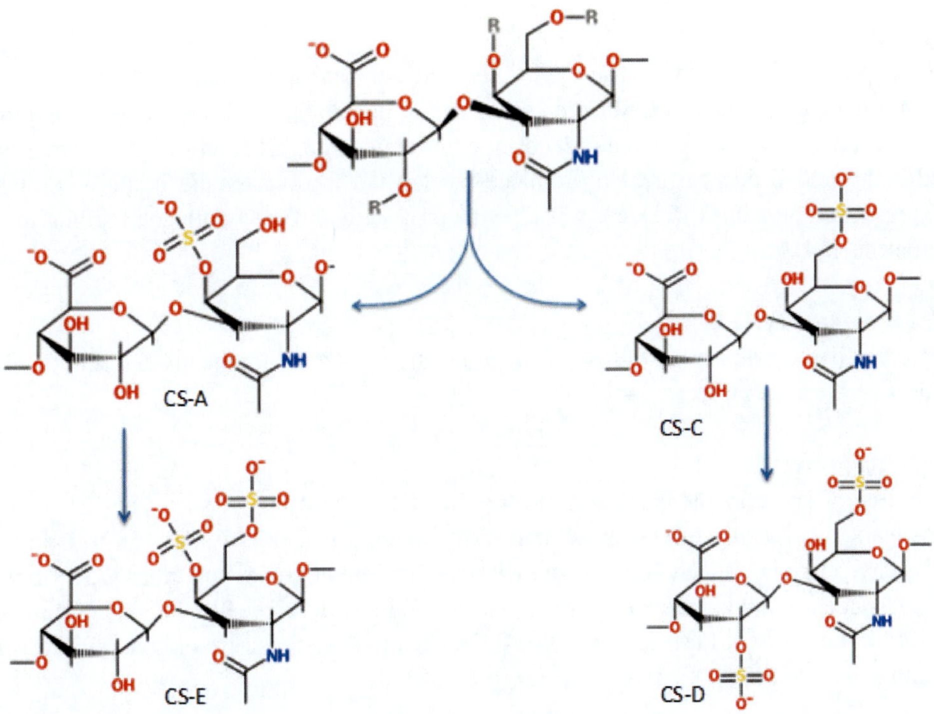

Figure 5. Abbreviated synthetic pathway for sulfated CS.

For the synthesis of CS-H, some of GlcA residues in the intact chondroitin disaccharide units are converted to IdoA by chondroitin-glucuronate C5 epimerase (C5-EP), and the resultant dermatan-type disaccharide unit serves as a substrate for stereocontrolled sulfation that catalyzed by dermatan 4-*O*-sulfotransferase-1 (D4ST-1) and 4,6-ST [4, 37].

The CS polymer chain can vary in length up to more than a hundred disaccharide units [32]. The polymer chain ends abruptly with either a GlcA or GalNAc, which can also vary in its sufation state [38, 39].

2.5. Tissue Distribution

The different subtypes of CS and different CSPGs are not equally distributed in the human body. The tissue location of CSPGs mainly depends on interactions between the core protein and molecules present on the cell surface or extracellular matrix (ECM) as well as controlled expression of particular CSPGs [40].

2.5.1. Cartilage

Articular cartilage is mostly composed of aggrecan, linked to unsulfated CS as well as C4S and C6S [24, 41, 42]. As one might expect, CSPGs are also found within the synovial fluid of joints. Similar to articular cartilage, the CSs expressed in the synovial fluid are mainly C4S and C6S [43]. Cartilage also includes biglycan [44].

2.5.2. Connective Tissues

Aggrecan and versican are the major PGs in connective tissues, and contribute to the mechanical properties of those tissues as well as hydration of the tissues [40].

2.5.3. Skin

Skin also contains CSPGs. The C6S CSPGs are present in the lamina densa of all skin basement membranes. Theses CSPGs present early in developing fetuses and are found in the skin of all ages [45]. Skin biglycan, decorin, and versican are all found in the dermal ECM [46-53]. The CSPG decorin is also produced by dermal fibroblasts, and is the predominant PG expressed in human skin. Decorin is also the predominant type I collagen binding protein in the dermis [54].

2.5.4. Ligament/ Tendon

Decorin is the primary CSPG of dense connective tissue of ligaments and tendons [55]. Tendon also contains biglycan [44].

2.5.5. Neural Tissues

All four CSPGs in the lectican class are found in neuronal tissues [56-57] although brevican and neurocan expression is mainly restricted to neural tissues [58-61]. Primary cultures have also shown that CSPGs are expressed by both neuronal cell types [25]. Brevican mRNA has been found in cerebellar astrocytes and oligodendrocytes [40, 58, 62]. Neurocan is expressed in neurons and astrocytes [63, 64]. Versican is produced by oligodendrocytes and Schwann cells [65, 66].

2.5.6. Eye

The ECM of the sclera contains the CSPGs; aggercan, biglycan, decorin, and lumican [67-71].

2.5.7. Blood Vessels

The umbilical veins and arteries were also found to contain CSPGs on the endothelial surface; however, while CSPGs are abundant in veins, they were not found to be expressed at a significant level in arteries. Versican is found in both veins and arteries, and a serglycin type

CSPG is found only in the veins [72]. Biglycan plays a major role on capillary endothelium [44].

2.5.8. Blood: Cells and Plasma

Blood cells contain CSPGs whose major components are C4S, C6S, C4,6S, and C2,4S [73]. Macrophages and monocytes also have CSPGs where the sulfation state of the associated CSs increases after these cells have been stimulated to induce the inflammation response [74]. Monocytes almost entirely express C4S; however, the proportion changes after the monocytes differentiate into macrophages. Macrophages contain a mixture of C4S and C6S. Cytolytic T lymphocytes and NK cells also contain C4S [75, 76]. C4S is also present in blood plasma. Here, C4S represents 75-80% of total GAGs in the plasma [77-80].

2.6. Physiological Role

CSPGs have many physiological functions such as cytokinesis, morphogenesis, and neuronal plasticity as well as growth development, which vary based on where in the body they are located as well as the developmental stage.

2.6.1. Role in Development

CSs show a high degree of spatiotemporal patterning in developing embryos [25]. These CSs, along with KSs, create a barrier to cell adhesion and axon outgrowth, and function as guidance cues for axon growth and neural crest cell migration [81-87]. While CS has a large effect on developmental processes, the core proteins themselves have also been shown to provide an impact on neurite outgrowth [25]. Phosphacan promotes neurite outgrowth [88-89]. Aggrecan has also been shown to inhibit adhesion and migration of neural crest cells [90]. Neurocan is involved in neural cell adhesion and outgrowth [91]. CSPGs can not only act as a barrier to cell motility, but also act as a pathway over which cells can migrate. This type of growth is called an addictive growth [92, 93].

2.6.2. Biological Function

Aggrecan makes up the majority of CSPGs found in articular cartilage. Aggrecan in cartilage forms aggregates with hyaluronan and a link protein through its N-terminal domain, which has a binding site for both the link protein and hyaluronan on it G1 domain [94-96]. The negative charges within the CSPG help to trap water molecules allowing the soft tissue of articular cartilage to be hydrated [6, 42]. This action enables the cartilage to produce an osmotic swelling pressure that allows the joints to withstand high compressive forces [6, 97]. The hydrated state of articular cartilage also provides synovial joints with a highly wear resistant, low friction surface, allowing for fluid, pain free movement [42]. Expression of several CSPGs (NG2, neurocan, brevican, versican) is upregulated in response to cortical injury [81, 98-102]. Brevican and neurocan are also upregulated in response to entorhinal lesions [84, 103, 104].

Skin CSPGs are necessary for maintaining the mechanical strength of the skin [54]. Decorin knockout mice have been used to show that this CSPG helps control the diameter of collagen fibers, as well as the organization of the ECM, helping to maintain the tensile strength of the skin [105].

Cell proliferation is also controlled by decorin which binds to epidermal growth factor receptors on the cell surface, inhibiting ligand binding [106]. Extracellular concentrations of transforming growth factor-β are also controlled by decorin. The protein core of decorin is able to trap the growth factor, lowering the extracellular concentration available for signaling [40, 107].

2.7. Age Dependency

2.7.1. Cartilage

The concentration and sulfation state of CSPGs in articular cartilage changes over time. The most pronounced result of this change is the loss of articular cartilage later in life, leading to osteoarthritis. Modifications to the concentration and sulfation state of CS stem from alterations in the chondrocyte regulation of CSPGs, which affect the aggregation of aggrecan with other ECM components of articular cartilage. Further, these changes in the regulation of CSPGs are not uniform across all joints or even within one joint [108-110]. It has also been shown in articular cartilage and intervertebral discs that non-glycated form of biglycan increases with age [111, 112]. Changes in CSPG regulation alter the proportions of C4S and C6S found in the articular cartilage. From birth to 20 years of age, the proportion of C6S rose, while during the same period, the proportion of C4S fell. After 20 years of age, the proportions of C4S and C6S remained constant [41].

The changes in articular cartilage over time can also be seen in CS expression in the synovial fluid. C4S and -6 are co-expressed in the synovial fluid before puberty; however, post puberty, only C6S is found [43]. The synovial fluid of osteoarthritis patients has also been shown to contain aggrecan fragments cleaved at distinctive sequences by aggrecanase, and it is thought that the degradation of aggrecan contributes to the pathogenesis of osteoarthritis [113-115].

2.7.2. Neuronal Tissues during Development

The levels of CSPGs in the rat brain also vary in an age-dependent manner. Neuronal tissues have spatiotemporal expression of CSPGs [116]. Aggrecan and brevican show the same expression patterns; the concentration of theses CSPGs increase from E (embryonic day) 14 to P (postnatal day) 150, when they reach their final concentrations. Similarly, neurocan increases in concentration by 5-fold during embryonic development, peaking at P2-6, followed by a sharp fall in concentration afterwards.

CSPGs also show different, developmentally regulated, alternative splicing. Neurocan can undergo developmentally regulated proteolytic cleavage of its core proteins [25, 117]. In embryonic and postnatal brains, the uncleaved form of neurocan predominates. However, in the adult brain, the majority of neurocan is in its cleaved form [25].

Versican also shows different, developmentally regulated, alternative splicing of the CS domain. The isoform, versican-CSα, shows increased expression from P10 to P100. Versican-CSβ has similar expression patterns to neurocan [118-119]. The tight regulation of the concentration of these CSPGs indicates that they play a critical role in the development of the nervous system.

The expression on individual PGs changes throughout life and sulfation pattern also varies with age while the expression of CS-sulfotransferases remains constant in neurogenic cells from mice [120].

2.7.3. Skin

Skin also undergoes changes in CSPG expression as a function of age. While CSPGs are found in all skin types of all ages, the expression of versican is age-dependent. Versican varies in two ways with respect to age. One-way is that fetal versican has different CS chains than adult versican. Another age-related difference is that age increases, the proportion of versican to total CSPG in the skin falls [121].

The polymer length on decorin also changes with age. The molecular weight of decorin extracted from aged (> 80 years) and young (21-30 years) skin varies, with CSPGs from young skin weighing more, indicating that polymer chain length attached to decorin shortens with age. Most likely this change contributes to the skin fragility seen in the elderly [54].

2.8. Pathological Conditions

2.8.1. Impairment of CS Degradation

Mucopolysaccharidoses (MPS) are a family of inheritable metabolic disorders caused by deficiency of the lysosomal enzymes needed for the complete degradation of glycosaminoglycans (GAGs). Each known type of MPS, 11 in total, involves a deficiency in a specific lysosomal enzyme required for the stepwise degradation of specific GAG(s) (CS, DS, HS, and KS). The enzymatic deficiency causes GAGs to accumulate in nearly all cell types. The accumulation disrupts normal cell function, leading to a multitude of secondary disease pathologies and unique phenotypes.

Failure in the degradation of the CS polymer is seen in three types of MPS; MPS IVA, MPS VI, and MPS VII, caused by a deficiency of GALNS, G4S, and β-glucuronidase, respectively. While MPS VII results in accumulation of C4,6S, MPS IVA and MPS VI result in accumulation of C6S and C4S, respectively [122-125].

All three types of MPS cannot fully degrade CS. MPS IVA also results in accumulation of KS, MPS VI accumulates DS and MPS VII accumulates HS and DS. Patients with these forms of MPS have the clinical features of skeletal abnormalities and joint restriction. Patients with MPS IVA have the most severe skeletal dysplasia with incomplete ossification in cartilage. Patients with MPS IVA and VI have no involvement of CNS, while patients with MPS VII have variable degrees of mental impairment [126]. Patients with MPS VI and VII have involvement of visceral organs, while patients with MPS IVA have a minimal involvement.

KS and CS are the major GAGs in articular and growth plate cartilage where bone grows [127, 128]. DS and HS, degraded by both GUSB and IDUA, enzymes deficient in MPS VII and MPS I, are minor components of the GAGs in both articular and epiphyseal cartilage [129-131]. Skeletal disease is reported to be mild in MPS IIIA [132] and MPS IIIB [133, 134] mice, which accumulate only HS. Bone pathology in MPS VII mice is more severe than in

MPS I mice. One potential reason for the difference is the accumulation of C4S in cartilage of MPS VII mice [128, 129]. MPS VI mice with a deficiency of G4S [135], which is critical for degradation of C4S, have a severe skeletal disease. Skeletal disease is severe in human MPS IVA patients who cannot degrade C6S and KS; however, MPS IVA mice have minimal skeletal disease [136, 137], even though these storage materials are evident in cartilage cells.

This finding was likely recently explained by the fact that C6S constitutes a minor component of cartilage [128, 138], and KS synthesis in mice is substantially limited when compared with human. Taken together, data from MPS mouse models are consistent with the hypothesis that the accumulation of C4S is responsible for a more severe chondrodysplasia in MPS VI and MPS VII than other MPS mouse models [128, 139].

Whether either C4S or C6S affects more the clinical manifestations on skeletal system in patients with MPS needs to be studied further.

2.8.2. Impairment of CS Sulfation

Uchimuma et al., first reported that C6ST-1 KO mice related to CS biosynthesis are born at the frequency expected from the Mendelian ratio and are viable through adulthood [140]. Except for significant reduction in the levels of CS-C and CS-D, and in the number of splenic naive T lymphocytes of the KO mice, no apparent clinical phenotype in mutant mice was observed. Apart from the mouse model, Thiele et al., [141] has demonstrated that a R304Q missense mutation in human C6ST-1 produces a severe spondyloepiphyseal dysplasia (SED) with progressive spinal involvement. Moreover, the congenital joint dislocations are unique clinical feature caused by missense mutations in the C6ST-1 gene [142, 143]. These findings suggest that 6-*O*-sulfated CS formed by C6ST-1 plays critical roles in skeletal development and maintenance in humans. The gene trap mutation in the C4ST-1 gene of mice showed that 4-*O*-sulfated structure of CS chains is reduced by more than 90% in the growth plates of the homozygous mutant mice and that the mutation causes severe chondrodysplasia characterized by a disorganized cartilage growth plate as well as specific alterations in the orientation of chondrocyte columns, leading to lethality in newborns [127].

This phenotype is associated with a chondroitin sulfation imbalance, mislocalization of chondroitin sulfation in the growth plate and an imbalance of apoptotic signals.

These results show that chondroitin 4-*O*-sulfation by C4ST-1 is required for proper CS localization, modulation of distinct signaling pathways and cartilage growth plate morphogenesis, demonstrating an important biological role of differential chondroitin sulfation in mammalian development. Partial overlapping but distinct phenotypes caused by defects in the initial "4-*O*-sulfation" and "6-*O*-sulfation" pathways support the notion of the functional importance of the specific sulfation in CS chains [4].

2.8.3. Neural Injury and Neuronal Plasticity

CS is the most abundant extracellular inhibitor of axon growth. The KO mice for ChGn-1, a key enzyme in CS biosynthesis, recover more completely from spinal cord injury than wild-type mice. Synthesis of heparan sulfate (HS), a glycosaminoglycan promoting axonal growth, is also upregulated in ChGn-1 KO mice [144]. Regulation of a single gene, ChGn-1, mediates excellent recovery from spinal cord injury by optimizing counteracting effectors of axon regeneration—an extracellular inhibitor of CS and intrinsic promoters, HS. Simultaneous upregulation of HS synthesis and downregulation of CS were essential to excellent recovery from spinal cord injury. Moreover, CS moieties of CSPGs are responsible for perineuronal net formation and control of the critical period plasticity. Sulfation pattern of CS chains in the plasticity plays a critical role while sulfation profiles of CS chains change dramatically during brain development [4, 145-148]. The proportion of 6-*O*-sulfation gradually decreases, whereas that of 4-*O*-sulfation progressively increases, which results in an incremental change in 4S/6S ratio during development with age [145].

C6ST-1 transgenic mice preserve ocular dominance plasticity even in adulthood, and have a reduced perineuronal net formation and decreased accumulation of Otx2 homeoprotein around parvalbumin+ interneurons [145]. Otx2 is known to regulate ocular dominance plasticity through its effect on maturation of parvalbumin+ interneurons [149, 150]. Therefore, the 4S/6S ratio of CSPGs may regulate the Otx2-mediated maturation of parvalbumin + interneurons that determines critical period of cortical plasticity.

2.9. Species Differences

Up to now the discussion of CS and CSPGs has focused on humans; however, which type of CS, which tissue, and which proportion vary between different species. The species differences are summarized in table 6 [41, 151-155].

Table 6. Proportions and tissue distribution of CS in animals

Source	Tissue	CS Sub-type (%)		
		C4S	C6S	Other
Human	Cartilage (24 yr)	~20%	~80%	
Shark	Cartilage-Fin	9.60%	8.80%	CS-D (~20%)
Squid	Cartilage	-	-	CS-E (~60%)
Hagfish	Notocord	-	-	CS-H (~68%)
Crocodile	Hyoid	14.80%	13.50%	
	Rib	5.60%	5.10%	
	Sternum	11.60%	10.50%	
	Trachea	9.50%	8.70%	
Ray	Cartilage	5.30%	4.80%	
Chicken	Keel	14.10%	13.40%	

2.10. Therapy

Supplements of CS are isolated from the cartilaginous tissues of assorted animals [156]. CS supplements administered orally are able to be absorbed as high molecular mass polysaccharides [157]. When CS is absorbed after oral administration, the vast majority of CS is broken down such that only 10% of CS remains by the time it reaches the blood compartment [158]. While CS appears to have a low bioavailability, it has been found that CS can accumulate in joint tissues [159]. CS has been shown to have both anti-inflammatory and anti-apoptotic effects both *in vitro* and *in vivo*. In a mouse model, it was shown that CS was able to promote synovitis and stimulate release of interleukin (IL)-6, a pro-inflammatory cytokine [160-161]. Jomphe et al. 2008 showed that 200ug/mL of CS was able to reduce the sensitivity of rabbit chondrocytes to apoptosis [162].

When CS is absorbed, it can stimulate the synthesis of the structural components of cartilage ECM; namely, the synthesis of GAGs as well as the synthesis of hyaluronic acid, which aggregate together to reform the cartilage ECM [163]. While it remains unknown why CS supplementation can stimulate the reconstruction of the cartilage ECM, it is most likely that CS can bind the signaling molecules, allowing time for such molecules to interact with surface receptors. The result is that supplementation with CS preserves articular cartilage by inhibiting the changes seen in C4S and C6S concentrations associated with osteoarthritis [164].

It had been reported that oral CS supplements provide a a positive effect for patients with osteoarthritis (OA). Therefore, oral CS is a registered drug for knee OA in most European countries. However, a recent and growing body of evidence has suggested that the previous reports of oral CS may has been misleading, and that oral CS has less impact in treating OA than described.

Recent clinical studies and meta-reviews have shown that oral CS has no significant effect in reducing joint pain [164-167] resulting in growing criticism of the use of CS in the treatment of pain. Meanhwhile, there have been several reports that CS can slow the progression of the disease.

Clinical studies have shown that oral CS is able to slow the change in minimum knee joint space width [164, 166, 168]. Oral CS administration could slow the mean change in total JSW [164, 168]. These findings indicate that there may be some effectiveness in reducing the degradation of articular cartilage, while CS might not be effective in reducing joint pain. Thus, therapeutic efficacy of oral CS for osteoarthritis remains unanswered and further investigations will be required. CS is also used for other pathological areas, including ophthalmologic diseases [169].

3. Keratan Sulfate (KS)

3.1. Historical Review

KS was identified in 1939 in bovine cornea mucoid [170]. The structure of KS was characterized as a linear polymer composed of alternating D-galactose (Gal) and N-acetyl-D-glucosamine (GlcNAc) residues linked β1→4 and β1→3 respectively, and can be sulfated at the C6 position of both hexose moieties (Figure 6) [171]. Studies for KS from the 1970s to the 1990s shed light mainly on the structure and biosynthesis of KS, and were successively focused on its biochemical function and as tools for clinical biomarkers. Many analytical methods including *thin-layer chromatography (TLC), high-performance liquid chromatography (HPLC),* ELISA, and tandem mass spectrometry (MS/MS) for KS have been developed.

3.2. Classification and Structure

The structures of KS are divided into three groups (KSI, KSII, and KSIII) based on its linkage structure to core protein [172]. KSI is *N*-linked to specific asparagine amino acids via *N*-acetylglucosamine, distributed mainly in cornea, cartilage, and bone [173,174]. KSII is *O*-linked to specific serine or threonine amino acids via *N*-acetylgalactosamine especially in cartilage tissue [175]. Novel techniques developing for studying glycosylation reveal that *O*-linked KS is also a significant component of the ovarian cancer ascites with lumican, agrin, versican, and dystroglycans [176].

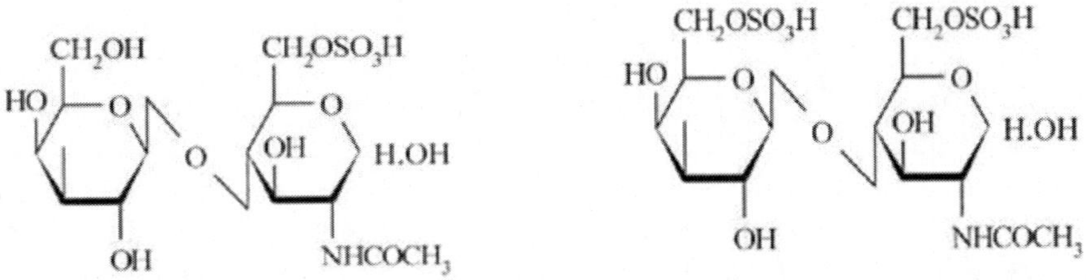

Gal β 1→4GlcNAc(6S) **Gal(6S) β 1→4GlcNAc(6S)**

Figure 6. Structure of keratan sulfate (KS).

A third type of KS (KSIII) has been isolated from brain tissue that is *O*-linked to specific serine or threonine amino acids via mannose [177]. The tissue-based classification of KS no longer exists since each type of KS has been shown to be non-tissue specific [172].

KS can be sulfated at the C6 of both hexose moieties. Since sulfation of C6 of GlcNAc proceeds simultaneously with enlongation of Gal to GlcNAc [178, 179], this moiety is generally sulfated, while C6 of Gal is sulfated to a variable extent, depending on the species, tissue, and age [180, 181]. The physiological functions of KS polymers depend on their molecular weight and sulfate pattern.

Sialic acid (*N*-acetylneuraminic acid, NeuAc) and fucose are contained in KS as minor structural components, and the expression patterns of these molecules are different depending upon tissues and age. In KSI of porcine corneal, the non-reducing ends of KS are capped with mainly neuraminic acid (70%), βGalNAc or αGal [182, 183]. The sialic acid residues extracted from cartilage KSII are linked to galactose as either α(2-3) or α(2-6) and occupy a non-reducing terminal position at C-3 or C-6 of GlcNAc [184]. The α-fucose residues are linked on C3 to sulfated GlcNAc within the main poly- N-acetyllactosamine repeat but not within four residues of the non-reducing terminus [185]. Meanwhile, KSII from tracheal cartilage is not fucosylated [175]. The function of sialic acid and fucose contained in KS is not understood well, although it is possible that these residues regulate degradation of KS. Sialic acid masks the galactose residues to which they are attached, preventing the removal of the glycoprotein from the circulation [186]. Sialidase treatment caused the disappearance of glycoprotein from circulation, suggesting that sialic acid chain caps, attached to KS, regulates the lifetime of the glycoprotein. The fucose residues showed an inhibition of cleavage in KS with keratanase and keratanase II, indicating that fucosylation reduces the sensitivity to glycosidases [187].

3.3. Tissue Distribution

KS is found mainly in cornea and cartilage. In the adult human corneal stroma, KS makes up 65% (by weight) of the total weight of GAGs [188]. Meanwhile, in the adult human eye, KS was detected in the sclera, but not retina and choroid by immunohistochemistry methods [189]. In 1986, Krusius first reported that the PGs in rat brain contain KS chains [177]. In brain, KS is distributed in thalamic nuclei and cortical areas to separate different cortical areas [190]. KS was identified in a wide range of human tissues from a 20-year-old autopsied case with MPS IVA including nucleus pulposus, vertebral disk, aorta, rib cartilage, cartilage, cornea, and heart valves [191]. KS is also detected in normal female genital tract, ectopic endometrium, lung, mammary gland, and thyroid; however, KS is not found in the digestive organs and urinary tract [192]. KS is the most abundant GAG in airway secretions and forms a major proportion of ECM in the heart valves [193, 194]. There are the differences of species- and sulfation- pattern in KS expression. The sulfation patterns of KS derived from bovine cornea and shark cartilage were confirmed using gel permeation chromatography (GPC)-HPLC. The proportion of mono-sulfated KS vs. di-sulfated KS in bovine cornea was 58% vs. 42%, while shark cartilage contained nearly 100% of di-sulfated KS [195]. Sulfation levels of KS in human blood and urine increases with age.

There has been no definite explanation as to why KS level and its sulfation level vary by age, species, and tissues.

3.4. Biosynthesis

KS is synthesized by glycosyltransferases and sulfotransferases (Figure 7) and is elongated via action of glycosyltransferases that alternately add Gal and GlcNAc to the growing polymer. The galactosyltranseferase activity in corneal cells due to a β-1,4-galactosyltranseferase enzyme (β4Gal-T1) which is a member of a family of at least seven β4Gal-T enzymes [196]. β4Gal-T1 is the most critical galactosyltranseferase for synthesis of KS chain, and β4Gal-T1 knockout mice show severe physiological problems and prenatal mortality [197]. Sulfation of the C-6 of GlcNAc is required for acceptor recognition by β4Gal-T1 prior to addition of Gal during extension of KS [198, 199].

The *N*-acetylglucosaminyltransferase (GnT) enzyme responsible for catalyzing addition of N-acetylglucosamine to Gal has not yet been clearly identified, although twoenzymes, i antigen-producing ß3Gn-T (iGnT) and β3GnTs (T2 to T7) that are active in the synthesis of linear polylactosamine, are candidate enzymes for KS synthesis [200-202]. ß1,3-*N*-acetylglucosaminyltransferase-7 (ß3Gn-T7) is the most efficient enzyme for catalyzing the *N*-acetylglucosaminyltransferase reaction in the biosynthesis of KS. ß3Gn-T2 and iGn-T also efficiently catalyze this transferase activity and are more efficient than ß3Gn-T3 to T6.

Sulfation in KS takes place at the C6-position of both Gal and GlcNAc. Gal sulfation in cornea and brain is accomplished by at least two sulfotransferase enzymes, chondroitin 6-sulfotransferase (C6ST) and keratan sulfate Gal-6-sulfotransferase (KS-Gal6ST), which add sulfate groups to the Gal of KS *in vitro* [203-207]. In C6ST knockout mice, 90% of sulfation of chondrotitin was lost; however, sulfation KS in spleen was unaffected, suggesting that C6ST is not responsible for sulfation of Gal during biosynthesis of KS *in vivo* [140]. KS-Gal6ST is expressed in brain and cornea and enhances KS synthesis [205]. Hoshino et al., reported that KS-Gal6ST is the primary enzyme responsible sulfation of Gal in KS in postnatal brain [208].

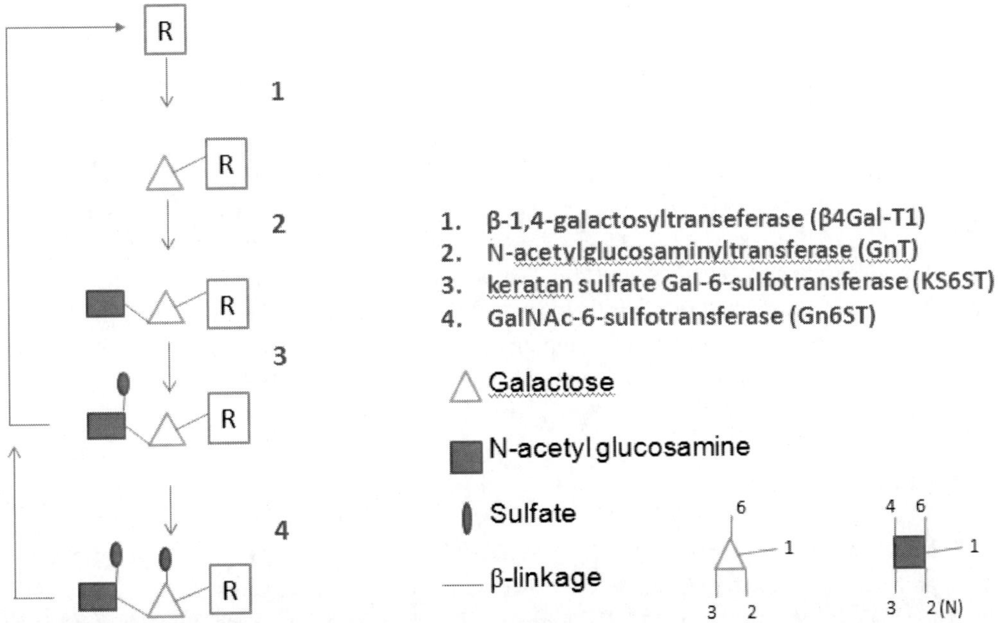

Figure 7. Biosynthesis of KS.

Nakazawa et al., reported that GlcNAc moieties are sulfated by a specific enzyme, GlcNAc-6-sulfotransferase (GlcNAc6ST) in keratocytes [209]. There are five human members of the GlcNAc6ST family, and four of them have murine orthologs. These enzymes show variations in substrate specificity and different tissue distributions [140, 210]. Mutation in the carbohydrate sulfotransferase gene, CHST6 (GlcNAc6ST5/c-GlcNAc6ST) causes macular corneal dystrophy [211] (see below). UDP-*N*-acetylglucosamine transporter (NGT) is also the key factor for the biosynthesis of KS since NGT may specifically supply ß3Gn-T7 to KS with UDP-GlcNAc [212]. Biosynthesis of KS is markedly altered in response to metabolic, pathologic, or developmental changes in tissues. In chicken cornea, KS is synthesized between embryonic day 5 (E5) and E7, and becomes highly sulfated between E12 and E15 [213, 214]. Liles et al. reported that after E15 highly sulfated KS serves to fine tune local matrix hydration and collagen fibril spacing during corneal growth [180]. KS in cartilage undergoes an age-dependent increase in KS chain length and sulfation of Gal [181, 215-217]. KS is present in small amounts in fetal rib cartilage, but levels increase with age [218]. Capping of KS with NeuAc and fucose also show age-dependent alterations, being absent in fetal articular cartilage and increasing with age [181, 217].

In vitro experiments on keratocyte and chondrocyte are limited by major factors of reduced KS biosynthesis, resulting in change of cell morphology like fibroblasts. Advanced development of a culture method for keratocytes with fibroblast growth factor (FGF) maintains biosynthesis of macromolecular, fully sulfated KS during the extended period *in vitro* [219]. KS levels in chondrocytes are maintained when these cells are cultured in 3D alginate beads [220]. It is more difficult to culture neuronal cells as most KS in the brain exhibits a unique, complex, developmentally-regulated pattern, suggesting a specialized developmental function for KS-linked molecules in the CNS [190]. HS regulates signal molecules during CNS development, and KS may play a similar role in the brain. KS expression in brain increases during spinal cord injury as mentioned below.

3.5. Degradation

Several enzymes regulate the catabolism of KS. Degradation of the KS chains occurs in a directional manner by removal or processing of the terminal sugar on the non-reducing end (NRE) of the GAG chain [122, 221] (Figure 8). *N*-acetylglucosamine 6-sulfatase, β-*N*-acetyl hexosaminidase A or S, GALNS, and β-galactosidase are involved in the degradation of KS. Mutations in these enzymes result in lysosomal storage of undegraded KS; the deficiency of GALNS and β-galactosidase causes mucopolysaccharidosis IVA and IVB (MPS IVA, Morquio A syndrome; MPS IVB, Morquio B syndrome), respectively (Table 2).

3.6. Proteoglycan

KS is attached to a number of core proteins: lumican, keratocan, mimecan, fibromodulin, aggrecan, PRELP, osteoadherin, PZP3, MUC1, CD44, cytokeratins, ABAKAN, SV2, phosphocan KS, claustrin, and PG-1000 (Table 7). Expression of these core proteins is an important determinant of KS biosynthesis [172]. The PGs associated with KS is shown in Table 7.

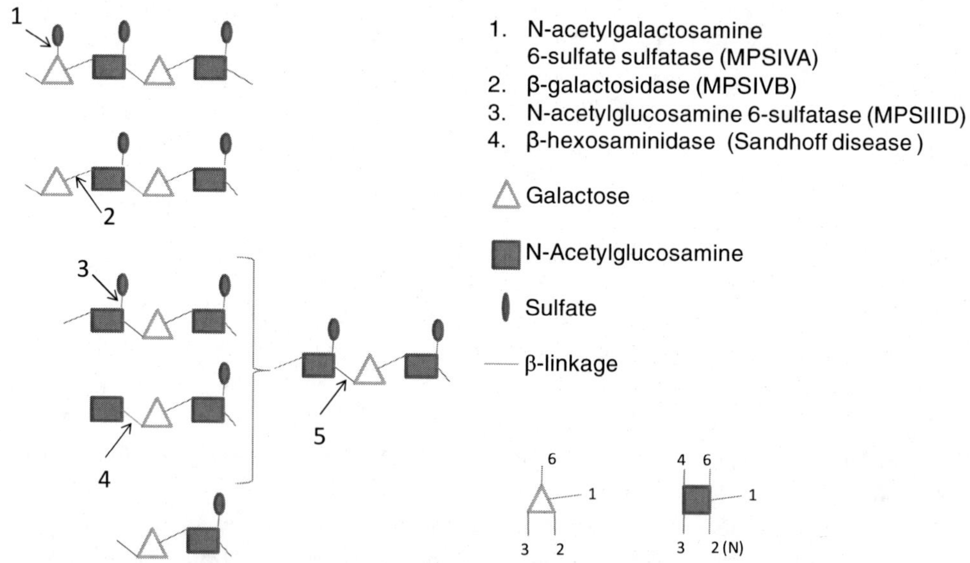

Figure 8. Degradation pathway of KS.

3.6.1. SLRPs

Small leucine-rich proteoglycans (SLRPs), consisting with a series of 24-amino acid leucine-rich repeats that make up the central portion, are the main core proteins for KS in lumican, keratocan, mimecan, fibromodulin, and PRELP [216-219, 227].

These proteins are typically the components of interstitial matrix. In cornea, KS is present on three SLRP proteins, lumican, keratocan, and mimecan [172, 223, 224]. Fibromodulin and PRELP are distributed in cartilage, and osteoadherin is in bone as SLRPs-family [172, 227]. In fibromodulin the KS chain is capped by α(2-3)-linked sialic acid [238].

Table 7. Tissue distribution of KS proteoglycans [172]

Protein/gene	KS-protein linkage	Tissue/ cell location	Reference
Lumican	KSI	Cornea, interstitial collagenous matrices	222
Keratocan	KSI	Cornea, sclera	223
Mimecan	KSI	Cornea	224
Fibromodulin	KSI	Cartilage	225
PRELP	KSI	Cartilage	227
Osteoadherin	KSI	Bone	174
Aggrecan	KSI and KSII	Cartilage	226
PZP3	KSI	Zoma pellucida	228
MUC1	NA	Endometrial epithelium	229
CD44	NA	Carcinoma line KM12L4	230
	NA	Epidermis	231
Cytokeratins	NA	Epidermal keratinocytes	232
ABAKAN	NA	Brain	233
SV2	NA	Synaptic vesicles	234
Phosphocan KS	KSIII	Brain	235
Claustrin KS	NA	Brain	236
PG-1000	NA	Electric organ	237

NA: not applicable.

3.6.2. Aggrecan

The major PG in cartilage comprises of the very large protein, aggrecan ($1x10^6 < M < 3x10^6$). KS is linked to aggrecan between the G2 and G3 regions of the protein [239], and the N-terminus of the protein contains a HA-binding domain [226]. The KS attachment sequence motif is highly preserved in different vertebrate species, but the numbers of repeated units varies, accounting for the differences in KS content of aggrecan in different species. In rodents, this motif sequence is not conserved, and in these species skeletal KS is greatly reduced or absent, compared to humans and other animals.

3.6.3. Cell Associated Proteoglycan

The endometrial protein MUC1, modified with KS [229], is a common component of the mucin layer associated with the apical surfaces of secretory epithelia. CD44 and SV2, other cell-surface molecules with KS, are integral membrane proteins [230, 240]. These GAGs are not restricted to interstitial connective tissues but modify various proteins with considerable diversity in location.

3.6.5.4. Brain Proteoglycan

The KS modified PGs ABAKAN [233], phosphacan-KS [235], SV2 [240], claustrin [236], and, PG-1000 [237] and are uniquely distributed in a limited populations of cells in the CNS.

3.7. Physiological Roles

3.7.1. Tissue Hydration

KS is a large, highly hydrated molecule which, in joints, can act as a cushion to absorb mechanical shock. Tissue hydration is a critical function of KS (and DS) in cornea to maintain corneal transparency [241]. Null mice for lumican show 20% reduction in KS and exhibit a loss in stromal hydration [242], suggesting that the corneal transparency depends on the presence of abundant KS. Aggrecan is a major component in ECM of cartilage that forms high molecular weight aggregates. The KS of aggrecan enables hydration of cartilage, supporting its weight-bearing mechanical properties.

3.7.2. Cell Biological Function

KS has similar cell biological function as other GAGs (hyaluronan, DS, and HS) and depends on its tissue location.

Linkage of KS to lumican diminishes the ability of the protein to contribute to cell migration and proliferation in embryonic development, tissue repair, and inflammatory response [243-247]. Macrophages contain a high-affinity cell surface receptor for lumican protein but do not bind when the KS is attached to the protein]. These characteristics are found in 1) the inhibition of direct axon growth during neuro-regeneration and development [236, 249] and 2) the suppression of osteoarthritis though blocking development of an immune response *in vivo* and *in vitro* with removal of KS with keratanase [250]. Lumican also promotes intestinal homeostasis by adding innate immune responses that are beneficial at the early stage of colitis.

The physiological functions of KS are regulated by the glycosylation and sulfation pattern, suggesting that molecular weight and sulfation level of KS are critical for physiological function, but not total KS levels.

Lumican with KS attached inhibits angiogenesis both *in vivo* and *in vitro* by interfering with α2 integrin receptor activity and down-regulating expression of matrix metalloproteinases [251]. The 6-*O*-sulfation of GlcNAc residues in KS reduces radiation-induced apoptosis of human Burkitt's lymphoma cells [252].

3.8. Pathological Conditions

3.8.1. Mucopolysaccharidoses

Mucopolysaccharidosis IV includes two subtypes, mucopolysaccharidosis IVA and IVB (MPS IVA, Morquio A syndrome, MPS IVB, Morquio B syndrome), caused by the deficiency of GALNS and β-galactosidase (β-Gal), respectively (Table 2). Either enzyme deficiency results in accumulation of undegraded KS especially in cells and extracellular matrix (ECM) of cartilage and cornea. In MPS IVA CS also accumulates. Patients with a classic (severe) form of MPS IVA have a unique systemic skeletal dysplasia including short trunk dwarfism, kyphoscoliosis, coxa valga, odontoid hypoplasia, abnormal gait, joint mobility problems, restriction of chest wall movement (Figure 9), and a life span of 20–30 years if an appropriate orthopedic intervention is unavailable. Patients with an attenuated form have a milder skeletal involvement, and have an almost normal life span [91, 253-263]. Postmortem tissues from a 20-year-old male case with MPS IVA demonstrated 1) systemic storage materials of CS and KS (vacuoles) in multiple tissues other than cartilage, 2) severely vacuolated and ballooned chondrocytes in trachea, humerus, vertebrae, and thyroid cartilage with disorganized extracellular matrix and poor ossification (Figure 10), 3) appearance of foam cells and macrophages in lung, aorta, heart valves, heart muscle, trachea, visceral organs, and bone marrow (Figure 10), and 4) storage of chondrotin-6-sulfate (C6S) in aorta.

Patients with MPS IVB show a milder phenotype of skeletal dysplasia than patients with a severe form of MPS IVA. Consequently, clinical features of MPS IVB are similar to those in patients with an attenuated form of MPS IVA. We reported that KS levels in blood and urine in patients with MPS IVA are associated with age and clinical severity [264], and decrease with enzyme replacement therapy (ERT) in a mouse model and bone marrow transplantation (BMT) in a patient with MPS IVA [265, 266]. Limitations of the current results are: 1) the mouse model does not have severe clinical features in bone, and therefore, we cannot assume that improvements in the mouse model will predict clinical improvement ly in human patients, and 2) pathological features in hyaline and fibrous cartilage in MPS IVA mice were not resolved despite ERT since birth. Thus additional unique strategies may be needed to improve impact on bone lesions, such as high dose administration, bone targeting, or infusion of long-circulating forms of enzyme. Since the bone lesion is already seen clinically in a newborn patient with MPS IVA and storage materials are observed in chondrocytes of an affected fetus and newborn MPS IVA mice, reversal of pathology and clinical improvement in bone lesions will need to be assessed carefully in future experiments and clinical trials.

Nevertheless, preclinical ERT studies with newborn and mature MPS IVA mice do show that improvements in bone pathology are better if ERT begins earlier [268].

ERT is available clinically for patients with MPS IVA. Patients with MPS IVA treated with ERT showed significant improvement in a 6-minute walk test than the group treated with placebo [267]. However, early diagnosis of newborns and early treatment should be recommended to ameliorate or remit devastating skeletal dysplasia.

3.8.2. Macular Corneal Dystrophy (MCD)

Macular corneal dystrophy (MCD is an autosomal recessive inherited disorder. The onset usually occurs in the first decade of life, and MCD patients have progressive development of corneal opacities, resulting in bilateral loss of vision, eventually need for corneal transplantation.

The defective gene in this disorder is carbohydrate sulfotransferase 6 (*CHST6*) [211]. The *CHST6* gene is a member of the GST-family (galactose/*N*-acetylgalactosamine/*N*-acetylglucosamine 6-*O*-sulfotransferases) that includes a group of Golgi enzymes that transfer sulfate to the 6-hydroxyl group of galactose, *N*-acetylgalactosamine (GalNac), glucose, or *N*-acetylglucosamine (GlcNAc) in nascent glycoproteins. The *CHST6* gene encodes a corneal *N*-acetylglucosamine-6-*O*-sulfotransferase (c-GlcNAc6ST) that initiates sulfation of KS chains on proteoglycans (PG).

MCD is classified into three immunophenotypes, MCD types I, IA, and II, based on the reactivity of serum and corneal tissue to an antibody against sulfated KS [269, 270]. In MCD type I, sulfated KS is undetectable in both serum and cornea. In MCD IA, KS is not detected from serum and extracellular stromal matrix, but is detected in keratocytes.

In MCD type II, KS is present at normal or subnormal levels in serum, and is eminent immunohistochemically in the cornea. All three subtypes of MCD have clinically indistinguishable phenotypes. Therefore, the immunophenotypes cannot be distinguished from each other clinically and are of no clinical significance.

MCD patients accumulate the deposits under the epithelium, within the corneal stroma, Descemet's membrane, and the corneal endothelium. These abnormal deposits are associated with a central stromal haze that gradually extends to the periphery of the cornea, leading to visual impairment. Previous biochemical studies have indicated that the pathologic corneal deposits contain accumulated GAGs. The histopathology of MCD shows a characteristic intracytoplasmic accumulation of GAGs (vacuoles) within keratocytes and the corneal endothelium [269].

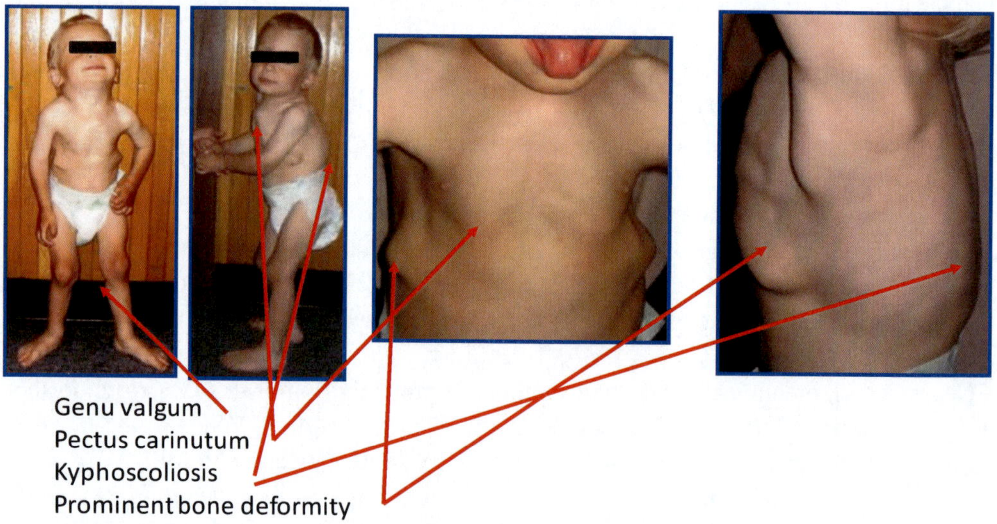

Figure 9. Clinical features of MPS IVA patient (copyright permission from International Morquio Organization and Carol Ann Foundation). A 3-year-old patient displays skeletal dysplasia such as short stature, pectus carinatum, kyphoscoliosis, genu valgum, prominent forehead, and abnormal gait; height 85 cm, 25 percentile of male Morquio A growth chart [259]; body weight 10 kg.

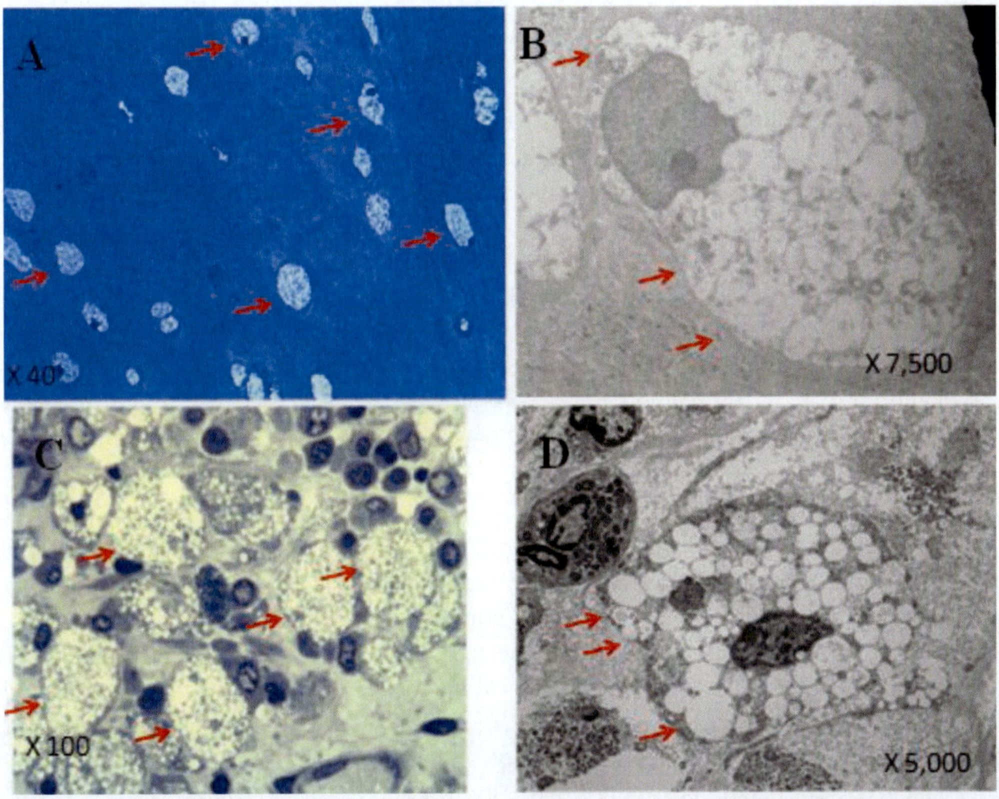

Figure 10. Pathology of chondrocytes and bone marrow from autopsy of a 20-year-old male with MPS IVA [191]. Vacuolated chondrocytes and foam cells appear in humerus and bone marrow. A) All chondrocytes in humerus are ballooned and contain numerous vacuoles (storage materials) (LM: x40); B) The chondrocyte in humerus shows numerous vacuoles in cytoplasm (EM: x7,500); C) Vacuolated foam cells in bone marrow are ballooned and enlarged (LM: x100); D) The vacuolated foam cell in bone marrow include lysosomal storage material in cytoplasm (EM: x5,000).

In MCD, vision can be restored by corneal grafting; however, the disease may recur in the graft after many years. Since the pathological status provides am impact on the entire corneal stroma, Descemet membrane, and the corneal endothelium, a lamellar keratoplasty will not excise all of the pathologic tissue.

3.8.3. Tumors

There are several reports that expression of KS in tumor tissues is regulated. Serum KS levels increased in pancreatic, breast, colorectal, and cervical cancers [271-273]. Lumican (with KS) inhibits cancer cell progression and migration, showing an anti-cancer function of this PG [274]. Lumican has been proposed as a valuable marker in prostate cancer staging [272]. Although KS levels are increased in several carcinomas, they are absent from carcinomas of the gastrointestinal and urinary tracts [192].

3.8.4. Osteoarthritis

Osteoarthritis is a joint disease caused by chronic inflammation and presents with musculoskeletal pain and disability in the elderly western population [275]. Patients with osteoarthritis have significantly higher levels of KS in serum and synovial fluid of the affected joint [276-277]. Serum KS was associated with progression of osteoarthritis [277] and correlated with the damage of the articular cartilage. KS level significantly increased even at an early stage after the injury [278]. As reported by Okumura et al. [279], early OA articular cartilage destruction begins with a loss of glycosaminoglycans (GAGs) from articular cartilage surfaces, followed by collagenolysis. However, other reports indicated no change of KS levels in the blood or synovial fluid of patients with osteoarthritis [280]. Since KS is invariably measured by ELISA methods that use monoclonal antibodies that recognize specific sulfation epitopes, the conflicting results might be caused by structural features of KS that vary in levels of sulfation and molecular size [181]. The current sandwich ELISA method only measures di-sulfated KS, Galβ1(6S)→4GlcNAc(6S), since the monoclonal antibody (Seikagaku, Tokyo, Japan) recognizes specifically this sulfated molecule [281-284]. Consequently this analytical method does not provide an accurate quantification of total KS. Serum KS level was also significantly higher in degenerative lumbar scoliosis [285].

3.8.5. Spinal Cord Injury

KS is a key factor in the development of neurons. KS is part of the molecular identity of thalamic nuclei and cortical domains [190]. Elevation of KS expression is related to spinal cord injury, unilateral nigrostriatal axotomy, amyotrophic lateral sclerosis [286-289], and recovery of axonal regeneration in GlcNAc6ST1 knock-out mice [290, 291] and keratanase II treated mice [249]. KS inhibits elongation of neural axons [249]. KS expression is especially enhanced in microglia/macrophages and ECM after spinal cord injury [249, 290], and amyotrophic lateral sclerosis [289]. KS plays a critical role and provides an impact comparable to that of CS, on post-injury plasticity. KS and CS are independent requirements for the proteoglycan-mediated inhibition of axonal regeneration/sprouting [249].

There is an opposite phenomenon that KS in microglia decreases in the spinal cord in experimental autoimmune neuritis [292-294]. The difference of KS regulation in CNS injury depends upon inflammatory milieu in the CNS [294]. Cerebral KS is reduced as a result of Alzheimer's disease, and the lack of highly sulfated KS structures may reflect a specific functional defect of the cells [295].

3.8.6. Diabetes and Obesity

KS has some relevance in lifestyle-related diseases. Ahn et al., reported that lumican, vasorin, and retinol binding protein-4 increase in diabetic nephropathy [296]. In obesity, degradative products of KS increased, compared with lean subjects [297]. There is a report that KS is expressed from stem cells in human subcutaneous adipose tissue and its expression is reduced during weight loss after exercise and diet [298].

3.8.7. Heart Disease

Lumican in heart increases in experimental and clinical heart failure. Cardiac fibroblasts produce lumican in response to inflammation and mechanical stimuli, and increased lumican regulates important factors that promote cardiac remodeling and fibrosis [299].

3.8.8. Lung Disease

Lumican regulates high tidal volume (Vt) mechanical stretch-induced lung damage through extracelluar signal-regulated kinase pathway [300].

3.8.9. Liver Disease

Lumican acts as a key factor for hepatic fibrosis. There are reports that lumican is upregulated with increasing severity of nonalcoholic steatohepatitis (NASH) [301, 302]. In lumican null mice, hepatic fibrosis was greatly reduced, and matrix turnover and proliferative response increased [302].

3.9. Therapy

Recent studies have shown the benefits of KS for several disorders. Shirato et al., showed that di-sulfated KS (L4) suppressed airway inflammatory responses by bacteria infections through inhibition of EGFR phosphorylation. KS is a candidate for cure of chronic obstructive pulmonary disease (COPD) [303].

Lumican is expressed by stromal keratocytes in normal unwounded cornea and promotes corneal epithelial wound healing, suggesting that KS may be useful method for therapies in epithelial defects of the cornea [304].

Similar to CS, KS could offer therapeutic efficacies for rheumatoid arthritis through suppression of cartilage degradation, suppression of joint damage and chondroprotection [305, 306].

4. Assay for KS and CS

The most commonly used methods for diagnosis of MPS are dye-spectrometric methods such as dimethylmethylene blue (DMB) [307-309] and alcian blue [310-312] on urine samples. The direct DMB method is automated for detection of MPS diseases [309]. However, these methods cannot be applied to blood specimens without prior protease, nuclease or hyaluronidase digestion and are impractical for screening newborn urine since collection is difficult and storage of urine samples in large numbers is costly. These spectrometric assays are also not adequate for screening: 1) proteins in the blood specimen block binding of the dyes to GAG, 2) the staining dyes are prone to decomposition leading to a low signal to noise ratio, and 3) the dye methods cannot predict MPS subtypes.

Thin layer chromatography, which is a semi-quantitative method for individual GAG including KS, is used to enable preliminary classification of patients among subgroups to reduce the number of enzyme assays necessary for final diagnosis [310, 311]. HPLC is a sensitive and specific method for diagnosis of MPS but not appropriate for mass screening since it is time-consuming and expensive [312-314]. Sandwich ELISA methods to measure specific GAGs such as HS and KS were devised for MPS detection [315-317]; however, ELISA does not cover all subtypes of each GAG including degree or position of sulfation.

We have established a quick, reproducible, sensitive, and specific assay technique to measure KS, HS, DS, and CS in blood, urine, dried blood spot (DBS), and tissue specimens using liquid chromatography tandem mass spectrometry (LC-MS/MS) [195, 256, 264, 318-323].

To digest "polymer" CS and KS to disaccharides, chondroitinase C (or alternatively, chondroitinase ABC) and keratanase II were used (Seikagaku Co, Tokyo, Japan).

Chondroitinase A produces disaccharides of C4S, while chondroitinase B and chondroitinase C produce disaccharides of DS and C6S, respectively (Figure 11).

In the C6S assay, we demonstrated 1) that the level of blood C6S in control subjects is age-dependent, being highest in newborns and declining with age, 2) that C6S levels in patients with MPS IVA and VII are significantly higher than that in age-matched control subjects, 3) that a combination of C6S and KS levels distinguishes patients with MPS IVA and control subjects more clearly than either C6S or KS level alone, and 4) that blood C6S and KS levels in patients with MPS IVB overlap with those in age-matched control subjects [323].

In the KS assay, we demonstrated 1) that the levels of blood and urine mono-sulfated and di-sulfated KS in control subjects are age-dependent, and decline with age, 2) that blood and urine di-sulfate KS levels in patients with MPS II and MPS IVA are significantly higher than those in age-matched control subjects, 3) that the level of blood and urine di-sulfated KS distinguishes control subjects and patients with MPS IVA more clearly than that of mono-sulfated KS, 4) that the proportion of blood di-sulfated KS increases with age and distinguishes young patients with MPS IVA from control subjects more clearly than the level of the individual sulfated KS, and 5) that levels of mono- and di- sulfated KS in urine provide a clearer differentiation between patients with MPS IVA and MPS IVB and age-matched controls than sulfated KS in blood.

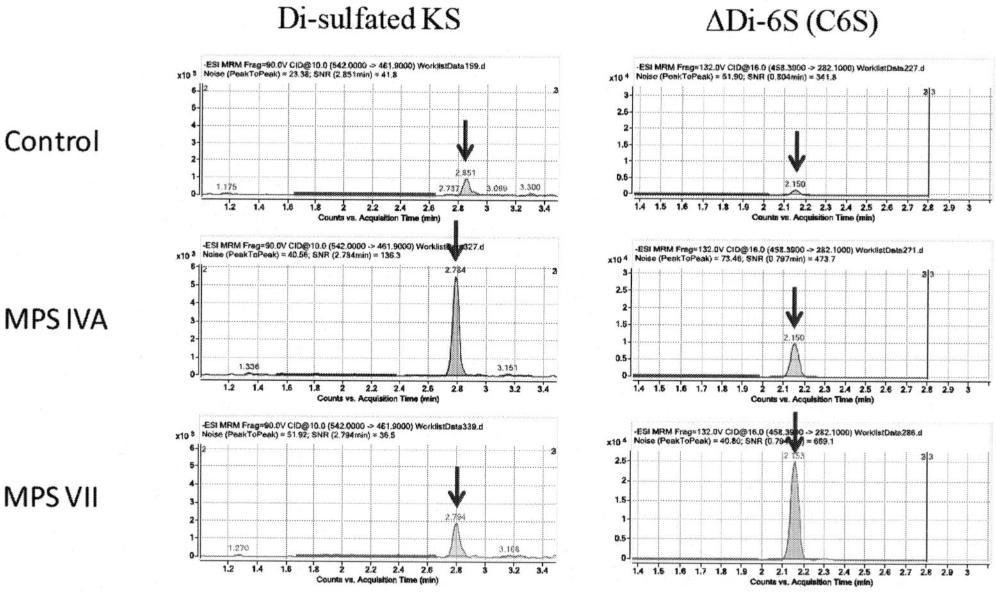

Figure 11. MRMs of di-sulfated KS and ΔDi-6S (C6S) in patients with MPS IVA and MPS VII.

In addition, total KS and levels of sulfation vary with different species and tissues [195]. Thus CS and KS levels in blood and/or urine should be a suitable biomarker for early diagnosis, screening, clinical severity, and therapeutic efficacy in not only MPS IVA, but also other type of MPS.

This CS and KS assay should be widely applied to a variety of biological specimens including serum/plasma, urine, *cerebrospinal fluid*, homogenated cells and tissues, and dried blood spot to understand physiological and pathological condition.

CONCLUSION

CS and KS play important physiological roles especially in connective tissue, cartilage and cornea. CS and/or KS accumulate in tissues due to deficiency of degradative enzymes and lead to MPS, while CS and KS are used for supplemental therapies for several common disorders. Newly developed analytical method for CS and KS by MS/MS can be used for prognosis, diagnosis, monitoring, and screening for MPS and other related disorders.

ACKNOWLEDGMENT

This work was supported by grants from the Austrian MPS Society, International Morquio Organization (Carol Ann Foundation) and Delaware Bioscience Center for Advanced Technology. This work was also supported by Japanese MPS Family Society. R.W.M. and S.T. were supported by an Institutional Development Award (IDeA) from the National Institute of General Medical Sciences of NIH under grant number P20GM103464. S.T. and A.M were supported by National Institutes of Health grant 1R01HD065767-02. The content of the article has not been influenced by the sponsors. F.K. was supported by Conselho Nacional de Desenvolvimento Científico e Tecnológico from Brazil (CNPq). Editorial assistance to the manuscript was kindly provided by Saori C. Tomatsu.

Conflicts of Interest

All the authors have contributed to this "Chapter" and have no conflict of interest with any other party.

REFERENCES

[1] Krukenberg, C. F. W. 1884. Die chemischen Bestandteile des Knorpels. *Z Biol.*, 20:307.

[2] Roseman, S. 2001. Reflections on glycobiology. *J. Biol. Chem.*, 276(45): 41527–41542.

[3] Levene P. A., La Forge F. B. 1915. On Chondrosamine. *Proc. Natl. Acad. Sci. USA*, 1(4):190-191.

[4] Mikami, T., and Kitagawa, H. 2013. Biosynthesis *Biochim. Biophys. Acta*, 1830(10):4719-4733.

[5] Lamari, F. N., and Karamanos, N. K. 2006. Structure of chondroitin sulfate. In Advances in Pharmacology. Chondroitin Sulfate: Structure, Role and Pharmacological Activity, Vol. 53, Volpi N (ed.). Elsevier: Oxford, 2005.

[6] Prydz, K., and Dalen, K. T. 2000. Synthesis and sorting of proteoglycans. *J. Cell Sci.*, 113(Pt 2):193-205.

[7] Lawrence, R., Brown, J. R., Lorey, F., Dickson, P. I., Crawford, B. E., and Esko, J. D. 2014. Glycan-based biomarkers for mucopolysaccharidoses. *Mol. Genet. Metab.*, 111(2):73-83.

[8] Hofinger, E. S., Bernhardt, G., and Buschauer, A. 2007. Kinetics of Hyal-1 and PH-20 hyaluronidases: comparison of minimal substrates *Glycobiology,* 17(9):963-971.

[9] Meyer, K., and Rapport, M. M. 1950. The hydrolysis of chondroitin sulfate by testicular hyaluronidase. *Arch. Biochem.*, 27(2):287-293.

[10] He, D., Xie, Q., Peng, H., Wei, W., Nie, L., and Yao, S. 2001. Use of a quartz crystal impedance technique to monitor the degradation of chondroitin sulfate by hyaluronidase. *Enzyme Microb. Technol.,* 29(1):84-89.

[11] Kaneiwa, T., Mizumoto, S., Sugahara, K., and Yamada, S. 2010. Identification of human hyaluronidase-4 as a novel chondroitin sulfate hydrolase that preferentially cleaves the galactosaminidic linkage in the trisulfated tetrasaccharide sequence. *Glycobiology,* 20(3):300–309.

[12] Meyer, K., Hoffman, P., and Linker, A. 1960. Hyaluronidases. In: Boyer, P.D., Lardy, H. and Myrback, K. (Eds.), *The Enzymes*, 2nd ed., vol. 4, Academic Press, New York.

[13] Rapport, M. M., Meyer, K., and Linker, A. 1951. Correlation of reductimetric and turbidimetric methods for hyaluronidase assay. *J. Biol. Chem.,* 186(2):615-623.

[14] Weissman, B. 1955. The transglycosylative action of testicular hyaluronidase. *J. Biol. Chem.*, 216(2):783-794.

[15] Stern, R. 2003. Devising a pathway for hyaluronan catabolism: are we there yet? *Glycobiology,* 13(12):105R-115R.

[16] Glössl, J., and Kresse, H. 1982. Impaired degradation of keratan sulphate by Morquio A fibroblasts. *Biochem. J.,* 203(1):335-338.

[17] Peters, C., Schmidt, B., Rommerskirch, W., Rupp, K., Zühlsdorf, M., Vingron, M., Meyer, H. E., Pohlmann, R., and von Figura, K. 1990. Phylogenetic conservation of arylsulfatases. cDNA cloning and expression of human arylsulfatase B. *J. Biol. Chem.*, 265(6):3374-3381.

[18] Schuchman, E. H., Jackson, C. E., and Desnick, R. J. 1990. Human arylsulfatase B: MOPAC cloning, nucleotide sequence of a full-length cDNA, and regions of amino acid identity with arylsulfatases A and C. *Genomics,* 6(1):149-158.

[19] Shaklee, P. N., Glaser, J. H., and Conrad, H. E. 1985. A sulfatase specific for glucuronic acid 2-sulfate residues in glycosaminoglycans. *J. Biol. Chem.,* 260(16):9146-9149.

[20] Freeman, C., and Hopwood, J. J. 1989. Human liver glucuronate 2-sulphatase. Purification, characterization and catalytic properties. *Biochem. J.,* 259(1):209-216.

[21] Esko, J. D., Kimata, K., and Lindahl, U. 2009. Proteoglycans and Sulfated Glycosaminoglycans. In: Varki A., Cummings R. D., Esko J. D., et al., editors. Essentials of Glycobiology. 2nd edition. Cold Spring Harbor (NY): Cold Spring Harbor Laboratory Press; Chapter 41.

[22] Tse, R., Vavougios, G., Hou, Y., and Mahuran, D. J. 1996. Identification of an active acidic residue in the catalytic site of beta-hexosaminidase. *Biochemistry*, 35(23):7599-7607.

[23] Hepbildikler, S. T., Sandhoff, R., Kolzer, M., Proia, R. L., and Sandhoff, K. 2002. Physiological substrates for human lysosomal beta -hexosaminidase S. *J. Biol. Chem.*, 277(4):2562-2572.

[24] Esko, J. D., Kimata, K., and Lindahl, U. 2009. Proteoglycans and Sulfated Glycosaminoglycans. In: Varki A., Cummings R. D., Esko J. D., et al., editors. Essentials of Glycobiology. 2nd edition. Cold Spring Harbor (NY): Cold Spring Harbor Laboratory Press; Chapter 16.

[25] Yamaguchi, Y. 2000. Lecticans: organizers of the brain extracellular matrix. *Cell Mol. Life Sci.*, 57(2):276-289.

[26] Ameye, L., and Young, M. F. 2002. Mice deficient in small leucine-rich proteoglycans: novel in vivo models for osteoporosis, osteoarthritis, Ehlers-Danlos syndrome, muscular dystrophy, and corneal diseases. *Glycobiology,* 12(9):107R-116R.

[27] Iozzo, R. V. 1999. The biology of the small leucine-rich proteoglycans. Functional network of interactive proteins. *J. Biol. Chem.,* 274(27):18843-18846.

[28] Hocking, A. M., Shinomura, T., and McQuillan, D. J. 1998. Leucine-rich repeat glycoproteins of the extracellular matrix. *Matrix Biol.,* 17(1):1-19.

[29] Chopra, R. K., Pearson, C. H., Pringle, G. A., Fackre, D. S., and Scott, P. G. 1985. Dermatan sulphate is located on serine-4 of bovine skin proteodermatan sulphate. Demonstration that most molecules possess only one glycosaminoglycan chain and comparison of amino acid sequences around glycosylation sites in different proteoglycans. *Biochem. J.,* 232:277–279.

[30] Lindahl, U. and Roden, L. 1966. The chondroitin 4-sulfate-protein linkage. *J. Biol. Chem.*, 241(9):2113–2119.

[31] Stern, E. L., Lindahl, B., and Roden, L. 1971. The linkage of dermatan sulfate to protein. II. Monosaccharide sequence of the linkage region. *J. Biol. Chem.*, 246(18):5707–5715.

[32] Silbert, J. E., and Sugumaran, G. 2002. Biosynthesis of chondroitin/ dermatan sulfate. *IUBMB Life,* 54(4):177-186.

[33] Richmond, M. E., DeLuca, S., and Silbert, J. E. 1973. Biosynthesis of chondroitin sulfate. Microsomal acceptors of sulfate, glucuronic acid and N- acetylgalactosamine. *Biochemistry*, 12(20):3898-3903.

[34] Rohrmann, K., Niemann, R., and Buddecke, E. 1985. Two N-acetylgalactosaminyl transferases are involved in the biosynthesis of chondroitin sulfate. *Eur. J. Biochem.*, 148:463-469.

[35] Sugumaran, G., Katsman, M., Sunthankar, P. and Drake, R. R. 1997. Biosynthesis of chondroitin sulfate. Purification of glucuronosyl transferase II and use of photoaffinity labeling for characterization of the enzyme as an 80-kDa protein. *J. Biol. Chem.*, 272(Pt 1):14399-14403.

[36] Nadanaka, S., Kitagawa, H., Goto, F., Tamura, J. I., Neumann, K. W., Ogawa, T., and Sugahara, K. 1999. Involvement of the core protein in the first β-N-acetylgalactosamine transfer to the glycosaminoglycan-protein linkage-region tetrasaccharide and in the subsequent polymerization: the critical determining step for chondroitin sulphate biosynthesis. *Biochem. J.*, 340:353-357.

[37] Maeda, N., Fukazawa, N., and Ishii, M. 2010. Chondroitin sulfate proteoglycans in neural development and plasticity. *Front. Biosci.,* 15:626-644.

[38] Silbert, J. E. 1978. Biosynthesis of chondroitin sulfate: chain termination. *J. Biol. Chem.*, 253(19):6888–6892.

[39] Otsu, K., Inoue, H., Tsuzuki, Y., Yonekura, H., Nakanishi, Y., and Suzuki, S. 1985. A distinct terminal structure in newly synthesized chondroitin sulphate chains. *Biochem. J.*, 227(1):37–48.

[40] Handley, C. J., Samiric, T., Ilic, M. Z. 2006. Structure, metabolism, and tissue roles of chondroitin sulfate proteoglycans. *Adv. Pharmacol.,* 53:219-232.

[41] Bayliss, M. T., Osborne, D., Woodhouse, S., and Davidson, C. 1999. Sulfation of chondroitin sulfate in human articular cartilage. The effect of age, topographical position, and zone of cartilage on tissue composition. *J. Biol. Chem.*, 274(22):15892-15900.

[42] Bathe, M., Rutledge, G. C., Grodzinsky, A. J., and Tidor, B. 2005. Osmotic Pressure of Aqueous Chondroitin Sulfate Solution: A Molecular Modeling Investigation. *Biophys. J.,* 89(4):2357–2371.

[43] Heimer, R., Sporer, R., Molinaro, L., Hansen, L., and Laposata, E. 1992. Normal human synovial fluid and articular cartilage contain similar intact proteoglycans. *Lab. Invest.,* 66(6):701-707.

[44] Fisher, L. W., Termine, J. D., and Young, M. F. 1989. Deduced protein sequence of bone small proteoglycan I (biglycan) shows homology with proteoglycan II (decorin) and several non-connective tissue proteins in a variety of species. *J. Biol. Chem.*, 264(8):4571–4576.

[45] Fine, J. D. and Couchman, J. R. 1988. Chondroitin-6-sulfate-containing proteoglycan: a new component of human skin dermoepidermal junction. *J. Invest. Dermatol.*, 90(3):283-288.

[46] Carrino, D. A., Calabro, A., Darr, A. B., Dours-Zimmermann, M. T., Sandy, J. D., Zimmermann, D. R., Sorrell, J. M., Hascall, V. C., and Caplan, A. I. 2011. Age-related differences in human skin proteoglycans. *Glycobiology*, 21(2):257-268.

[47] Seidler, D. G. 2012. The galactosaminoglycan-containing decorin and its impact on diseases. *Curr. Opin. Struct. Biol.*, 22(5):578–582.

[48] Carrino, D. A., Mesiano, S., Barker, N. M., Hurd, W. W., and Caplan, A. I. 2012. Proteoglycans of uterine fibroids and keloid scars: similarity in their proteoglycan composition. *Biochem. J.,* 443(2):361–368.

[49] Fleischmajer, R., Fisher, L. W., MacDonald, E. D., Jacobs, L. Jr., Perlish, J. S., and Termine, J. D. 1991. Decorin interacts with fibrillar collagen of embryonic and adult human skin. *J. Struct. Biol.*, 106(1):82-90.

[50] Hasegawa, K., Yoneda, M., Kuwabara, H., Miyaishi, O., Itano, N., Ohno, A., Zako, M., and Isogai, Z. 2007. Versican, a major hyaluronan-binding component in the dermis, loses its hyaluronan-binding ability in solar elastosis. *J. Invest. Dermatol.*, 127(7):1657–1663.

[51] Honardoust, D., Varkey, M., Marcoux, Y., Shankowsky, H. A., and Tredget, E. E. 2012. Reduced decorin, fibromodulin, and transforming growth factor-beta3 in deep dermis leads to hypertrophic scarring. *J. Burn Care Res.,* 33(2):218–227.

[52] Lundqvist, K. and Schmidtchen, A. 2001. Immunohistochemical studies on proteoglycan expression in normal skin and chronic ulcers. *Br. J. Dermatol.*, 144(2):254–259.

[53] Malgouries, S., Thibaut, S. and Bernard, B. A. 2008. Proteoglycan expression patterns in human hair follicle. *Br. J. Dermatol.*, 158(2):234–242.

[54] Li, Y., Liu, Y., Xia, W., Lei, D., Voorhees, J. J., and Fisher, G. J. 2013. Age-dependent alterations of decorin glycosaminoglycans in human skin. *Sci. Rep.,* 3, 2422. DOI: 10.1038/srep02422.

[55] Vogel, K. G., and Heinegard, D. 1985. Characterisation of proteoglycans from adult bovine tendon. *J. Biol. Chem.*, 260(16):9298–9306.

[56] Schwartz, N. B., Domowicz, M., Krueger, R. C. Jr., Li, H., and Mangoura, D. 1996. Brain aggrecan. Perspect. *Dev. Neuro-biol.,* 3(4):291–306.

[57] Milev, P., Maurel, P., Chiba, A., Mevissen, M., Popp, S., Yamaguchi, Y., Margolis, R. K., and Margolis, R. U. 1998. Differential regulation of expression of hyaluronan-binding proteoglycans in developing brain: aggrecan, versican, neurocan and brevican. *Biochem. Biophys. Res. Commun.,* 247(2):207 – 212.

[58] Yamada, H., Watanabe, K., Shimonaka, M., and Yamaguchi, Y. 1994. Molecular cloning of brevican, a novel brain proteoglycan of the aggrecan/versican family. *J. Biol. Chem.*, 269(13):10119 – 10126.

[59] Seidenbecher, C. I., Richter, K., Rauch, U., Fa¨ssler, R., Garner, C. C., and Gundelfinger, E. D. 1995. Brevican, a chondroitin sulfate proteoglycan of rat brain, occurs as secreted and cell surface glycosylphosphatidylinositol-anchored isoforms. *J. Biol. Chem.*, 270 (45): 27206 – 27212.

[60] Jaworski, D. M., Kelly, G. M., and Hockfield, S. 1995. The CNS-specific hyaluronan-binding protein BEHAB is expressed in ventricular zones coincident with gliogenesis. *J. Neurosci.*, 15(2):1352–1362.

[61] Jaworski, D. M., Kelly. G. M., Piepmeier. J. M., and Hockfield, S. 1996. BEHAB (brain enriched hyaluronan binding) is expressed in surgical samples of glioma and in intracranial grafts of invasive glioma cell lines. *Cancer Res.,* 56:2293– 2298.

[62] Seidenbecher, C. I., Gundelfinger, E. D., Böckers, T. M., Trotter. J., and Kreutz, M. R. 1998. Transcripts for secreted and GPI-anchored brevican are differentially distributed in rat brain. *Eur. J. Neurosci.*, 10(5):1621–1630.

[63] Oohira, A., Matsui, F., Watanabe, E., Kushima, Y., and Maeda, N. 1994. Developmentally regulated expression of a brain specific species of chondroitin sulfate proteoglycan, neurocan, identified with a monoclonal antibody IG2 in the rat cerebrum. *Neuroscience*, 60:145–157.

[64] Margolis, R. U., and Margolis, R. K. 1994. Aggrecan-versican-neurocan family of proteoglycans. *Methods Enzymol.*, 245:105–126.

[65] LeBaron, R. G. 1996. Versican. Perspect. *Dev. Neurobiol.*, 3(4):261–271.

[66] Courel, M. N., Marret, S., Girard, N., Chauzy, C., Olivier, A., Bertrand, P., Delpech, A., Laquerriere, A., Asou, H., and Delpech, B. 1998. Hyaluronectin is produced by oligodendrocytes and Schwann cells in vitro. *J. Neurocytol.*, 27:27–32.

[67] Michelacci, Y. M. 2003. Collagens and proteoglycans of the corneal extracellular matrix. *Braz. J. Med. Biol. Res.*, 36(8):1037–1046.

[68] Rada, J. A., Achen, V. R., Perry, C. A., and Fox, P. W. 1997. Proteoglycans in the human sclera. Evidence for the presence of aggrecan. *Invest. Ophthalmol. Vis. Sci.*, 38(9):1740–1751.

[69] Austin, B. A., Coulon, C., Liu, C. Y., Kao, W. W., and Rada, J. A. 2002. Altered collagen fibril formation in the sclera of lumican-deficient mice. *Invest. Ophthalmol. Vis. Sci.,* 43(6):1695–1701.

[70] Dunlevy, J. R., and Rada, J. A. 2004. Interaction of lumican with aggrecan in the aging human sclera. *Invest. Ophthalmol. Vis. Sci.,* 45(11):3849–3856.

[71] Chakravarti, S., Paul, J., Roberts, L., Chervoneva, I., Oldberg, A., and Birk, D. E. 2003. Ocular and scleral alterations in gene-targeted lumican-fibromodulin double-null mice. *Invest. Ophthalmol. Vis. Sci.,* 44(6):2422–2432.

[72] Valiyaveettil, M., Achur, R. N., Muthusamy, A., and Gowda, D. C. 2004 Chondroitin sulfate proteoglycans of the endothelia of human umbilical vein and arteries and assessment for the adherence of Plasmodium falciparum infected erythrocytes. *Mol. Biochem. Parasitol.,* 134(1):115-126.

[73] Kolset, S. O. and Gallagher, J. T. 1990. Proteoglycans in haemopoietic cells. *Biochim. Biophys. Acta*, 1032(2-3):191–211.

[74] Uhlin-Hansen, L., Eskeland, T., and Kolset, S. O. 1989. Modulation of the expression of chondroitin sulfate proteoglycan in stimulated human monocytes. *J. Biol. Chem.*, 264(25):14916-14922.

[75] Kolset, S. O., Kjellén, L., Seljelid, R., and Lindahl, U. 1983. Changes in glycosaminoglycan biosynthesis during differentiation in vitro of human monocytes. *Biochem. J.,* 210(3):661–667.

[76] Kolset, S. O. 1986. Oversulfated chondroitin sulfate proteoglycan in cultured human peritoneal macrophages. *Biochem. Biophys. Res. Commun.*, 139(2):377 – 382.

[77] Schiller, S. 1958. The isolation of chondroitin sulfuric acid from normal human plasma. *Biochim. Biophys. Acta*, 28(2):413-416.

[78] Calatroni, A., Donnelly, P. V., and Di Ferrante, N. 1969. The glycosaminoglycans of human plasma. *J. Clin. Invest.,* 48(2):332-343.

[79] Murata, K., Ogura, T., and Okuyama, T. 1972. Studies on the disaccharide subunits of chondroitin sulfates in human plasma. *Biochemical Medicine*, 6(3):223–230.

[80] Melet, J., Hooghwinkel, G. J., Giesberts, M. A., van Gelderen. H. H. 1980. A semi-quantitative micromethod for the determination of free glycosaminoglycans in serum. Results from studies on serum of healthy children of various age and patients affected by mucopolysaccharidosis. *Clin. Chim. Acta*, 108(2):179-188.

[81] Jones, L. L., Margolis, R. U., and Tuszynski, M. H. 2003. The chondroitin sulfate proteoglycans neurocan, brevican, phosphacan, and versican are differentially regulated following spinal cord injury. *Exp. Neurol.*, 182(2):399–411.

[82] Braunewell, K. H., Pesheva, P., McCarthy, J. B., Furcht, L. T., Schmitz, B., and Schachner, M. 1995. Functional involvement of sciatic nerve-derived versican- and decorin-like molecules and other chondroitin sulphate proteoglycans in ECM-mediated cell adhesion and neurite outgrowth. *Eur. J. Neurosci.*, 7(4):805–814.

[83] Fidler, P. S., Schuette, K., Asher, R. A., Dobbertin, A., Thornton, S. R., Calle-Patino, Y., Muir, E., Levine, J. M., Geller, H. M., Rogers, J. H., Faissner, A., and Fawcett, J. W. 1999. Comparing astrocytic cell lines that are inhibitory or permissive for axon

growth: the major axon-inhibitory proteoglycan is NG2. *J. Neurosci.*, 19(20):8778–8788.

[84] Yamada, H., Fredette, B., Shitara, K., Hagihara, K., Miura, R., Ranscht, B., Stallcup, W. B., and Yamaguchi, Y. 1997. The brain chondroitin sulfate proteoglycan brevican associates with astrocytes ensheathing cerebellar glomeruli and inhibits neurite outgrowth from granule neurons. *J. Neurosci.*, 17(20):7784–7795.

[85] Perris, R., Krotoski, D., Lallier, T., Domingo, C., Sorrel, J. M., and Bronner-Fraser, M. 1991. Spatial and temporal changes in the distribution of proteoglycans during avian neural crest development. *Development*, 111(2):583–599.

[86] Oakley, R. A. and Tosney, K. W. 1991. Peanut agglutinin and chondroitin-6-sulfate are molecular markers for tissues that act as barriers to axon advance in the avian embryo. *Dev. Biol.*, 147(1):187–206.

[87] Brittis, P. A., Canning, D. R., and Silver, J. 1992. Chondroitin sulfate as a regulator of neuronal patterning in the retina. *Science,* 225(5045):733–736.

[88] Sakurai, T., Lustig, M., Nativ, M., Hemperly, J. J., Schlessinger, J., Peles, E., and Grumet, M. 1997. Induction of neurite outgrowth through contactin and Nr-CAM by extracellular regions of glial receptor tyrosine phosphatase beta. *J. Cell Biol.*, 136(4):907-918.

[89] Peles, E., Nativ, M., Campbell, P. L., Sakurai, T., Martinez, R., Lev, S., Clary, D. O., Schilling, J., Barnea, G., Plowman, G. D., Grumet, M., and Schlessinger, J. 1995. The carbonic anhydrase domain of receptor tyrosine phosphatase beta is a functional ligand for the axonal cell recognition molecule contactin. *Cell,* 82(2):251-60.

[90] Perris, R. and Johansson, S. 1990. Inhibition of neural crest cell migration by aggregating chondroitin sulfate proteoglycans is mediated by their hyaluronan-binding region. *Dev. Biol.*, 137(1):1–12.

[91] Margolis, R. U. and Margolis, R. K. 1997. Chondroitin sulfate proteoglycans as mediators of axon growth and pathfinding. *Cell Tissue Res.*, 290(2):343–348.

[92] Thomas, L. B., Gates, M. A., and Steindler, D. A. 1996. Young neurons from the adult subependymal zone proliferate and migrate along an astrocyte, extracellular matrix-rich pathway. *Glia*, 17(1):1-14.

[93] Bicknese, A. R., Sheppard, A. M., O'Leary, D. D., and Pearlman, A. L. 1994. Thalamocortical axons extend along a chondroitin sulfate proteoglycan-enriched pathway coincident with the neocortical subplate and distinct from the efferent path. *J. Neurosci.*, 14:3500-3510.

[94] Watanabe, H., Cheung, S. C., Itano. N., Kimata. K., and Yamada Y. 1997. Identification of hyaluronan-binding domains of aggrecan. *J. Biol. Chem.*, 272(44):28057–28065.

[95] Perkins, S. J., Nealis, A. S., Dudhia, J., and Hardingham, T. E. 1989. Immunoglobulin fold and tandem repeat structures in proteoglycan N-terminal domains and link protein. *J. Mol. Biol.,* 206(4):737–753.

[96] Grover, J. and Roughley P. J. 1994. The expression of functional link protein in a baculovirus system: analysis of mutants lacking the A, B and B' domains. *Biochem. J.*, 300:317–324.

[97] Maroudas, A. 1979. in Adult Articular Cartilage, 2nd Ed., Pitman, Tunbridge Wells, United Kingdom.

[98] Levine, J. M. 1994. Increased expression of the NG2 chondroitin-sulfate proteoglycan after brain injury. *J. Neurosci.*, 14(8):4716–4730.

[99] Asher, R. A., Morgenstern, D. A., Fidler, P. S., Adcock, K. H., Oohira, A., Braistead, J. E., Levine, J. M., Margolis, R. U., Rogers, J. H., and Fawcett, J. W. 2000. Neurocan is upregulated in injured brain and in cytokine- treated astrocytes. *J. Neurosci.*, 20(7):2427–2438.

[100] Asher, R. A., Morgenstern, D. A., Shearer, M. C., Adcock, K. H., Pesheva, P., and Fawcett, J. W. 2002. Versican is upregulated in CNS injury and is a product of oligodendrocyte lineage cells. *J. Neurosci.,* 22(6):2225–2236.

[101] McKeon, R. J., Jurynec, M. J., and Buck, C. R. 1999. The chondroitin sulfate proteoglycans neurocan and phosphacan are expressed by reactive astrocytes in the chronic CNS glial scar. *J. Neurosci.,* 19(24):10778–10788.

[102] Jaworski, D. M., Kelly, G. M., and Hockfield, S. 1999. Intracranial injury acutely induces the expression of the secreted isoform of the CNS- specific hyaluronan-binding protein BEHAB/brevican. *Exp. Neurol.*, 157(2):327–337.

[103] Haas, C. A., Rauch, U., Thon, N., Merten, T., and Deller, T. 1999. Entorhinal cortex lesion in adult rats induces the expression of the neuronal chondroitin sulfate proteoglycan neurocan in reactive astrocytes. *J. Neurosci*., 19(22):9953–9963.

[104] Thon, N., Haas, C. A., Rauch, U., Merten, T., Fassler, R., Frotscher, M., Deller, T. 2000. The chondroitin sulphate proteoglycan brevican is upregulated by astrocytes after entorhinal cortex lesions in adult rats. *Eur. J. Neurosci*., 12(7):2547–2558.

[105] Danielson, K. G., Baribault, H., Holmes, D. F., Graham, H., Kadler, K. E., and Iozzo, R. V. 1997. Targeted disruption of decorin leads to abnormal collagen fibril morphology and skin fragility. *J. Cell Biol.,* 136(3):729–743.

[106] Moscatello, D. K., Santra, M., Mann, D. M., McQuillan, D. J., Wong, A. J., and Iozzo, R. V. 1998. Decorin suppresses tumor cell growth by activating the epidermal growth factor receptor. *J. Clin. Invest.,* 101(2):406–412.

[107] Yamaguchi, Y., Mann, D. M., and Ruoslahti, E. 1990. Negative regulation of transforming growth factor-beta by the proteoglycan decorin. *Nature,* 346(6281):281–284.

[108] Ali, S. Y., and Bayliss, M. T. 1981. Cathepsin B and other proteases in human articular cartilage. In Seminars in *Arthritis and Rheumatism,* 11(1):56-58.

[109] Roughley, P. J., and White, R. J. 1980. Age-related changes in the structure of the proteoglycan subunits from human articular cartilage. *J. Biol. Chem.,* 255(1):217-224.

[110] Hardingham, T. and Bayliss, M. 1990. Proteoglycans of articular cartilage: changes in aging and in joint disease. *Semin. Arthritis Rheum.*, 20(3 Suppl. 1):12-33.

[111] Johnstone, B., Markopoulos, M., Neame, P., and Caterson, B. 1993. Identification and characterization of glycanated and non-glycanated forms of biglycan and decorin in the human intervertebral disc. *Biochem. J.*, 292(Pt 3):661–666.

[112] Roughley, P. J., White, R. J., Magny, M. C., Liu, J., Pearce, R. H., and Mort, J. S. 1993. Non-proteoglycan forms of biglycan increase with age in human articular cartilage. *Biochem. J.,* 295(Pt 2):421–426.

[113] Sandy, J. D., Flannery, C. R., Neame, P. J., and Lohmander, L. S. 1992. The structure of aggrecan fragment in human synovial fluid: evidence for the involvement in osteoarthritis of a novel protease which cleaves the Glu 373-Ala 374 bond of the interglobular domain. *J. Clin. Invest.,* 89(5):1512–1516.

[114] Sandy, J. D., Neame, P. J., Boynton, R. E., and Flannery, C. R. 1991. Catabolism of aggrecan in cartilage explants: identification of a major cleavage site within the interglobular domain. *J. Biol. Chem.*, 266(14):8683–8685.

[115] Ilic, M. Z., Handley, C. J., Robinson, H. C., and Mok M. T. 1992. Mechanism of catabolism of aggrecan by articular cartilage. *Arch. Biochem. Biophys.*, 294(1):115–122.

[116] Bandtlow, C. E., and Zimmermann, D. R. 2000. Proteoglycans in the developing brain: New conceptual insights for old proteins. *Physiol. Rev.*, 80(4):1267–1290.

[117] Rauch, U., Gao, P., Janetzko, A., Flaccus, A., Hilgenberg, L., Tekotte, H., Margolis, R. K., and Margolis, R. U. 1991. Isolation and characterization of developmentally regulated chondroitin sulfate and chondroitin/keratan sulfate proteoglycans of brain identified with monoclonal antibodies. *J. Biol. Chem.*, 266(22):14785-14801.

[118] Schmalfeldt, M., Dours-Zimmermann, M. T., Winterhalter, K. H., and Zimmermann D.R. 1998. Versican V2 is a major extracellular matrix component of the mature bovine brain. *J. Biol. Chem.*, 273(25):15758–15764.

[119] Yamada, H., Fredette, B., Shitara, K., Hagihara, K., Miura, R., Ranscht, B., Stallcup, W. B., and Yamaguchi, Y. 1997. The brain chondroitin sulfate proteoglycan brevican associates with astrocytes ensheathing cerebellar glomeruli and inhibits neurites outgrowth from granule neurons. *J. Neurosci.*, 17(20):7784–7795.

[120] Akita, K., von Holst, A., Furukawa, Y., Mikami, T., Sugahara, K., and Faissner, A. 2008. Expression of multiple chondroitin/ dermatan sulfotransferases in the neurogenic regions of the embryonic and adult central nervous system implies that complex chondroitin sulfates have a role in neural stem cell maintenance. *Stem Cells*, 26(3):798-809.

[121] Carrino, D. A., Sorrell, J. M., Caplan, A. I. 2000. Age-related changes in the proteoglycans of human skin. *Arch. Biochem. Biophys.*, 373(1):91-101.

[122] Neufeld E. F., Muenzer J., The Mucopolysaccharidoses, in: C. R. Scriver, A. L. Beaudet, W. S. Sly, D. Valle, B. Childs, K. W. Kinzler, B. Vogelstein (Eds.), *Metabolic and Molecular Basis of Inherited Disease*, MacGraw-Hill, San Francisco, 2001, pp. 3421–3452.

[123] Coutinho, M. F., Lacerda, L., Alves, S. (2012). Glycosaminoglycan Storage Disorders: A Review. *Biochem. Res. Int.*, 2012, 471325.

[124] Tomatsu, S., Montaño, A. M., Oikawa, H., Smith, M., Barrera, L., Chinen, Y., Thacker, M. M., Mackenzie, W. G., Suzuki, Y., and Orii, T. 2011. Mucopolysaccharidosis type IVA (Morquio A disease): clinical review and current treatment. *Curr. Pharm. Biotechnol.*, 12(6):931-945.

[125] Northover, H., Cowie, R. A., and Wraith, J. E. 1996. Mucopolysaccharidosis type IVA (Morquio syndrome): a clinical review. *J. Inherit. Metab. Dis.*, 19(3):357-365.

[126] Shipley, J. M., Klinkenberg, M., Wu, B. M,. Bachinsky, D. R., Grubb, J. H., and Sly, W. S. 1993. Mutational analysis of a patient with mucopolysaccharidosis type VII, and identification of pseudogenes. *Am. J. Hum. Genet.*, 52(3):517–526.

[127] Klüppel, M., Wight, T. N., Chan, C., Hinek, A., and Wrana, J. L. 2005. Maintenance of chondroitin sulfation balance by chondroitin-4-sulfotransferase 1 is required for chondrocyte development and growth factor signaling during cartilage morphogenesis. *Development,* 132(17):3989-4003.

[128] Metcalf, J. A., Zhang, Y., Hilton, M. J., Long, F., Ponder, K. P. 2009. Mechanism of shortened bones in mucopolysaccharidosis VII. *Mol. Genet. Metab.,* 97:202–211.

[129] Herati, R. S., Knox, V. W., O'Donnell, P., D'Angelo, M., Haskins, M. E., Ponder, K. P. 2008. Radiographic evaluation of bones and joints in mucopolysaccharidosis I and VII dogs after neonatal gene therapy. *Mol. Genet. Metab.*, 95:142–151.

[130] Clarke, L. A. Russell, C. S., Pownall, S., Warrington, C. L., Borowski, A., Dimmick, J. E., Toone, J., Jirik, F. R. 1997. Murine mucopolysaccharidosis type I: targeted disruption of the murine alpha-L-iduronidase gene. *Hum. Mol. Genet.*, 6:503–511.

[131] Liu, Y., Xu, L., Hennig, A. K., Kovacs, A., Fu, A., Chung, S., Lee, D., Wang, B., Herati, R. S., Ogilvie, J. M., Cai, S., Ponder, K. P. 2005. Liver-directed neonatal gene therapy prevents cardiac, bone, ear, and eye disease in mucopolysaccharidosis I mice. *Mol. Ther.*, 11:35–47.

[132] Bhaumik, M., Mulle, V. J., Rozaklis, T., Johnson, L., Dobrenis, K., Bhattacharyya, R., Wurzelmann, S., Finamore, P., Hopwood, J. J., Walkley, S. U., Stanley, P. 1999. A mouse model for mucopolysaccharidosis type III A (Sanfilippo syndrome). *Glycobiology,* 9(12):1389–1396.

[133] Ohmi, K., Greenberg, D. S., Rajavel, D. S, Ryazantsev, S., Li, H. H., Neufeld, E. F. 2003. Activated microglia in cortex of mouse models of mucopolysaccharidoses I and IIIB. *Proc. Natl. Acad. Sci. USA,* 100 (4): 1902–1907.

[134] Li, H. H., Yu, W. H., Rozengurt, N., Zhao, H. Z., Lyons, K. M., Anagnostaras, S., Fanselow, M. S., Suzuki, K., Vanier, M. T., Neufeld, E. F. 1999. Mouse model of Sanfilippo syndrome type B produced by targeted disruption of the gene encoding alpha- N-acetylglucosaminidase. *Proc. Natl. Acad. Sci. USA,* 96(25):14505–14510.

[135] Evers, M., Saftig, P., Schmidt, P., Hafner, A., McLoghlin, D. B., Schmahl, W., Hess, B., von Figura, K., Peters, C. 1996. Targeted disruption of the arylsulfatase B gene results in mice resembling the phenotype of mucopolysaccharidosis VI. *Proc. Natl. Acad. Sci. USA,* 93(3):8214–8219.

[136] Tomatsu, S., Vogler, C., Montaño, A. M., Gutierrez, M., Oikawa, H., Dung, V. C., Orii, T., Noguchi, A., Sly, W. S. 2007. Murine model (Galns(tm(C76S)slu)) of MPS IVA with missense mutation at the active site cysteine conserved among sulfatase proteins. *Mol. Genet. Metab.,* 91(24):251–258.

[137] Tomatsu, S., Orii, K. O., Vogler, C., Nakayama, J., Levy, B., Grubb, J. H., Gutierrez, M. A., Shim, S., Yamaguchi, S., Nishioka, T., Montaño, A. M., Noguchi, A., Orii, T., Kondo, N., Sly, W. S. 2003. Mouse model of N-acetylgalactosamine-6-sulfate sulfatase deficiency (Galns-/-) produced by targeted disruption of the gene defective in Morquio A disease. *Hum. Mol. Genet.*, 12:3349-3358.

[138] Rowan, D. J., Tomatsu, S., Grubb, J. H., Montaño, A. M., and Sly, W. S. 2013. Assessment of bone dysplasia by micro-CT and glycosaminoglycan levels in mouse models for mucopolysaccharidosis type I, IIIA, IVA, and VII. *J. Inherit. Metab. Dis.,* 36(2):235-246.

[139] Byers, S., Caterson, B., Hopwood, J. J, Foster, B. K. 1992. Immunolocation analysis of glycosaminoglycans in the human growth plate. *J. Histochem. Cytochem.*, 40(2):275-282.

[140] Uchimura, K., Kadomatsu, K., Nishimura, H., Muramatsu, H., Nakamura, E., Kurosawa, N., Habuchi, O., El-Fasakhany, F. M., Yoshikai, Y., and Muramatsu. T. 2002. Functional analysis of the chondroitin 6-sulfotransferase gene in relation to

lymphocyte subpopulations, brain development, and oversulfated chondroitin sulfates. *J. Biol. Chem.*, 277(2):1443–1450.

[141] Thiele, H., Sakano, M., Kitagawa, H., Sugahara, K., Rajab, A., Höhne, W., Ritter, H., Leschik, G., Nürnberg, P., and Mundlos, S. 2004. Loss of chondroitin 6-O-sulfotransferase-1 function results in severe human chondrodysplasia with progressive spinal involvement. *Proc. Natl. Acad. Sci. USA,* 101(27):10155–10160.

[142] Hermanns, P., Unger, S., Rossi, A., Perez-Aytes, A., Cortina, H., Bonafé, L., Boccone, L., Setzu, V., Dutoit, M., Sangiorgi, L., Pecora, F., Reicherter, K., Nishimura, G., Spranger, J., Zabel, B., and Superti-Furga. A. 2008. Congenital joint dislocations caused by carbohydrate sulfotransferase 3 deficiency in recessive Larsen syndrome and humero-spinal dysostosis. *Am. J. Hum. Genet.,* 82(6):1368–1374.

[143] Unger, S., Lausch, E., Rossi, A., Mégarbané, A., Sillence, D., Alcausin, M., Aytes, A., Mendoza-Londono, R., Nampoothiri, S., Afroze, B., Hall, B., Lo, I. F., Lam, S. T., Hoefele, J., Rost, I., Wakeling, E., Mangold, E., Godbole, K., Vatanavicharn, N., Franco, L. M., Chandler, K., Hollander, S., Velten, T., Reicherter, K., Spranger, J., Robertson, S., Bonafé, L., Zabel, B., and Superti-Furga. A. 2010. Phenotypic features of carbohydrate sulfotransferase 3 (CHST3) deficiency in 24 patients: congenital dislocations and vertebral changes as principal diagnostic features. *Am. J. Med. Genet.*, 152A(10):2543–2549.

[144] Takeuchi, K., Yoshioka, N., Higa Onaga, S., Watanabe, Y., Miyata, S., Wada, Y., Kudo, C., Okada, M., Ohko, K., Oda, K., Sato, T., Yokoyama, M., Matsushita, N., Nakamura, M., Okano, H., Sakimura, K., Kawano, H., Kitagawa, H., and Igarashi, M. 2013. Chondroitin sulphate N-acetylgalactosaminyl-transferase-1 inhibits recovery from neural injury. *Nat. Commun.*, 4:2740.

[145] Miyata, S., Komatsu, Y., Yoshimura, Y., Taya, C., and Kitagawa, H. 2012. Persistent cortical plasticity by upregulation of chondroitin 6-sulfation. *Nat. Neurosci.*, 15(3):414-422.

[146] Properzi, F., Carulli, D., Asher, R. A., Muir, E., Camargo, L. M., van Kuppevelt, T. H., ten Dam, G. B., Furukawa, Y., Mikami, T., Sugahara, K., Toida, T., Geller, H. M., Fawcett, J. W. 2005. Chondroitin 6-sulphate synthesis is up-regulated in injured CNS, induced by injury-related cytokines and enhanced in axon-growth inhibitory glia. *Eur. J. Neurosci.,* 21(2):378-390.

[147] Kitagawa, H., Tsutsumi, K., Tone, Y., and Sugahara, K. 1997. Developmental regulation of the sulfation profile of chondroitin sulfate chains in the chicken embryo brain. *J. Biol. Chem.*, 272(50):31377-31381.

[148] Mitsunaga, C., Mikami, T., Mizumoto, S., Fukuda, J., and Sugahara, K. 2006. Chondroiti sulfate/dermatan sulfate hybrid chains in the development of cerebellum. Spatiotemporal regulation of the expression of critical disulfated disaccharides by specific sulfotransferases. *J. Biol. Chem.,* 281(28):18942-18952.

[149] Sugiyama, S., Di Nardo, A. A., Aizawa, S., Matsuo, I., Volovitch, M., Prochiantz, A., Hensch, T. K. 2008. Experience-dependent transfer of Otx2 homeoprotein into the visual cortex activates postnatal plasticity. *Cell,* 134(3):508–520.

[150] Beurdeley, M., Spatazza, J., Lee, H. H., Sugiyama, S., Bernard, C., Di Nardo, A. A., Hensch, T. K., Prochiantz, A. 2012. Otx2 binding to perineuronal nets persistently regulates plasticity in the mature visual cortex. *J. Neurosci.,* 32(27):9429–9437.

[151] Garnjanagoonchorn, W., Wongekalak, L., Engkagul, A. 2007. Determination of chondroitin sulfate from different sources of cartilage. *Chem. Eng. Process.*, 46:465–471.

[152] Suzuki, S., Saito, H., Yamagata, T., Anno, K., Seno, N., Kawai, Y., Furuhashi, T. 1968. Formation of three types of disulfated disaccharides from chondroitin sulfates by chondroitinase digestion. *J. Biol. Chem.*, 243(7):1543–1550.

[153] Sugahara, K. and Yamada, S. 2000. Structure and function of oversulfated chondroitin sulfate variants: Unique sulfation patterns and neuroregulatory activities. *Trends Glycosci. Glycotechnol.*, 12(67):321–349.

[154] Anno, K., Seno, N., Mathews, M. B., Yamagata, T., Suzuki, S. 1971. A new dermatan polysulfate, chondroitin sulfate H, from hagfish notochord. *Biochim. Biophys. Acta*, 237(1):173–177.

[155] Ueoka, C., Nadanaka, S., Seno, N., Khoo, K. H., Sugahara, K. 1999. Structural determination of novel tetra- and hexasaccharide sequences isolated from chondroitin sulfate H (oversulfated dermatan sulfate) of hagfish notochord. *Glycoconj. J.*, 16(6):291–305.

[156] Lamari, F. N., Theocharis, A. D., Asimakopoulou, A. P., Malavaki, C. J., Karamanos, N. K. 2006. Metabolism and biochemical/physiological roles of chondroitin sulfates: analysis of endogenous and supplemental chondroitin sulfates in blood circulation. *Biomed. Chromatogr.*, 20(6-7): 539-550.

[157] Volpi, N. 2003. Oral absorption and bioavailability of ichthyic origin chondroitin sulfate in healthy male volunteers. *Osteoarthritis Cartilage,* 11(6):433-441.

[158] Henrotin, Y., Mathy, M., Sanchez, C., Lambert, C. 2010. Chondroitin sulfate in the treatment of osteoarthritis: from in vitro studies to clinical recommendations. *Ther. Adv. Musculoskelet. Dis.*, 2(6):335-348.

[159] Ronca, F., Palmieri, L., Panicucci, P., Ronca, G. 1998. Anti-inflammatory activity of chondroitin sulfate. *Osteoarthritis Cartilage,* 6(Suppl. A):14-21.

[160] Omata, T., Itokazu, Y., Inoue, N., Segawa, Y. 2000. Effects of chondroitin sulfate-C on articular cartilage destruction in murine collagen-induced arthritis. *Arzneimittelforschung,* 50(2):148-153.

[161] Cho, S. Y., Sim, J. S., Jeong, C. S., Chang, S. Y., Choi, D. W., Toida, T., Kim, Y. S. 2004. Effects of low molecular weight chondroitin sulfate on type II collagen-induced arthritis in DBA/1J mice. *Biol. Pharm. Bull.,* 27(1):47-51.

[162] Jomphe, C., Gabriac, M., Hale, T. M., Héroux, L., Trudeau, L. E., Deblois, D., Montell, E., Vergés, J., du Souich, P. 2008. Chondroitin sulfate inhibits the nuclear translocation of nuclear factor-kappaB in interleukin-1beta-stimulated chondrocytes. *Basic Clin. Pharmacol. Toxicol.,* 102(1):59-65.

[163] Bali, J. P., Cousse, H., Neuzil, E. 2001. Biochemical basis of the pharmacologic action of chondroitin sulfates on the osteoarticular system. *Semin. Arthritis Rheum.,* 31(1): 58-68.

[164] Michel, B. A., Stucki, G., Frey, D., De Vathaire, F., Vignon, E., Bruehlmann, P., Uebelhart, D. 2005. Chondroitins 4 and 6 sulfate in osteoarthritis of the knee: a randomized, controlled trial. *Arthritis Rheum.,* 52(3):779-786.

[165] Clegg, D. O., Reda, D. J., Harris, C. L., Klein, M. A., O'Dell, J. R., Hooper, M. M., Bradley, J. D., Bingham, C. O. 3rd., Weisman, M. H., Jackson, C. G., Lane, N. E., Cush, J. J., Moreland, L. W., Schumacher, H. R. Jr., Oddis, C. V., Wolfe, F., Molitor, J.

A., Yocum, D. E., Schnitzer, T. J., Furst, D. E., Sawitzke, A. D., Shi, H., Brandt, K. D., Moskowitz, R. W., Williams, H. J. 2006. Glucosamine, chondroitin sulfate, and the two in combination for painful knee osteoarthritis. *N. Engl. J. Med.,* 354(8):795-808.

[166] Kahan, A., Uebelhart, D., De Vathaire, F., Delmas, P. D., Reginster, J. Y. 2009. Long-term effects of chondroitins 4 and 6 sulfate on knee osteoarthritis: the study on osteoarthritis progression prevention, a two-year, randomized, double-blind, placebo-controlled trial. *Arthritis Rheum.,* 60(2). 524-533.

[167] Sawitzke, A. D., Shi, H., Finco, M. F., Dunlop, D. D., Harris, C. L., Singer, N. G., Bradley, J. D., Silver, D., Jackson, C. G., Lane, N. E., Oddis, C. V., Wolfe, F., Lisse, J., Furst, D. E., Bingham, C. O., Reda, D. J., Moskowitz, R. W., Williams, H. J., Clegg, D. O. 2010. Clinical efficacy and safety of glucosamine, chondroitin sulphate, their combination, celecoxib or placebo taken to treat osteoarthritis of the knee: 2-year results from GAIT. *Ann. Rheum. Dis.*, 69(8):1459-1464.

[168] Uebelhart, D. 2008. Clinical review of chondroitin sulfate in osteoarthritis. *Osteoarthritis Cartilage*, 16(Suppl. 3):S19-S21.

[169] Ledbetter, E. C, Munger, R. J, Ring, R. D, Scarlett, J. M. 2006. Efficacy of two chondroitin sulfate ophthalmic solutions in the therapy of spontaneous chronic corneal epithelial defects and ulcerative keratitis associated with bullous keratopathy in dogs. *Vet. Ophthalmol.*, 9(2):77-87.

[170] Suzuki, M. 1939. Prosthetic group of cornea mucoid. *J. Biochem.,* 30:185-191.

[171] Meyer, K., Linker, A., Davidson, E. A., Weissmann, B. 1953. The mucopolysaccharides of bovine cornea. *J. Biol. Chem.*, 205(2):611-616.

[172] Funderburgh, J. L. 2000. Keratan sulfate: structure, biosynthesis, and function. *Glycobiology,* 10(10):951-958.

[173] Antonsson, P., Heinegård, D., and Oldberg, A. 1991. Posttranslational modifications of fibromodulin. *J. Biol. Chem.*, 266(25):16859-16861.

[174] Sommarin, Y., Wendel, M., Shen, Z., Hellman, U., and Heinegârd, D. 1998. Osteoadherin, a cell-binding keratan sulfate proteoglycan in bone, belongs to the family of leucine-rich repeat proteins of the extracellular matrix. *J. Biol. Chem.*, 273(27):16723-16729.

[175] Nieduszynski, I. A., Huckerby, T. N., Dickenson, J. M., Brown, G. M., Tai, G. H., Morris, H. G., and Eady, S. 1990. There are two major types of skeletal keratan sulphates. *Biochem. J.,* 271(1):243-245.

[176] Karlsson, N. G., and McGuckin, M. A. 2012. O-Linked glycome and proteome of high-molecular-mass proteins in human ovarian cancer ascites: Identification of sulfation, disialic acid and O-linked fucose. *Glycobiology,* 22(7):918-929.

[177] Krusius,T., Finne, J., Margolis, R. K., and Margolis, R. U. 1986. Identification of an O-glycosidic mannose-linked sialylated tetrasaccharide and keratan sulfate oligosaccharides in the chondroitin sulfate proteoglycan of brain. *J. Biol. Chem.*, 261(18):8237–8242.

[178] Degroote, S., Lo-Guidice, J. M., Strecker, G., Ducourouble, M. P., Roussel, P., Lamblin, G. 1997. Characterization of an N-acetylglucosamine-6-O-sulfotransferase from human respiratory mucosa active on mucin carbohydrate chains. *J. Biol. Chem.*, 272(47):29493-29501.

[179] Uchimura, K., Muramatsu, H., Kaname, T., Ogawa, H., Yamakawa, T., Fan, Q. W., Mitsuoka, C., Kannagi, R., Habuchi, O., Yokoyama, I., Yamamura, K., Ozaki, T.,

Nakagawara, A., Kadomatsu, K., and Muramatsu, T. 1998. Human N-acetylglucosamine-6-O-sulfotransferase involved in the biosynthesis of 6-sulfo sialyl Lewis X: molecular cloning, chromosomal mapping, and expression in various organs and tumor cells. *J. Biochem.*, 124(3):670-678.

[180] Liles, M., Palka, B. P., Harris, A., Kerr, B., Hughes, C., Young, R. D., Meek, K. M., Caterson, B., and Quantock, A. J. 2010. Differential relative sulfation of Keratan sulfate glycosaminoglycan in the chick cornea during embryonic development. *Invest. Ophthalmol. Vis. Sci.,* 51:1365-1372.

[181] Brown, G. M., Huckerby, T. N., Bayliss, M. T., and Nieduszynski, I. A. 1998. Human aggrecan keratan sulfate undergoes structural changes during adolescent development. *J. Biol. Chem.*, 273(41):26408-26414.

[182] Tai, G. H., Huckerby, T. N., and Nieduszynski, I. A. Multiple non-reducing chain termini isolated from bovine corneal keratan sulfates. 1996. *J. Biol. Chem.*, 271(38):23535-23546.

[183] Tai, G. H., Nieduszynski, I. A., Fullwood, N. J., and Huckerby, T. N. 1997. Human corneal keratan sulfates. *J. Biol. Chem.*, 7;272(45):28227-28231.

[184] Brown, G. M., Huckerby, T. N., Morris, H. G., Abram, B. L., and Nieduszynski, I. A. 1994. Oligosaccharides derived from bovine articular cartilage keratan sulfates after keratanase II digestion: implications for keratan sulfate structural fingerprinting. *Biochemistry,* 33(16):4836-4846.

[185] Brown, G. M., Huckerby, T. N., Morris, H. G., and Nieduszynski, I. A. 1992. Degradation of articular cartilage keratan sulphates using hydrazinolysis and nitrous acid. Environment of fucose residues. *Biochem. J.,* 286(Pt 1):235-241.

[186] Ashwell, G., and Morell, A. 1974. The dual role of sialic acid in the hepatic recognition and catabolism of serum glycoproteins. *Biochem. Soc. Symp.*, 40:117-124.

[187] Tai, G. H., Huckerby, T. N., and Nieduszynski, I. A. 1994. 600 MHz 1H NMR study of a fucose-containing heptasaccharide derived from a keratanase digestion of bovine articular cartilage keratan sulphate. *Carbohydr. Res.,* 255:303-309.

[188] Gipson, I. K., Joyce, N. C., and Zieske, J. D. 2004. The anatomy and cell biology of the human cornea, limbus, conjunctiva and adnexa. In: Smolin G., Foster C. S., Azar D. T., Dohman C. H., editors. (Eds.), Smolin and Thoft's The Cornea, Scientific Foundations and Clinical Practice. (4th ed. Pp3-37). By Lippincott, Williams, and Wilkins.

[189] Clark, S. J., Keenan, T. D., Fielder, H. L., Collinson, L. J., Holley, R. J., Merry, C. L., van Kuppevelt, T. H., Day, A. J., and Bishop, P. N. 2011. Mapping the differential distribution of glycosaminoglycans in the adult human retina, choroid, and sclera. *Invest. Ophthalmol. Vis. Sci.,* 52(9):6511-6521.

[190] Miller, B., Sheppard, A. M., and Pearlman, A. L. 1997. Developmental expression of keratan sulfate-like immunoreactivity distinguishes thalamic nuclei and cortical domains. *J. Comp. Neurol.*, 380(4):533-552.

[191] Yasuda, E., Fushimi, K., Suzuki, Y., Shimizu, K., Takami, T., Zustin, J., Patel, P., Ruhnke, K., Shimada, T., Boyce, B., Kokas, T., Barone, C., Theroux, M., Mackenzie, W., Nagel, B., Ryerse, J. S., Orii, K. E., Iida, H., Orii, T., and Tomatsu, S. 2013. Pathogenesis of Morquio A syndrome: an autopsied case reveals systemic storage disorder. *Mol. Genet. Metab.,* 109(3):301-311.

[192] Miyamoto, T., Ishii, K., Asaka, R., Suzuki, A., Takatsu, A., Kashima, H., and Shiozawa, T. 2011. Immunohistochemical expression of keratan sulfate: a possible

diagnostic marker for carcinomas of the female genital tract. *J. Clin. Pathol.*, 64(12):1058-1063.

[193] Souza-Fernandes, A. B., Pelosi, P., and Rocco, P. R. 2006. Bench-to-bedside review: the role of glycosaminoglycans in respiratory disease. *Crit. Care*, 10(6):237.

[194] Latif, N., Sarathchandra, P., Taylor, P. M., Antoniw, J., and Yacoub, M. H. 2005. Localization and pattern of expression of extracellular matrix components in human heart valves. *J. Heart Valve Dis.*, 14(2):218-227.

[195] Shimada, T., Tomatsu S., Mason R. W., Yasuda, Y., Mackenzie, W. G., Hossain, J., Shibata, Y., Montaño, A. M., Kubaski, F., Giugliani, R., Yamaguchi, S., Suzuki, Y,. Orii, K. E. Fukao, T., and Orii, T. 2014. Di-sulfated keratan sulfate as a novel biomarker for mucopolysaccharidosis IVA. *J. Inherit. Metab. Dis. Rep.,* (in press).

[196] Amado, M., Almeida, R., Schwientek, T., and Clausen, H. 1999. Identification and characterization of large galactosyltransferase gene families: galactosyltransferases for all functions. *Biochim. Biophys. Acta,* 1473(1):35-53.

[197] Kotani, N., Asano, M., Iwakura, Y., and Takasaki, S. 2001. Knockout of mouse beta 1,4-galactosyltransferase-1 gene results in a dramatic shift of outer chain moieties of N-glycans from type 2 to type 1 chains in hepatic membrane and plasma glycoproteins. *Biochem. J.,* 357(Pt 3): 827-834.

[198] Seko, A., Hara-Kuge, S., Yonezawa, S., Nagata, K., and Yamashita, K. 1998. Identification and characterization of N-acetylglucosamine-6-O-sulfate-specific beta1,4-galactosyltransferase in human colorectal mucosa. *FEBS Lett.,* 440:307-310.

[199] Seko, A., Dohmae, N., Takio, K., and Yamashita, K. 2003. Beta 1,4-galactosyltransferase (beta 4GalT)-IV is specific for GlcNAc 6-O-sulfate. Beta 4GalT-IV acts on keratan sulfate-related glycans and a precursor glycan of 6-sulfosialyl-Lewis X. *J. Biol. Chem.,* 278(11): 9150-9158.

[200] Sasaki, K., Kurata-Miura, K., Ujita, M., Angata, K., Nakagawa, S., Sekine, S., Nishi, T., and Fukuda, M. 1997. Expression cloning of cDNA encoding a human beta-1,3-N-acetylglucosaminyltransferase that is essential for poly-N-acetyllactosamine synthesis. *Proc. Natl. Acad. Sci. USA,* 94(26): 14294-14299.

[201] Zhou, D., Dinter, A., Gutiérrez Gallego, R., Kamerling, J. P., Vliegenthart, J. F., Berger, E. G., and Hennet, T. 1999. A beta-1,3-N-acetylglucosaminyltransferase with poly-N-acetyllactosamine synthase activity is structurally related to beta-1,3-galactosyltransferases. *Proc. Natl. Acad. Sci. USA,* 96(2):406-411.

[202] Seko, A., and Yamashita, K. 2004. beta1,3-N-Acetylglucosaminyltransferase-7 (beta3Gn-T7) acts efficiently on keratan sulfate-related glycans. *FEBS Lett.,* 556(1-3):216-220.

[203] Rüter, E. R., and Kresse, H. 1984. Partial purification and characterization of 3'-phosphoadenylylsulfate:keratan sulfate sulfotransferases. *J. Biol. Chem.*, 259(19):11771-11776.

[204] Habuchi, O., Hirahara, Y., Uchimura, K., and Fukuta, M. 1996. Enzymatic sulfation of galactose residue of keratan sulfate by chondroitin 6-sulfotransferase. *Glycobiology,* 6(1):51-57.

[205] Fukuta, M., Inazawa, J., Torii, T., Tsuzuki, K., Shimada, E., and Habuchi, O. 1997. Molecular cloning and characterization of human keratan sulfate Gal-6-sulfotransferase. *J. Biol. Chem.*, 272(51):32321-32328.

[206] Torii, T., Fukuta, M., and Habuchi, O. 2000. Sulfation of sialyl N-acetyllactosamine oligosaccharides and fetuin oligosaccharides by keratan sulfate Gal-6-sulfotransferase. *Glycobiology*, 10(2):203-211.

[207] Uchimura, K., Kadomatsu, K., Nishimura, H., Muramatsu, H., Nakamura, E., Kurosawa, N., Habuchi, O., El-Fasakhany, F. M., Yoshikai, Y., and Muramatsu, T. 2002. Functional analysis of the chondroitin 6-sulfotransferase gene in relation to lymphocyte subpopulations, brain development, and oversulfated chondroitin sulfates. *J. Biol. Chem.*, 277(2):1443-1450.

[208] Hoshino, H., Foyez, T., Ohtake-Niimi, S., Takeda-Uchimura, Y., Michikawa, M., Kadomatsu, K., and Uchimura, K. 2014. KSGal6ST is essential for the 6-sulfation of galactose within keratan sulfate in early postnatal brain. *J. Histochem. Cytochem.*, 62(2):145-156.

[209] Nakazawa, K., Takahashi, I., and Yamamoto, Y. 1998. Glycosyltransferase and sulfotransferase activities in chick corneal stromal cells before and after in vitro culture. *Arch. Biochem. Biophys.,* 359(2):269-282.

[210] Uchimura, K., and Rosen, S. D. 2006. Sulfated L-selectin ligands as a therapeutic target in chronic inflammation. *Trends Immunol.,* 27(12):559-565.

[211] Akama, T. O., Nishida, K., Nakayama, J., Watanabe, H., Ozaki, K., Nakamura, T., Dota, A., Kawasaki, S., Inoue, Y., Maeda, N., Yamamoto, S., Fujiwara, T., Thonar, E. J., Shimomura, Y., Kinoshita, S., Tanigami, A., and Fukuda, M. N. 2000. Macular corneal dystrophy type I and type II are caused by distinct mutations in a new sulphotransferase gene. *Nat. Genet.,* 26(2):237-241.

[212] Maszczak-Seneczko, D., Sosicka, P., Olczak, T., Jakimowicz, P., Majkowski, M., and Olczak, M. 2013. UDP-N-acetylglucosamine transporter (SLC35A3) regulates biosynthesis of highly branched N-glycans and keratan sulfate. *J. Biol. Chem.,* 288(30):21850-21860.

[213] Cornuet, P. K., Blochberger, T. C., and Hassell, J. R. 1994. Molecular polymorphism of lumican during corneal development. *Invest. Ophthalmol. Vis. Sci.*, 35(3):870–877.

[214] Hart, G. W. 1976. Biosynthesis of glycosaminoglycans during corneal development. *J. Biol. Chem.*, 251(21):6513–6521.

[215] Funderburgh, J. L., Caterson, B., Conrad, and G. W. 1986. Keratan sulfate proteoglycan during embryonic development of the chicken cornea. *Dev. Biol.*, 116(2):267-277.

[216] Ying, S., Shiraishi, A., Kao, C. W., Converse, R. L., Funderburgh, J. L., Swiergiel, J., Roth, M. R., Conrad, G. W., and Kao, W. W. 1997. Characterization and expression of the mouse lumican gene. *J. Biol. Chem.,* 272(48):30306-30313.

[217] Lauder, R. M., Huckerby, T. N., Nieduszynski, I. A., and Plaas, A. H. 1998. Age-related changes in the structure of the keratan sulphate chains attached to fibromodulin isolated from articular cartilage. *Biochem. J.,* 330(Pt 2):753-757.

[218] Kaplan, D., and Meyer, K. 1959. Ageing of human cartilage. *Nature,* 183(4670):1267-1268.

[219] Long, C. J., Roth, M. R., Tasheva, E. S., Funderburgh, M., Smit, R., Conrad, G. W., and Funderburgh, J. L. 2000. Fibroblast growth factor-2 promotes keratan sulfate proteoglycan expression by keratocytes in vitro. *J. Biol. Chem.,* 275(18):13918-13923.

[220] Domm, C., Schünke, M., Christesen, K., and Kurz, B. 2002. Redifferentiation of dedifferentiated bovine articular chondrocytes in alginate culture under low oxygen tension. *Osteoarthritis Cartilage,* 10(1):13-22.

[221] Tomatsu, S., Shimada, T., Mason, R. W., Kelly, J., LaMarr, W. A., Yasuda, E., Shibata, Y., Futatsumori, H., Montaño, A. M., Yamaguchi, S., Suzuki, Y. and Orii, T. 2014. Assay for glycosaminoglycans by tandem mass spectrometry and its applications. *J. Anal. Bioanal. Tech.,* S2, 006. doi:10.4172/2155-9872.S002-S006.

[222] Blochberger, T. C., Vergnes, J. P., Hempel, J., and Hassell, J. R. 1992. cDNA to chick lumican (corneal keratan sulfate proteoglycan) reveals homology to the small interstitial proteoglycan gene family and expression in muscle and intestine. *J. Biol. Chem.*, 267:347-352.

[223] Corpuz, L. M., Funderburgh, J. L., Funderburgh, M. L., Bottomley, G. S., Prakash, S., and Conrad, G. W. 1996. Molecular cloning and tissue distribution of keratocan. Bovine corneal keratan sulfate proteoglycan 37A. *J. Biol. Chem.*, 271(16):9759-9763.

[224] Funderburgh, J. L., Corpuz, L. M., Roth, M. R., Funderburgh, M. L., Tasheva, E. S., and Conrad, G. W. 1997. Mimecan, the 25-kDa corneal keratan sulfate proteoglycan, is a product of the gene producing osteoglycin. *J. Biol. Chem.*, 272(44):28089-28095.

[225] Oldberg, A., Antonsson, P., Lindblom, K., and Heinegård, D. 1989. A collagen-binding 59-kd protein (fibromodulin) is structurally related to the small interstitial proteoglycans PG-S1 and PG-S2 (decorin). *EMBO J.*, 8(9):2601-2604.

[226] Barry, F. P., Rosenberg, L. C., Gaw, J. U., Gaw, J. U., Koob, T. J., and Neame, P. J. 1995. N- and O-linked keratan sulfate on the hyaluronan binding region of aggrecan from mature and immature bovine cartilage. *J. Biol. Chem.*, 270(35):20516-20524.

[227] Bengtsson, E., Neame, P. J., Heinegård, D., and Sommarin, Y. 1995. The primary structure of a basic leucine-rich repeat protein, PRELP, found in connective tissues. *J. Biol. Chem.*, 270(43):25639-25644.

[228] Noguchi, S., and Nakano, M. 1992. Structure of the acidic N-linked carbohydrate chains of the 55-kDa glycoprotein family (PZP3) from porcine zona pellucida. *Eur. J. Biochem.*, 209(3):883-894.

[229] Aplin, J. D., Hey, N. A., and Graham, R. A. 1998. Human endometrial MUC1 carries keratan sulfate: characteristic glycoforms in the luminal epithelium at receptivity. *Glycobiology*, 8(3):269-276.

[230] Takahashi, K., Stamenkovic, I., Cutler, M., Dasgupta, A., and Tanabe, K. K. 1996. Keratan sulfate modification of CD44 modulates adhesion to hyaluronate. *J. Biol. Chem.*, 271(16):9490-9496.

[231] Tuhkanen, A. L., Tammi, M., and Tammi, R. 1997. CD44 substituted with heparan sulfate and endo-beta-galactosidase-sensitive oligosaccharides: a major proteoglycan in adult human epidermis. *J. Invest. Dermatol.,* 109(2):213-218.

[232] Schafer, I. A., and Sorrell, J. M. 1993. Human keratinocytes contain keratin filaments that are glycosylated with keratan sulfate. *Exp. Cell Res.,* 207(2):213-219.

[233] Seo, H., and Geisert, E. E. Jr. 1995. A keratan sulfate proteoglycan marks the boundaries in the cortical barrel fields of the adult rat. *Neurosci. Lett.,* 197(1):13-16.

[234] Carlson, S. S. 1996. SV2proteoglycan: a potential synaptic vesicle transporter and nerve terminal extracellular matrix receptor. *Perspect. Dev. Neurobiol.,* 3(4):373-386.

[235] Margolis, R. K., Rauch, U., Maurel, P., and Margolis, R. U. Neurocan and phosphacan: two major nervous tissue-specific chondroitin sulfate proteoglycans. *Perspect. Dev. Neurobiol.*, 3(4):273-290.

[236] Burg, M. A., and Cole, G. J. 1994. Claustrin, an antiadhesive neural keratan sulfate proteoglycan, is structurally related to MAP1B. *J. Neurobiol.*, 25(1):1-22.

[237] Carlson, S. S., Iwata, M., and Wight, T. N. 1996. A chondroitin sulfate/keratan sulfate proteoglycan, PG-1000, forms complexes which are concentrated in the reticular laminae of electric organ basement membranes. *Matrix Biol.,* 15(4):281-292.

[238] Lauder, R. M., Huckerby, T. N., and Nieduszynski, I. A. 2011. Lectin affinity chromatography of articular cartilage fibromodulin: Some molecules have keratan sulphate chains exclusively capped by α(2-3)-linked sialic acid. *Glycoconj J.*, 28:453-461.

[239] Flannery, C. R., Little, C. B., and Caterson, B. 1998. Molecular cloning and sequence analysis of the aggrecan interglobular domain from porcine, equine, bovine and ovine cartilage: comparison of proteinase-susceptible regions and sites of keratan sulfate substitution. *Matrix Biol.*, 16:507-511.

[240] Snow, A. D., Nochlin, D., Sekiguichi, R., and Carlson, S. S. 1996. Identification in immunolocalization of a new class of proteoglycan (keratan sulfate) to the neuritic plaques of Alzheimer's disease. *Exp. Neurol.*, 138(2):305-317.

[241] Bettelheim, F. A., and Plessy, B. 1975.The hydration of proteoglycans of bovine cornea. *Biochim. Biophys. Acta,* 381(1):203-214.

[242] Chakravarti, S., Magnuson, T., Lass, J. H., Jepsen, K. J., LaMantia, C., and Carroll, H. 1998. Lumican regulates collagen fibril assembly: skin fragility and corneal opacity in the absence of lumican. *J. Cell Biol.,* 141(5):1277-1286.

[243] Saika, S., Miyamoto, T., Tanaka, S., Tanaka, T., Ishida, I., Ohnishi, Y., Ooshima, A., Ishiwata, T., Asano, G., Chikama, T., Shiraishi, A., Liu, C. Y., Kao, C. W., and Kao, W. W. 2003. Response of lens epithelial cells to injury: role of lumican in epithelial-mesenchymal transition. *Invest. Ophthalmol. Vis. Sci.*, 44(5):2094-2102.

[244] Yole, M., and Gibson, J. G. 2005. Organic farmers are required to medicate sick animals. *Can. Vet. J.,* 46(6):479.

[245] Kao, W. W., Funderburgh, J. L., Xia, Y., Liu, C. Y., and Conrad, G. W. 2006. Focus on molecules: lumican. *Exp. Eye Res.*, 82(1):3-4.

[246] DeLoia, J. A., Krasnow, J. S., Brekosky, J., Babaknia, A., Julian, J., and Carson, D. D. 1998. Regional specialization of the cell membrane-associated, polymorphic mucin (MUC1) in human uterine epithelia. *Hum. Reprod.*, 13(1O):2902-2909.

[247] Davies, Y., Lewis, D., Fullwood, N. J., Nieduszynski, I. A., Marcyniuk, B., Albon, J., and Tullo, A. 1999. Proteoglycans on normal and migrating human corneal endothelium. *Exp. Eye Res.*, 68(3):303-311.

[248] Funderburgh, J. L., Mitschler, R. R., Funderburgh, M. L., Roth, M. R., Chapes, S. K., and Conrad, G. W. 1997. Macrophage receptors for lumican. A corneal keratan sulfate proteoglycan. *Invest. Ophthalmol. Vis. Sci.,* 38(6):1159-1167.

[249] Imagama, S., Sakamoto, K., Tauchi, R., Shinjo, R., Ohgomori, T., Ito, Z., Zhang, H., Nishida, Y., Asami, N., Takeshita, S., Sugiura, N., Watanabe, H., Yamashita, T., Ishiguro, N., Matsuyama, Y., and Kadomatsu, K. 2011. Keratan sulfate restricts neural plasticity after spinal cord injury. *J. Neurosci.,* 31(47):17091-17102.

[250] Guerassimov, A., Zhang, Y., Cartman, A., Rosenberg, L. C., Esdaile, J., Fitzcharles, M. A., and Poole, A. R. 1999. Immune responses to cartilage link protein and the G1 domain of proteoglycan aggrecan in patients with osteoarthritis. *Arthritis Rheum.*, 42(3):527-533.

[251] Niewiarowska, J., Brézillon, S., Sacewicz-Hofman, I., Bednarek, R., Maquart, F. X., Malinowski, M., Wiktorska, M., Wegrowski, Y., and Cierniewski, C. S. 2011. Lumican inhibits angiogenesis by interfering with α2β1 receptor activity and downregulating MMP-14 expression. *Thromb Res.*, 128:452-457.

[252] Nakayama, F., Umeda, S., Ichimiya, T., Kamiyama, S., Hazawa, M., Yasuda, T., Nishihara, S., and Imai, T. 2013. Sulfation of keratan sulfate proteoglycan reduces radiation-induced apoptosis in human Burkitt's lymphoma cell lines. *FEBS Lett.*, 587:231-237.

[253] Dũng, V. C., Tomatsu, S., Montaño, A. M., Gottesman, G., Bober, M. B., Mackenzie, W., Maeda, M., Mitchell, G. A., Suzuki, Y., and Orii, T. 2013. Mucopolysaccharidosis IVA: correlation between genotype, phenotype and keratan sulfate levels. *Mol. Genet. Metab.*, 110(1-2):129-138.

[254] Tomatsu, S., Montaño, A. M., Oikawa, H., Smith, M., Barrera, L., Chinen, Y., Thacker, M. M., Mackenzie, W. G., Suzuki, Y., and Orii, T. 2011. Mucopolysaccharidosis type IVA (Morquio A disease): clinical review and current treatment. *Curr. Pharm. Biotechnol.*, 12(6):931-945.

[255] Tomatsu, S., Mackenzie, W. G., Theroux, M. C., Mason, R. W., Thacker, M. M., Shaffer, T. H., Montaño, A. M., Rowan, D., Sly, W., Alméciga-Díaz, C. J., Barrera, L. A., Chinen, Y., Yasuda, E., Ruhnke, K., Suzuki, Y., and Orii, T. 2012. Current and emerging treatments and surgical interventions for Morquio A Syndrome: A review. *Res. Rep. End. Dis.*, 2012(2):65-77.

[256] Tomatsu, S., Fujii, T., Fukushi, M., Oguma, T., Shimada, T., Maeda, M., Kida, K., Shibata, Y., Futatsumori, H., Montaño, A. M., Mason, R. W., Yamaguchi, S., Suzuki, Y., and Orii, T. 2013. Newborn screening and diagnosis of mucopolysaccharidoses. *Mol. Genet. Metab.*, 110(1-2): 42-53.

[257] Tomatsu, S., Alméciga-Díaz, C. J., Barbosa, H., Montaño, A. M., Barrera, L. A., Shimada, T., Yasuda, E., Mackenzie, W. G., Mason, R., Suzuki, Y., Orii, K. E., and Orii, T. 2013. Therapies of Mucopolysaccharidosis IVA (Morquio A Syndrome). *Expert Opin.Orphan Drugs,* 1:805-818.

[258] Montaño, A. M., Tomatsu, S., Gottesman, G. S., Smith, M., and Orii, T. 2007. International Morquio A Registry: clinical manifestation and natural course of Morquio A disease. *J. Inherit. Metab. Dis.*, 30(2):165-174.

[259] Montaño, A. M., Tomatsu, S., Brusius, A., Smith, M., and Orii, T. 2008. Growth charts for patients affected with Morquio A disease. *Am. J. Med. Genet. A.,* 146A(10):1286-1295.

[260] Suzuki, Y., Oshima, A., and Namba, E. 2001. β-Galactosidase deficiency (β-galactosidosis) GM1 gangliosidosis and Morquio B disease. In: Scriver C. R., Beaudet A. L., Sly W. S., Valle D., editors. The Metabolic and Molecular Bases of Inherited Disease. (pp. 3775–3809). New York: McGraw-Hill.

[261] Hendriksz, C. J., Harmatz, P., Beck, M., Jones, S., Wood, T., Lachman, R., Gravance, C. G., Orii, T., and Tomatsu, S. 2013. Review of clinical presentation and diagnosis of mucopolysaccharidosis IVA. *Mol. Genet. Metab.,* 110(1-2):54-64.

[262] Möllmann, C., Lampe, C. G., Müller-Forell, W., Scarpa, M., Harmatz, P., Schwarz, M., Beck, M., and Lampe, C. 2013. Development of a Scoring System to Evaluate the Severity of Craniocervical Spinal Cord Compression in Patients with Mucopolysaccharidosis IVA (Morquio A Syndrome). *JIMD Rep.*, 11:65-72.

[263] Harmatz, P., Mengel, K. E., Giugliani, R., Valayannopoulos, V., Lin, S. P., Parini, R., Guffon, N., Burton, B. K., Hendriksz, C. J., Mitchell, J., Martins, A., Jones, S., Guelbert, N., Vellodi, A., Hollak, C., Slasor, P., and Decker, C. 2013. The Morquio A Clinical Assessment Program: baseline results illustrating progressive, multisystemic clinical impairments in Morquio A subjects. *Mol. Genet. Metab.*, 109(1):54-61.

[264] Tomatsu, S., Montaño, A. M., Oguma, T., Dung, V. C., Oikawa, H., de Carvalho, T. G., Gutiérrez, M. L., Yamaguchi, S., Suzuki, Y., Fukushi, M., Kida, K., Kubota, M., Barrera, L., and Orii, T. 2010. Validation of keratan sulfate level in mucopolysaccharidosis type IVA by liquid chromatography-tandem mass spectrometry. *J. Inherit. Metab. Dis.,* 33(Suppl. 3):S35-S42.

[265] Tomatsu, S., Montaño, A. M., Ohashi, A., Gutierrez, M. A., Oikawa, H., Oguma, T., Dung, V. C., Nishioka, T., Orii, T., and Sly, W. S. 2008. Enzyme replacement therapy in a murine model of Morquio A syndrome. *Hum. Mol. Genet.*, 17(6):815-824.

[266] Chinen, Y., Higa, T., Tomatsu, S., Suzuki, Y., Orii, T., Hyakuna, N. 2014. Long-term therapeutic efficacy of allogenic bone marrow transplantation in a patient with mucopolysaccharidosis IVA. *Mol. Genet. Metab. Rep.,* 1:31–41.

[267] Hendriksz, C. J., Burton, B., Fleming, T. R., Harmatz, P., Hughes, D., Jones, S. A., Lin, S. P., Mengel, E., Scarpa, M., Valayannopoulos, V., Giugliani, R.; STRIVE Investigators, Slasor, P., Lounsbury, D., and Dummer, W. 2014. Efficacy and safety of enzyme replacement therapy with BMN 110 (elosulfase alfa) for Morquio A syndrome (mucopolysaccharidosis IVA): a phase 3 randomised placebo-controlled study. *J. Inherit. Metab. Dis.*, [Epub ahead of print].

[268] Tomatsu, S., Montaño, A. M., Oikawa, H., Dũng, V. C., Hashimoto, A., Oguma, T., Gutiérrez, M. L., Takahashi, T., Shimada, T., Orii, T., Sly W. S. 2014. Enzyme replacement therapy in newborn mucopolysaccharidosis IVA mice: early treatment rescues bone lesions? *Mol. Genet. Metab.*, [Epub ahead of print].

[269] T. Saito, K. Nishida, J. Nakayama, T. O. Akama, M. N. Fukuda, K. Watanabe, A. J. Quantock, N. Maeda, H. Watanabe, and Y. Tano. 2008. Sulfation patterns of keratan sulfate in different macular corneal dystrophy immunophenotypes using three different probes. *Br. J. Ophthalmol.*, 92(10):1434–1436.

[270] Klintworth, G. K. 2009. Corneal dystrophies. *Orphanet J. Rare Dis.,* 23;4:7. doi: 10.1186/1750-1172-4-7.

[271] Pan, S., Chen, R., Brand, R. E., Hawley, S., Tamura, Y., Gafken, P. R., Milless, B. P., Goodlett, D. R., Rush, J., and Brentnall, T. A. 2012. Multiplex targeted proteomic assay for biomarker detection in plasma: a pancreatic cancer biomarker case study. *J. Proteome. Res.,* 11(3):1937-1948.

[272] Coulson-Thomas, V. J., Coulson-Thomas, Y. M., Gesteira, T. F., Andrade de Paula, C. A., Carneiro, C. R., Ortiz, V., Toma, L., Kao, W. W., and Nader, H. B. 2013. Lumican expression, localization and antitumor activity in prostate cancer. *Exp. Cell Res.*, 319(7):967-981.

[273] Brézillon, S., Venteo, L., Ramont, L., D'Onofrio, M. F., Perreau, C., Pluot, M., Maquart, F. X., and Wegrowski, Y. 2007. Expression of lumican, a small leucine-rich

proteoglycan with antitumour activity, in human malignant melanoma. *Clin. Exp. Dermatol.*, 32(4):405-416.

[274] Brézillon, S., Radwanska, A., Zeltz, C., Malkowski, A., Ploton, D., Bobichon, H., Perreau, C., Malicka-Blaszkiewicz, M., Maquart, F. X., and Wegrowski, Y. 2009. Lumican core protein inhibits melanoma cell migration via alterations of focal adhesion complexes. *Cancer Lett.,* 283(1):92-100.

[275] Lawrence, R. C., Helmick, C. G., Arnett, F. C., Deyo, R. A., Felson, D. T., Giannini, E. H., Heyse, S. P., Hirsch, R., Hochberg, M. C., Hunder, G. G., Liang, M. H., Pillemer, S. R., Steen, V. D., and Wolfe, F. 1998. Estimates of the prevalence of arthritis and selected musculoskeletal disorders in the United States. *Arthritis Rheum.*, 41(5):778–799.

[276] Leipold, H. R., Goldberg, R. L., and Lust, G. (1989). Canine serum keratan sulfate and hyaluronate concentrations. Relationship to age and osteoarthritis. *Arthritis Rheum.*, 32(3), 312-321.

[277] Tang, T., Muneta, T., Ju, Y. J., Nimura, A., Miyazaki, K., Masuda, H., Mochizuki, T., and Sekiya, I. 2008. Serum keratan sulfate transiently increases in the early stage of osteoarthritis during strenuous running of rats: protective effect of intraarticular hyaluronan injection. *Arthritis Res. Ther.*, 10(1):R13.

[278] Wakitani, S., Nawata, M., Kawaguchi, A., Okabe, T., Takaoka, K., Tsuchiya, T., Nakaoka, R., Masuda, H., and Miyazaki, K. 2007. Serum keratan sulfate is a promising marker of early articular cartilage breakdown. *Rheumatology,* (Oxford). 46(11):1652-1656.

[279] Okumura, M., Tagami, M., and Fujinaga, T. 1998. Measurement of serum and synovial fluid keratan sulphate and antibody to collagen type II in equine osteoarthritis. *J. Vet. Med. A.*, 45(8):513–516.

[280] Fawthrop, F., Yaqub, R., Belcher, C., Bayliss, M., Ledingham, J., and Doherty, M. 1997. Chondroitin and keratan sulphate epitopes, glycosaminoglycans, and hyaluronan in progressive versus non-progressive osteoarthritis. *Ann. Rheum. Dis.*, 56(2):119-122.

[281] Caterson, B., Christner, J. E., and Baker, J. R. 1983. Identification of a monoclonal antibody that specifically recognizes corneal and skeletal keratan sulfate. Monoclonal antibodies to cartilage proteoglycan. *J. Biol. Chem.,* 25(14):8848–8854.

[282] Cappelletti, R., Del Rosso, M., Chiarugi, V. P. 1979. A new electrophoretic method for the complete separation of all known animal glycosaminoglycans in a monodimensional run. *Anal. Biochem.,* 99: 311-315.

[283] Hopwood, J. J., Harrison, J. R. 1982. High-resolution electrophoresis of urinary glycosaminoglycans: an improved screening test for the mucopolysaccharidoses. *Anal. Biochem.*, 119:120-127.

[284] Mehmet, H., Scudder, P., Tang, P. W., Hounsell, E. F., Caterson, B., and Feizi, T. 1986. The antigenic determinants recognized by three monoclonal antibodies to keratan sulphate involve sulphated hepta- or larger oligosaccharides of the poly (N-acetyllactosamine) series. *Eur. J. Biochem.,* 157(2):385–391.

[285] Hosogane, N., Watanabe, K., Tsuji, T., Miyamoto, T., Ishii, K., Niki, Y., Nakamura, M., Toyama, Y., Chiba, K., and Matsumoto, M. 2012. Serum cartilage metabolites as biomarkers of degenerative lumbar scoliosis. *J. Orthop. Res.*, 30(8):1249-1253.

[286] Zhang, H., Uchimura, K., Kadomatsu, K. 2006. Brain keratan sulfate and glial scar formation. *Ann. N Y Acad. Sci.*, 1086:81-90.

[287] Jones, L. L., and Tuszynski, M. H. 2002. Spinal cord injury elicits expression of keratan sulfate proteoglycans by macrophages, reactive microglia, and oligodendrocyte progenitors. *J. Neurosci.*, 22(11):4611-4624.

[288] Krautstrunk, M., Scholtes, F., Martin, D., Schoenen, J., Schmitt, A. B., Plate, D., Nacimiento, W., Noth, J., and Brook, G. A. 2002. Increased expression of the putative axon growth-repulsive extracellular matrix molecule, keratan sulphate proteoglycan, following traumatic injury of the adult rat spinal cord. *Acta Neuropathol.*, 104(6):592-600.

[289] Hirano, K., Ohgomori, T., Kobayashi, K., Tanaka, F., Matsumoto, T., Natori, T., Matsuyama, Y., Uchimura, K., Sakamoto, K., Takeuchi, H., Hirakawa, A., Suzumura, A., Sobue, G., Ishiguro, N., Imagama, S., and Kadomatsu, K. 2013. Ablation of keratan sulfate accelerates early phase pathogenesis of ALS. *PLoS One*, 8(6):e66969.

[290] Ito, Z., Sakamoto, K., Imagama, S., Matsuyama, Y., Zhang, H., Hirano, K., Ando, K., Yamashita, T., Ishiguro, N., and Kadomatsu, K. 2010. N-acetylglucosamine 6-O-sulfotransferase-1-deficient mice show better functional recovery after spinal cord injury. *J. Neurosci.*, 30:5937-5947.

[291] Zhang, H., Muramatsu, T., Murase, A., Yuasa, S., Uchimura, K., and Kadomatsu, K. 2006. N-Acetylglucosamine 6-O-sulfotransferase-1 is required for brain keratan sulfate biosynthesis and glial scar formation after brain injury. *Glycobiology,* 16(8):702-710.

[292] Jander, S., and Stoll, G. 1996. Downregulation of microglial keratan sulfate proteoglycans coincident with lymphomonocytic infiltration of the rat central nervous system. *Am. J. Pathol.*, 148(1):71-78.

[293] Jander, S., Schroeter, M., Fischer, J., and Stoll, G. 2000. Differential regulation of microglial keratan sulfate immunoreactivity by proinflammatory cytokines and colony-stimulating factors. *Glia,* 30(4):401-410.

[294] Matsui, H., Ohgomori, T., Natori, T., Miyamoto, K., Kusunoki, S., Sakamoto, K., Ishiguro, N., Imagama, S., and Kadomatsu, K. 2013. Keratan sulfate expression in microglia is diminished in the spinal cord in experimental autoimmune neuritis. *Cell Death Dis.,* 4:e946.

[295] Lindahl, B., Eriksson, L., Spillmann, D., Caterson, B., and Lindahl, U. 1996. Selective loss of cerebral keratan sulfate in Alzheimer's disease. *J. Biol. Chem.*, 271(29):16991-16994.

[296] Ahn, J. M., Kim, B. G., Yu, M. H., Lee, I. K., and Cho, J. Y. 2010. Identification of diabetic nephropathy-selective proteins in human plasma by multi-lectin affinity chromatography and LC-MS/MS. *Proteomics Clin. Appl.*, 4(6-7):644-653.

[297] Mardinoglu, A., Agren, R., Kampf, C., Asplund, A., Nookaew, I., Jacobson, P., Walley, A. J., Froguel, P., Carlsson, L. M., Uhlen, M., and Nielsen, J. 2013. Integration of clinical data with a genome-scale metabolic model of the human adipocyte. *Mol. Syst. Biol.*, 9:649.

[298] Messier, S. P., Loeser, R. F., Mitchell, M. N., Valle, G., Morgan, T. P., Rejeski, W. J., and Ettinger, W. H. 2000. Exercise and weight loss in obese older adults with knee osteoarthritis: A preliminary study. *J. Am. Geriatr. Soc.,* 48(9):1062–1072.

[299] Engebretsen, K. V., Lunde, I. G., Strand, M. E., Waehre, A., Sjaastad, I., Marstein, H. S., Skrbic, B., Dahl, C. P., Askevold, E. T., Christensen, G., Bjørnstad, J. L., and Tønnessen, T. 2013. Lumican is increased in experimental and clinical heart failure,

and its production by cardiac fibroblasts is induced by mechanical and proinflammatory stimuli. *FEBS J.*, 280(10):2382-2398.

[300] Li, L. F., Chu, P. H., Hung, C. Y., Kao, W. W., Lin, M. C., Liu, Y. Y., and Yang, C. T. 2013. Lumican regulates ventilation-induced epithelial-mesenchymal transition through extracelluar signal-regulated kinase pathway. *Chest*, 143:125212-125260.

[301] Charlton, M., Viker, K., Krishnan, A., Sanderson, S., Veldt, B., Kaalsbeek, A. J., Kendrick, M., Thompson, G., Que, F., Swain, J., and Sarr, M. 2009. Differential expression of lumican and fatty acid binding protein-1: new insights into the histologic spectrum of nonalcoholic fatty liver disease. *Hepatology*, 49(4):1375-1384.

[302] Krishnan, A., Li, X., Kao, W. Y., Viker, K., Butters, K., Masuoka, H., Knudsen, B., Gores, G., and Charlton, M. 2012. Lumican, an extracellular matrix proteoglycan, is a novel requisite for hepatic fibrosis. *Lab. Invest.*, 92:1712-1725.

[303] Shirato, K., Gao, C., Ota, F., Angata, T., Shogomori, H., Ohtsubo, K., Yoshida, K., Lepenies, B., and Taniguchi, N. 2013. Flagellin/Toll-like receptor 5 response was specifically attenuated by keratan sulfate disaccharide via decreased EGFR phosphorylation in normal human bronchial epithelial cells. *Biochem. Biophys. Res. Commun.*, 435(3): 460-465.

[304] Liu, C. Y., and Kao, W. W. Lumican promotes corneal epithelial wound healing. *Methods Mol. Biol.*, 836:85-90.

[305] Hayashi, M., Kadomatsu, K., and Ishiguro, N. Keratan sulfate suppresses cartilage damage and ameliorates inflammation in an experimental mice arthritis model. *Biochem. Biophys. Res. Commun.*, 401(3):463-468.

[306] Hayashi, M., Kadomatsu, K., Kojima, T., and Ishiguro, N. 2011. Keratan sulfate and related murine glycosylation can suppress murine cartilage damage in vitro and in vivo. *Biochem. Biophys. Res. Commun.*, 409(4):732-737.

[307] Farndale, R. W., Buttle, D. J. and Barrett, A. J. 1986. Improved quantitation and discrimination of sulphated glycosaminoglycans by use of dimethylmethylene blue. *Biochim. Biophys. Acta*, 883(2):173-177.

[308] de Jong, J. G., Wevers, R. A., and Liebrand-van Sambeek, R. 1992. Measuring urinary glycosaminoglycans in the presence of protein: an improved screening procedure for mucopolysaccharidoses based on dimethylmethylene blue. *Clin. Chem.*, 38(6):803-807.

[309] Whitley, C. B., Spielmann, R. C., Herro, G. and Teragawa, S. S. 2002. Urinary glycosaminoglycan excretion quantified by an automated semimicro method in specimens conveniently transported from around the globe. *Mol. Genet. Metab.*, 75(1):56-64.

[310] Shinmei, M., Miyauchi, S., Machida, A., Miyazaki, K. 1992. Quantitation of chondroitin 4-sulfate and chondroitin 6-sulfate in pathologic joint fluid. *Arthritis Rheum.*, 35(11):1304-1308.

[311] Yoshida, K., Miyauchi, S., Kikuchi, H., Tawada, A. and Tokuyasu, K. 1989. Analysis of unsaturated disaccharides from glycosaminoglycuronan by high-performance liquid chromatography. *Anal. Biochem.*, 177(2):327-332.

[312] Toyoda, H., Demachi, Y., Komoriya, S., Furuya, N., Toida, T. and Imanari, T. 1998. Characterization and determination of human urinary keratan sulfate. *Chem. Pharm. Bull.*, (Tokyo). 46(1):97-101.

[313] Kinoshita, A. and Sugahara, K. 1999. Microanalysis of glycosaminoglycan-derived oligosaccharides labeled with a fluorophore 2-aminobenzamide by high-performance

liquid chromatography: application to disaccharide composition analysis and exosequencing of oligosaccharides. *Anal. Biochem.*, 269:367-378.

[314] Yamada, H., Miyauchi, S., Morita, M., Yoshida, Y., Yoshihara, Y., Kikuchi, T., Washimi, O., Washimi, Y., Terada, N., Seki, T. and Fujikawa, K. 2000. Content and sulfation pattern of keratan sulfate in hip osteoarthritis using high performance liquid chromatography. *J. Rheumatol.,* 27(7):1721-1724.

[315] Tomatsu, S., Okamura, K., Taketani, T., Orii, K. O., Nishioka, T., Gutierrez, M. A., Velez-Castrillon, S., Fachel, A. A., Grubb, J. H., Cooper, A., Thornley, M., Wraith, E., Barrera, L. A., Giugliani, R., Schwartz, I. V., Frenking, G. S., Beck, M., Kircher, S. G., Paschke, E., Yamaguchi, S., Ullrich, K., Isogai, K., Suzuki, Y., Orii, T., Kondo, N., Creer, M. and Noguchi, A. 2004. Development and testing of new screening method for keratan sulfate in mucopolysaccharidosis IVA. *Pediatr. Res.*, 55(4):592-597.

[316] Tomatsu, S., Okamura, K., Maeda, H., Taketani, T., Castrillon, S. V., Gutierrez, M. A., Nishioka, T., Fachel, A. A., Orii, K. O., Grubb, J. H., Cooper, A., Thornley, M., Wraith, E., Barrera, L. A., Laybauer, L. S., Giugliani, R., Schwartz, I. V., Frenking, G. S., Beck, M., Kircher, S. G., Paschke, E., Yamaguchi, S., Ullrich, K., Haskins, M., Isogai, K., Suzuki, Y., Orii, T., Kondo, N., Creer, M., Okuyama, T., Tanaka, A. and Noguchi, A. 2005. Keratan sulphate levels in mucopolysaccharidoses and mucolipidoses. *J. Inherit. Metab. Dis.,* 28(2):187-202.

[317] Tomatsu, S., Gutierrez, M. A., Ishimaru, T., Pena, O. M., Montano, A. M., Maeda, H., Velez-Castrillon, S., Nishioka, T., Fachel, A. A., Cooper, A., Thornley, M., Wraith, E., Barrera, L. A., Laybauer, L. S., Giugliani, R., Schwartz, I. V., Frenking, G. S., Beck, M., Kircher, S. G., Paschke, E., Yamaguchi, S., Ullrich, K., Isogai, K., Suzuki, Y., Orii, T. and Noguchi, A. 2005. Heparan sulfate levels in mucopolysaccharidoses and mucolipidoses. *J. Inherit. Metab. Dis.,* 28:743-757.

[318] Oguma, T., Tomatsu, S., Montano, A. M. and Okazaki, O. 2007. Analytical method for the determination of disaccharides derived from keratan, heparan, and dermatan sulfates in human serum and plasma by high-performance liquid chromatography/turbo ionspray ionization tandem mass spectrometry. *Anal. Biochem.*, 368(1):79-86.

[319] Oguma, T., Tomatsu, S. and Okazaki, O. 2007. Analytical method for determination of disaccharides derived from keratan sulfates in human serum and plasma by high-performance liquid chromatography/turbo-ionspray ionization tandem mass spectrometry. *Biomed. Chromatogr.,* 21(4):356-362.

[320] Tomatsu, S., Montaño, A. M., Oguma, T., Dung, V. C., Oikawa, H., Gutiérrez, M. L., Yamaguchi, S., Suzuki, Y., Fukushi, M., Barrera, L. A., Kida, K., Kubota, M. and Orii, T. 2010. Validation of disaccharide compositions derived from dermatan sulfate and heparan sulfate in mucopolysaccharidoses and mucolipidoses II and III by tandem mass spectrometry. *Mol. Genet. Metab.*, 99(2):124-131.

[321] Tomatsu, S., Montaño, A. M., Oguma, T., Dung, V. C., Oikawa, H., de Carvalho, T. G., Gutiérrez, M. L., Yamaguchi, S., Suzuki, Y., Fukushi, M., Sakura, N., Barrera, L., Kida, K., Kubota, M. and Orii, T. 2010. Dermatan sulfate and heparan sulfate as a biomarker for mucopolysaccharidosis I. *J. Inherit. Metab. Dis.,* 33(2):141-150.

[322] Zhang, Y., Kariya, Y., Conrad, A. H., Tasheva, E. S. and Conrad, G. W. 2005. Analysis of keratan sulfate oligosaccharides by electrospray ionization tandem mass spectrometry. *Anal. Chem.*, 77(3):902-910.

[323] Shimada T., Tomatsu S., Yasuda E., Mason R. W., Mackenzie W. G., Shibata Y., Kubaski F., Giugliani R., Yamaguchi S., Suzuki Y., Orii K., Orii T. Chondroitin 6-Sulfate as a Novel Biomarker for Mucopolysaccharidosis IVA and VII. *JIMD Rep.*, 2014 May 22. [Epub ahead of]

In: Sulfated Polysaccharides
Editors: M. Gama, H. Nader, H. de Oliveira Rocha

ISBN: 978-1-63482-975-5

Chapter 3

CHARACTERIZATION AND CHEMICAL MODIFICATION OF GLYCOSAMINOGLYCANS OF THE EXTRACELLULAR MATRIX

***J. Schiller*[1*]*, J. Becher*[2]*, S. Möller*[2]*, K. Lemmnitzer*[1]*, T. Riemer*[1] *and M. Schnabelrauch*[2]**

[1]University of Leipzig, Medical Faculty,
Institute of Medical Physics and Biophysics, Leipzig, Germany
[2]INNOVENT e. V., Biomaterials Department, Jena, Germany

ABSTRACT

Glycosaminoglycans (GAGs) occur in many connective tissues such as cartilage and skin. Due to their significant water-binding capacity, GAGs are essential for the biomechanical properties of these tissues. In addition, GAGs participate in many biochemical processes of the cells, mostly through binding to proteins and regulating, thus, their biological activities. Certain GAGs (such as heparin) have also interesting pharmacological properties.

There is nowadays growing evidence that the sulfation of GAGs does not occur at random, but a "sulfation code" exists that mediates the physiological functions of GAGs. Thus, the biochemical properties of GAGs are influenced by (a) the monosaccharide composition, (b) the degree of sulfation, and (c) the sulfate group distribution along the polymer chain. Finally, it is commonly accepted that GAGs represent promising biomaterials in the field of tissue engineering. Overall, GAGs are experiencing increasing (bio)medical interest.

In this book chapter we will focus on (chemically) sulfated GAGs and provide a survey of these compounds on three different levels. First, the structures and peculiarities of the native GAGs (with the focus on CS and DS) will be discussed. Second, we will provide an overview on chemical functionalization strategies of naturally occurring GAGs with special emphasis on derivatives, which have gained attention as promising biomaterials in tissue regeneration. Third, chemical and biochemical methods to

[*] All authors contributed equally to this work. Email address: juergen.schiller@medizin.uni-leipzig.de.

characterize GAGs will be introduced with the focus on methods based on nuclear magnetic resonance (NMR) and mass spectrometry (MS).

1. Introduction

The "ExtraCellular Matrix" (ECM) is a biological tissue that occurs to a significant extent in all vertebrates. Connective tissues such as skin, bone, cartilage and the intervertebral discs are particularly rich in ECM but differ in the collagen/glycosaminoglycan (GAG) ratio as well as the alignments of the individual compounds [1]. There are numerous (often inflammatory) diseases such as rheumatoid arthritis that affect the structure of the related biopolymers and/or the composition of the ECM [2]. Although these diseases are not acute life-threatening, they cause immense costs for the health and social insurance systems. Therefore, recent attempts of "regenerative medicine" [3] are often focused on the treatment of ECM-related diseases.

It is remarkable that the regenerative capacities of the individual tissues are very different. While bone is characterized by a reasonable self healing capacity (provided the defect does not exceed a critical size limit), skin and particularly (articular) cartilage has only a limited regeneration potential [4] - most probably due to the relatively small cell number and the fact that these tissues are bradytroph, i.e., there are no blood vessels and, thus, transport processes (of nutrients into and metabolic waste products out of the tissue) are exclusively mediated by diffusion.

Nevertheless, the organism often requires external "help" to overcome damages of the ECM: the transplantation of tissue cells (e.g., chondrocytes in the case of cartilage) [5] or stem cells [6] that can be differentiated into the respective tissue cells are currently the most promising approaches to treat ECM-related diseases.

Additionally, there is also considerable interest in the enhancement of the long-term stability of implants (such as titanium in dental implants) [7]. The improvement of implant stability requires, on the one hand, attraction of tissue cells and, on the other hand, the reduction of inflammatory processes mediated particularly by granulocytes and macrophages. There is sufficient evidence that coating of the implant by selected, chemically modified components of the ECM (such as sulfated GAGs) may improve the long-term stabilities of the implants [8]. It is interesting to note that sulfated GAGs occur exclusively in the animal kingdom. GAGs presumably appeared with the cnidarians, the earliest animals assumed to have muscles and a nervous system [9]. In mice, the complete elimination of GAG biosynthesis results in embryonic lethality. Knockout of specific GAG-modifying enzymes results in a variety of organ-specific phenotypes of varying severity [10].

It is reasonable to assume that the physiological functions of GAGs are mediated by protein binding: due to the anionic character of GAGs, these polysaccharides have the tendency to bind basic (cationic) protein domains that are found in numerous classes of growth factors and the related receptors [11]. Therefore, one might argue that protein-GAG interactions are comparatively nonspecific.

However, since a few years there is increasing evidence that the interaction between GAGs and proteins is much more specific than initially assumed [12] and this may also explain the different effects of GAGs on different cell lines [13].

This review consists of three different major parts. First, the structures and peculiarities of the native GAGs (with the focus on CS and DS) will be discussed. Second, we will provide an overview on chemical functionalization strategies of naturally occurring GAGs with special emphasis on GAG derivatives that gained attraction as promising biomaterials in tissue regeneration. Third, chemical and biochemical methods to characterize GAGs will be introduced with the focus on methods based on nuclear magnetic resonance (NMR) and mass spectrometry (MS).

2. Architecture of the Native Extracellular Matrix

ECM is a complex, interlocking meshwork formed by fibrous proteins. In this meshwork, proteoglycans in the form of hydrated gels fill the majority of the interstitial space [1]. A sketch of the ECM structure and a typical proteoglycan (PG) aggregate is shown in Figure 1.

As typical for biological tissues, the most abundant constituent of the ECM is water which constitutes about 80 % of the wet weight of the ECM and is very important regarding the nutrition of the cells in the ECM (and also the removal of metabolic waste products) because these processes rely exclusively on diffusion [14].

The main fibrous proteins in the ECM are collagens contributing significantly to the integrity and mechanical properties of the individual tissues [15]. Further fibrous proteins of the ECM include elastin, fibronectin and laminin [16] but will not be discussed here due to the limited space. Proteoglycans consist of a core protein to which one ore more sulfated GAG chains are covalently attached via an oligosaccharide linker. The GAG chains on the core protein are unbranched polysaccharides which can be further classified into sulfated (chondroitin sulfate (CS), dermatan sulfate (DS), keratan sulfate (KS), heparan sulfate (HS), heparin (HE)) and the non-sulfated GAG hyaluronan (HA). These compounds (which we will focus on in this review) are very hydrophilic macromolecules able to adopt highly extended conformations that are essential for hydrogel formation - a putative prerequisite to induce healing processes.

Among native GAGs the highly sulfated ones, heparin and heparan sulfate, occur only in very low quantities in biological tissues [17]. However, at least since the discovery of the so-called "sulfation code" those derivatives and their chemically sulfated analogs have attracted significant attention [4, 18,].

The term "sulfation code" implies that GAGs with a specific sulfation pattern exhibit direct effects on the metabolism of cells which are involved either in tissue regeneration (e.g., fibroblasts, chondrocytes, etc.) [19] or the autoimmune response (e. g. macrophages or neutrophils) [20].

Furthermore, there is an increasing evidence that GAG/PG are involved in cancer growth and progression [21] and play also a significant role in the human brain [22]. In the following chapter the main structural features of native GAGs are briefly discussed.

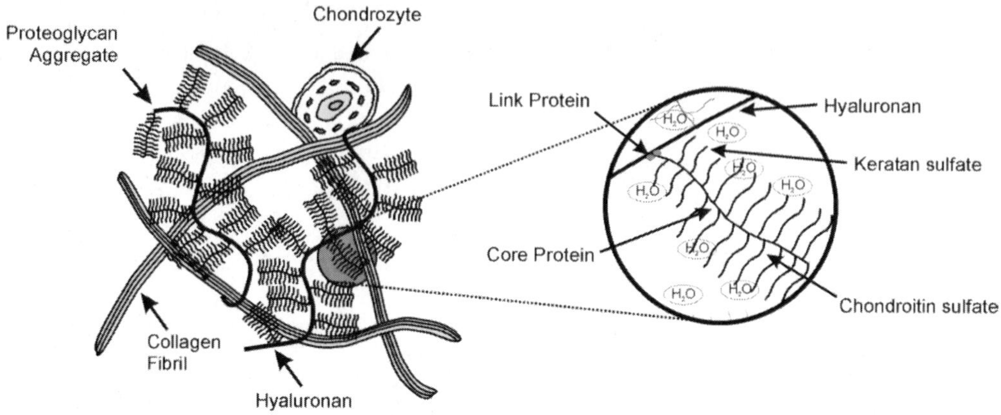

Figure 1. Scheme of the molecular organization of articular cartilage (left). The architecture of a proteoglycan aggregate is emphasized at the right. Note the considerable water content of the ECM.

3. Structural Aspects of Native Glycosaminoglycans (GAGs)

GAGs are natural, anionic, non-branched and multifunctional heteropolysaccharides composed of characteristic disaccharide repeating units. Normally, the disaccharide unit is formed from D-glucuronic acid (GlcA) or L-iduronic acid (IdoA) (only keratan sulfate (KS) has a galactose instead of an uronic acid moiety) linked to a D-*N*-acetylglucosamine (GlcNAc) or D-*N*-acetylgalactosamine (GalNAc) residue. The characteristic repeating units of the different GAGs are schematically illustrated in Figure 2.

HA is the only non-sulfated GAG consisting of repeating disaccharide units of D-*N*-acetylglucosamine (GlcNAc) and D-glucuronic acid (GlcA) linked by alternating β-1→3 and β-1→4 glycosidic linkages. All other GAGs are modified by post-biosynthetic modifications, such as *O*-sulfation (sulfuric acid half-ester), C-5-epimerization to form iduronic acid (IdoA) residues and de-*N*-acetylation to produce glucosamine *N*-sulfate GlcN-sulfate) residues, which are particularly typical of heparin [23, 24]. These modifications play a key role in many biological and pharmacological processes (as it was particularly seen during the heparin contamination crises [25]).

HA is widely used in medicine and cosmetics because it is capable of binding huge amounts of water and possesses excellent gel-forming properties [26]. As an exception in the GAG family, native HA may reach an extremely high molecular weight (MW): for instance, the molecular weight of HA from articular cartilage and healthy joint (synovial) fluid is about 10^7 Dalton [27], while the MW of the residual GAGs are much smaller. At physiological pH (about 7.4), the majority of the carboxyl groups is deprotonated and, therefore, HA (pK ≈ 3.2) is negatively charged. To indicate all the different charge states, "hyaluronan" is the most commonly used term instead of "hyaluronic acid" (free acid) or "hyaluronate" (salt form). HA is omnipresent in the body of all vertebrates - an adult human of an average body weight of 70 kg contains approximately 15 g of HA. Whereas HA was formerly isolated from animal sources, nearly all of the commercially available HA is nowadays produced from bacteria in fermentation processes [28, 29].

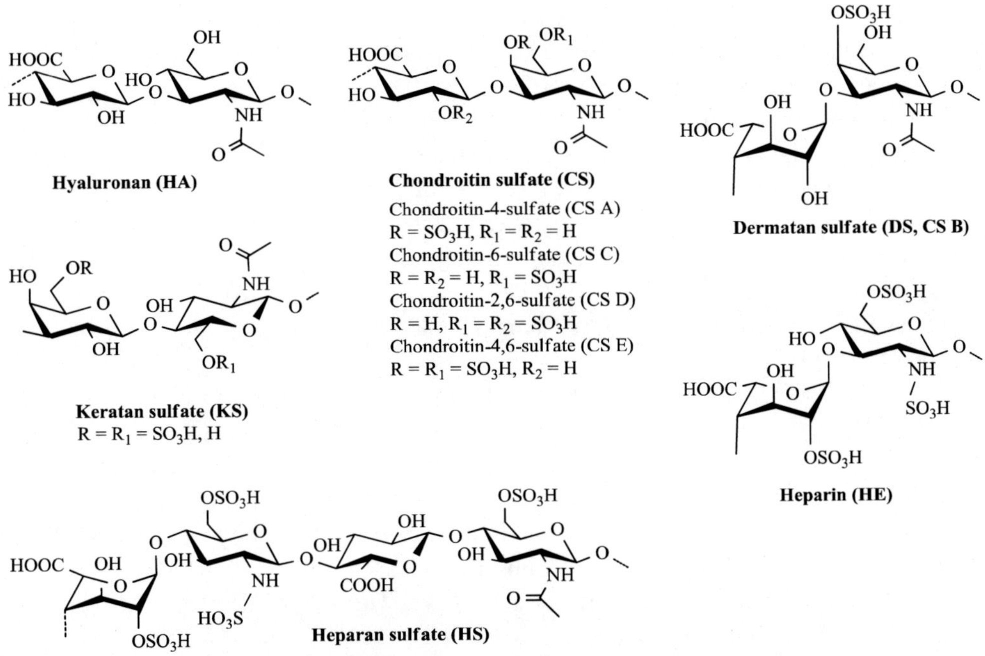

Figure 2. Major di- or tetrasaccharide repeating units of natural GAGs. Based on the biological source of the GAGs, structural variations may occur to a minor extent (e.g., differences of the sulfation patterns) in low-sulfated GAGs (CS, DS, KS) but to a larger extent (variations of sugars in the repeating unit) in high-sulfated HE and HS.

CS comprises the disaccharide unit [GlcA (β1→3) GalNAc (β1→4)] which can be variously sulfated in different positions of the hexosamine unit and/or the uronic acid residue. The regular disaccharide sequence of chondroitin-4-sulfate (formerly termed as CS A) is sulfated in position C-4 of the GalNAc unit, while chondroitin-6-sulfate (CS C) is sulfated in position C-6 of GalNAc. Disaccharides with varying numbers and positions of the sulfate groups can be also located along the polysaccharide chains, such as the disulfated disaccharides with two sulfate groups *O*-linked in position C-2 of GlcA and C-6 of GalNAc (chondroitin-2,6-disulfate, CS D) or in position C-4 and C-6 of GalNAc (chondroitin-4,6-sulfate, CS E) [30]. Therefore, it is important to note that extremely homogeneous CS is not available from biological materials - even if suppliers sometimes claim that their products have a defined sulfation pattern. A comprehensive review on the biosynthesis of GAGs has been published recently [31].

In the case of DS, further enzymatic modifications, in particular C-5 epimerization of GlcA to IdoA as well as *O*-sulfation at C-2 of IdoA occur. Therefore, GAG chains of DS feature a prevailing disaccharide unit [IdoA (α1→3) GalNAc (β1→4)] sulfated at the C-4 position of GalNAc with minor amounts of disulfated disaccharides sulfated in position C-4 of GalNAc and C-2 of the IdoA unit (not shown in Figure 2) [30]. These heterogeneous structures are responsible for the different and more specialized functions of DS. One important aspect related to those GAGs is the "heparin contamination crisis": this means that severe adverse events (even deaths) occurred when patients were treated with heparin that contained "oversulfated" CS. This has forced, for instance, the Baxter Healthcare Corp. (as well as many other companies) to recall many preparations of contaminated heparin [32] and significantly enhanced the interest in detailed heparin analysis [25] to monitor the potential presence of harmful components [33]. This is a convincing proof of the pharmacological activities regarding the different GAGs [34].

The insertion of IdoA residues imparts conformational flexibility to the DS chain altering the shape and spatial orientation of the sulfate residues and endowing the chain with a higher negative charge density compared to the presence of GlcA. Although the biosynthesis of DS has not yet been elucidated in detail, it is already known that the DS chain has a hybrid co-polymeric structure consisting of low modified (CS) and highly modified (DS) domains [35]. This is important because these structural differences are accompanied by different affinities to proteins [36]. To our knowledge 24 different CS/DS disaccharide repeating units have been identified so far [37].

Keratan sulfate (KS) is less abundant in the ECM in comparison to CS or DS [1]. It is commercially only rarely available and this hampers applications of this GAG. There are also much less studies about the structural features and chemical modifications of KS. The repeating disaccharide unit of KS [Gal (β1→4) GlcNAc (β1→3)] contains a galactose instead of an uronic acid residue and the glycosidic linkages are reversed in comparison to HA and CS/DS (Figure 2). Sulfate esters are normally present at the C-6 of one or both monosaccharide units of KS but only rarely at C-4 [38]. The MW of KS ranges between 4 and 19 kDa and is slightly lower compared to the one of CS [39].

Heparin (HE) and heparan sulfate (HS) possess a quite different repeating disaccharide structure compared to the GAGs discussed before. Both high-sulfated GAGs have variably sulfated disaccharide repeating units. In HE, the most common disaccharide repeating unit is based on a [IdoA (β1→4) GlcNAc (α1→4)] unit, where the C-2 position of IdoA and the C-6 position of GlcNAc are sulfated and, additionally, the *N*-acetyl group of GlcNAc is replaced by a *N*-sulfate moiety (Figure 2). In contrast, HS is characterized by a [GlcA((β1→4) GlcNAc (α1→4)] repeating unit sulfated at C-6 position of GlcNAc. However, major variations may also occur: up to 50 % of GlcA can be replaced by IdoA and 40-60 % of the GlcNAc units can be replaced by GlcNS sugars. This is the reason why the HS structure is often depicted as a tetrasaccharide unit as shown in Figure 2. Both compounds have been shown to differ regarding their degree of sulfation, with HE being stronger negatively charged and displaying higher *O*- and particularly *N*-sulfation than HS. It must be emphasized that HE has the highest charge density of any known biological macromolecule, while HS is generally less sulfated and possesses lower IdoA contents. This extreme negative charge density makes the structural characterization of these GAGs very difficult and the term "heparanomics" has been recently introduced to define the related discipline of analytical chemistry (*vide infra*) [40].

Both, HE and HS are polydisperse with broad (Gaussian-shaped) molecular weight distributions and average molecular weights of the order of 30 kDa for the native polymers and 12-15 kDa for the most common commercial HE preparations.

HE and HS exhibit diverse biological functions and participate in many interactions with proteins and cells. Many excellent reviews about the physiological roles of heparin and heparan sulfate are available [41]. Anyway, the charge density of GAGs is very important regarding their biological activity. Therefore, and because of the already mentioned difficulties to isolate these GAGs in larger quantities from natural sources, several different experimental approaches have been investigated to provide high-sulfated GAGs by chemical synthesis routes from readily available starting materials. An overview on these approaches and further chemical modifications will be given now.

4. Chemical Modifications of Glycosaminoglycans

4.1. General Remarks

The chemical derivatization of native GAGs represents a powerful method to generate new biopolymer products with unique physical and biological properties. Recent work has shown that certain GAGs can attract or defeat selected cells under *in vivo* conditions and that highly sulfated GAGs are able to interact with growth factors to stabilize them against proteolysis (for review see [4,36]). This opens up a new window for the application of GAGs as implant or cell scaffold coatings as well as in growth factor formulations.

Considering the multi-functionality of all GAG macromolecules, their limited solubility in organic solvents suitable for chemical reactions and their sensitivity against enzymatic, thermal or oxidative degradation are real challenges. Even in aqueous solutions reactions on GAGs are often possible exclusively at high dilution due to the high viscosity of HA and other GAG solutions. The careful selection of modifying reagents and reaction conditions is essential to obtain products with the desired structure and structural uniformity.

4.2. Sulfation and Desulfation of GAGs

First attempts to sulfate HA or "over"sulfate the already sulfated GAGs (CS, DS, KS, HE, HS) have been reported in the 1930s and 1940s [42,43]. In most of these early studies SO_2/SO_3 mixtures, sulfuric acid or chlorosulfonic acid were used as sulfating agents resulting in a considerable depolymerization of the GAG educts [44,45,46]. In more recent attempts, complexes of SO_3 with organic amines (trimethylamine, triethylamine, pyridine) [47,48,49] or amides (DMF) [50] are the preferred reagents which cause less polymer degradation. Due to their different reactivity in aprotic solvents, SO_3 complexes allow the synthesis of GAG derivatives with a gradually adjustable degree of sulfation (ds) in the range between $0 < ds \leq 3.5$ with regard to the used starting GAGs [51,52]. The "ds" value refers to the average number of substituents related to the saccharide repeating unit of the GAG. Concerning the controlled synthesis of HA or CS derivatives with low or medium degree of sulfation ($ds \leq 2$), SO_3-pyridine has been recently identified as the most suitable sulfating agent, whereas SO_3-DMF can be preferentially used for the preparation of highly sulfated GAGs ($ds \geq 2$, see Figure 3) [53].

The sulfation reactions are usually performed in aprotic solvents such as dimethylformamide (DMF) at room temperature. To increase the solubility of the starting polymers in DMF and by this means to accomplish homogeneous reaction conditions, GAGs were transferred into their corresponding tetrabutyl ammonium (TBA) salts before sulfation.

Studies on the regioselectivity of these sulfation reactions using ^{13}C NMR spectroscopy revealed a preferred sulfation of the primary hydroxy group at the C-6' (i.e., at the amino sugar) position of HA [53,54]. Low-sulfated HAs ($ds \leq 1$) are exclusively sulfated at this primary hydroxy group. In HA sulfates with higher degrees of sulfation (ds > 1) the order of sulfation was recently examined to be in the order C_6' >> C_2 / C_3 > C_4' [55]. In contrast to HA, native, commercially available CSs are usually mixtures of chondroitin-4-sulfate (C-4'-sulfated) and chondroitin-6-sulfate (C-6'-sulfated). Therefore, further sulfation attempts do not lead to products exclusively sulfated at C-6'. However, also in this case, NMR studies have evidenced a favored sulfation of the primary hydroxy group of CS [55].

Similar results regarding the regioselectivity have been reported for the sulfation of DS (Figure 4). With SO_3-trimethylamine in DMF as sulfating agent substitution occurred initially at the primary hydroxyl group at C-6' of the *N*-acetyl-galactosamine-4-sulfate unit to produce the 4,6-disulfate (method A). After achieving a substitution level higher than about 50 % by increasing the temperature, sulfation took place equally at C-2 and C-3 of the iduronic acid residue (method B) [57].

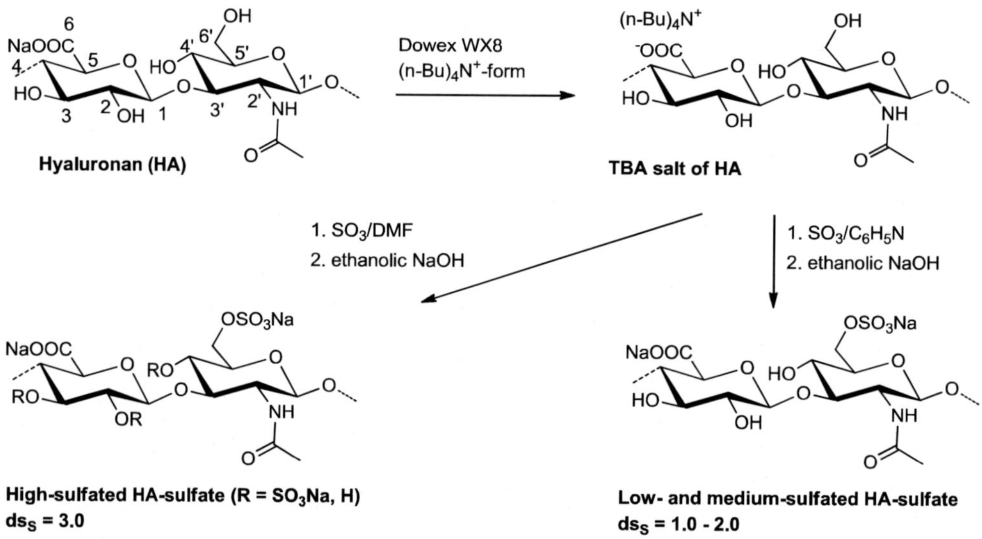

Figure 3. Sulfation of HA using SO_3-DMF and SO_3-pyridine complexes, respectively, as sulfating agents. For further details see [56].

The "over"sulfation of the native high-sulfated GAGs, HE und HS, has also been reported using both chlorosulfonic acid [46] and SO_3 complexes [58,59] as sulfating agents. A hog-mucosal HE which was desulfated and *N*-resulfated with SO_3-pyridine in DMF gave a heparin derivative with the following order of reactivity of the hydroxyl groups: C-6-HO (GlcN) >> C-2-OH-2 (IdoA) > other available hydroxyl groups (HO-3 in GlcNAc, HO-3 in IdoA, and HO-2 or HO-3 in GlcA) [59]. Obviously, the primary OH-group is preferentially sulfated also in the case of those more complex GAGs. Considering the order of reactivity of the secondary OH-groups in HS, a preferred sulfation of the HO-2 position of GlcA compared to the other secondary OH-positions was found performing sulfation with SO_3-triethylamine [60].

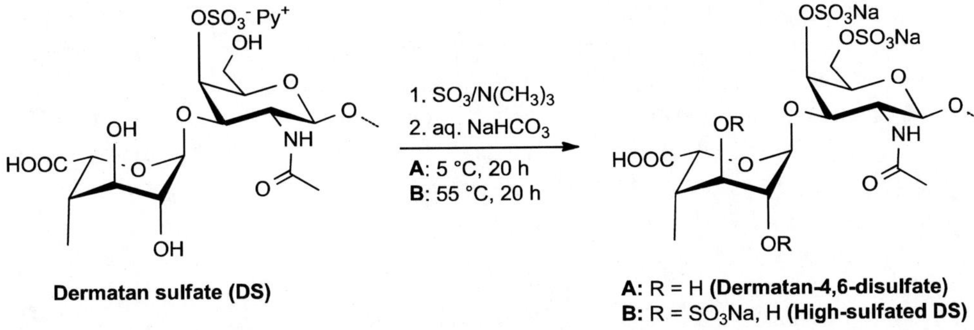

Figure 4. Influence of temperature on DS sulfation with SO_3/trimethylamine as sulfating agent.

GAG derivatives already bearing other previously introduced functionalities (such as carboxymethyl groups) have also been sulfated using SO_3 complexes. HA-based hydrogels prepared by crosslinking HA with different diamines have been subsequently sulfated to generate stimuli-sensitive materials for biomedical applications [61]. Regarding carboxymethyl HA sulfate, both carboxymethylation and sulfation were found to take place preferentially at the primary hydroxyl groups of HA (Figure 5) [62]. The introduced carboxymethyl groups can be used as potential "anchor" groups to immobilize marker molecules, bioactive agents or peptide ligands to HA.

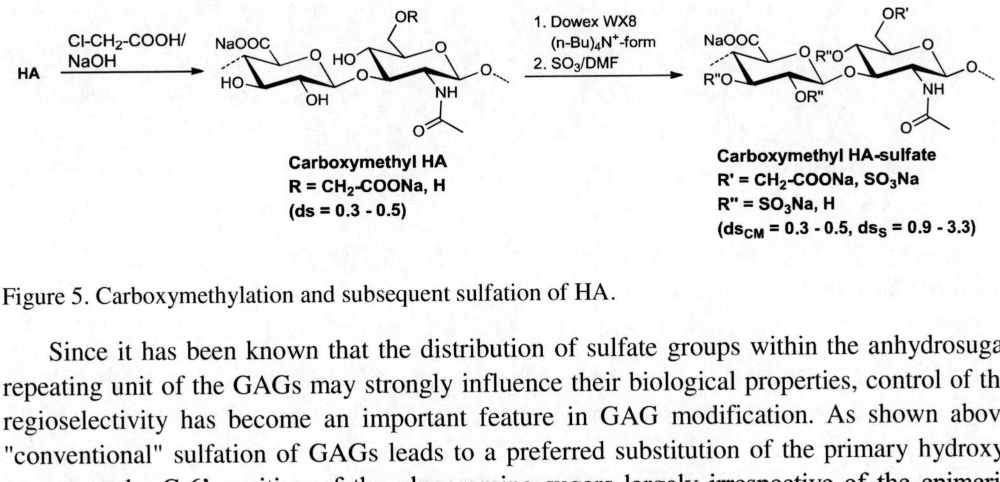

Figure 5. Carboxymethylation and subsequent sulfation of HA.

Since it has been known that the distribution of sulfate groups within the anhydrosugar repeating unit of the GAGs may strongly influence their biological properties, control of the regioselectivity has become an important feature in GAG modification. As shown above "conventional" sulfation of GAGs leads to a preferred substitution of the primary hydroxyl groups at the C-6' position of the glucosamine sugars largely irrespective of the epimeric sugar configuration and the used sulfation conditions. Synthetic routes directing sulfate groups into other positions of the GAGs are therefore of particular interest to study the influence of the sulfation pattern on biological properties of the GAGs in more detail. Due to the lower reactivity of secondary hydroxyl positions towards esterifications and the multi-functional character of all GAGs, the generation of sulfated GAGs with alternative substitution patterns than those discussed above is challenging and requires new synthesis strategies.

One approach to GAGs with sulfation patterns differing from the one obtained by conventional sulfation is based on the use of protecting groups prior to sulfation. It could recently be demonstrated that the benzoyl ester group is most suitable to protect the primary hydroxy group at C-6' of the HA backbone from a subsequent sulfation. This applies because the benzoyl ester group is sterically demanding enough to allow regioselective introduction into HA. Additionally, it is also stable under the acidic conditions of sulfation [63]. After sulfation, the protecting group can easily be cleaved under basic conditions leading to sulfated HA derivatives with ds values between 1.5 and 3 (Figure 6). As confirmed by ^{13}C-NMR studies, the primary OH-group in these HA sulfates are not sulfated at these conditions.

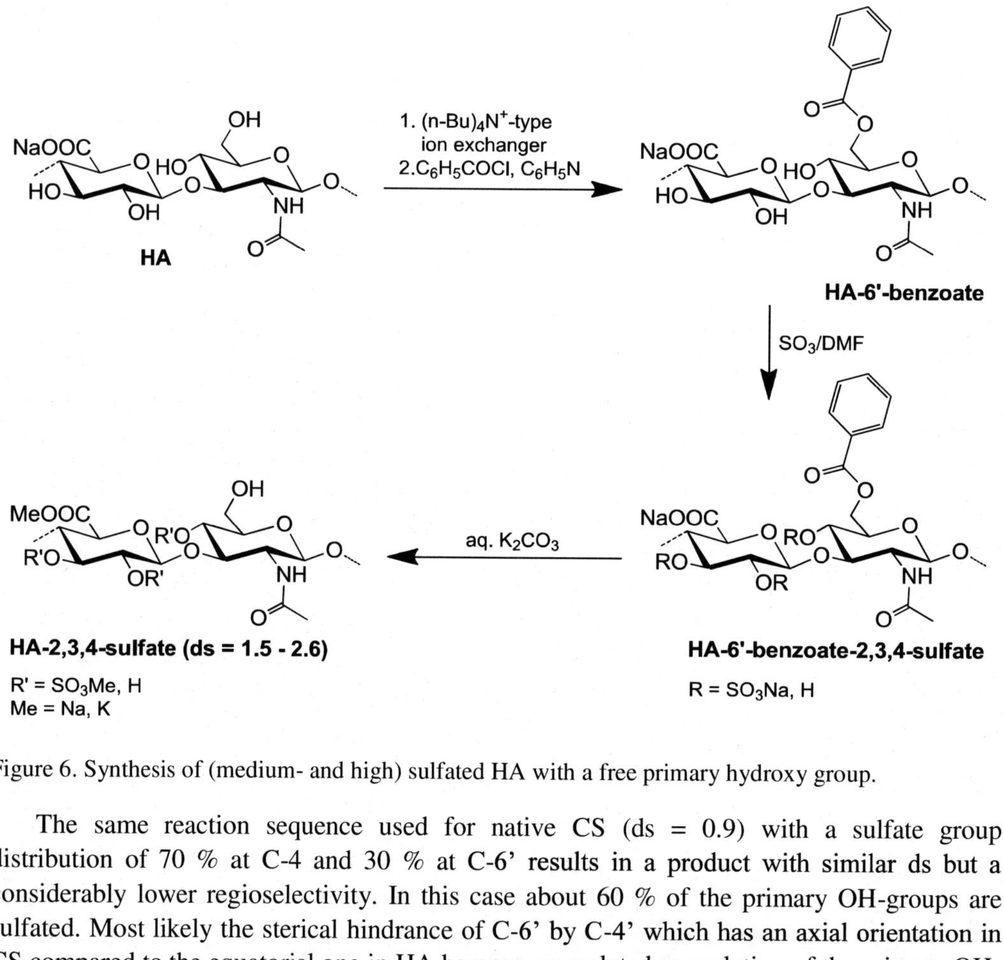

Figure 6. Synthesis of (medium- and high) sulfated HA with a free primary hydroxy group.

The same reaction sequence used for native CS (ds = 0.9) with a sulfate group distribution of 70 % at C-4 and 30 % at C-6' results in a product with similar ds but a considerably lower regioselectivity. In this case about 60 % of the primary OH-groups are sulfated. Most likely the sterical hindrance of C-6' by C-4' which has an axial orientation in CS compared to the equatorial one in HA hampers complete benzoylation of the primary OH-group [55]. Use of the less bulky acetyl protecting group instead of the benzoyl one gave sulfated CS with ds values of about 2 and a sulfate group content of only about 30 % at the primary hydroxyl group. This corresponds to the sulfate group content at C-6' of the starting CS at a total ds value which is two times higher compared to the native CS. Choosing a suitable protecting strategy for the primary hydroxyl group makes thus regioselective sulfation possible even for CS.

Another approach to polysaccharides bearing sulfate moieties mainly at their secondary OH-groups is regioselective desulfation. Normally, sulfated sugars can be desulfated in the presence of acid [64] or alkali [65], by heating of their pyridinium salts in polar aprotic solvents (solvolytic desulfation) [66] or by treatment with silyl amides [67]. A serious problem of these treatments is to avoid remarkable depolymerization of the GAGs. A variety of methods has been established for the *N*-desulfation of HE or HS (see [68] for a more detailed review). As a relatively mild procedure only slightly affecting *O*-sulfate groups and glycosidic bonds, solvolytic desulfation can be achieved by treating the pyridinium salts of the GAG with dimethyl sulfoxide containing 5 % of water or methanol at elevated temperatures [69].

Besides *N*-desulfation, two reactions resulting in regioselective *O*-desulfation have been explored for high-sulfated GAGs. Treatment of HE under defined alkaline conditions leads to a 2,3-desulfated derivative that is 2-*O*-desulfated on the α-L-iduronic acid residue and 3-*O*-desulfated at the D-*N*-glucosamine unit [65,70]. Whereas the C-6-sulfate of the *N*-hexosamine remains widely unchanged during the latter reaction, a regioselective C-6-desulfation can be achieved by treating the pyridinium salts of HE or HS with silyl acetamides such as *N,O*-bis(trimethylsilyl)acetamide (BTSA) and *N*-methyl-*N*-(trimethylsilyl)-trifluoro acetamide (MTSTFA) [71,72]. Despite the occurrence of C-2-desulfation to a smaller extent, the reaction is highly regioselective.

O-Desulfation reactions have also been reported for the moderately sulfated CS, DS, and KS [67,73,74]. Using silyl amide derivatives as already introduced for high-sulfated GAGs, the *O*-desulfation was regioselective regarding the C-6'-OSO_3-position [67].

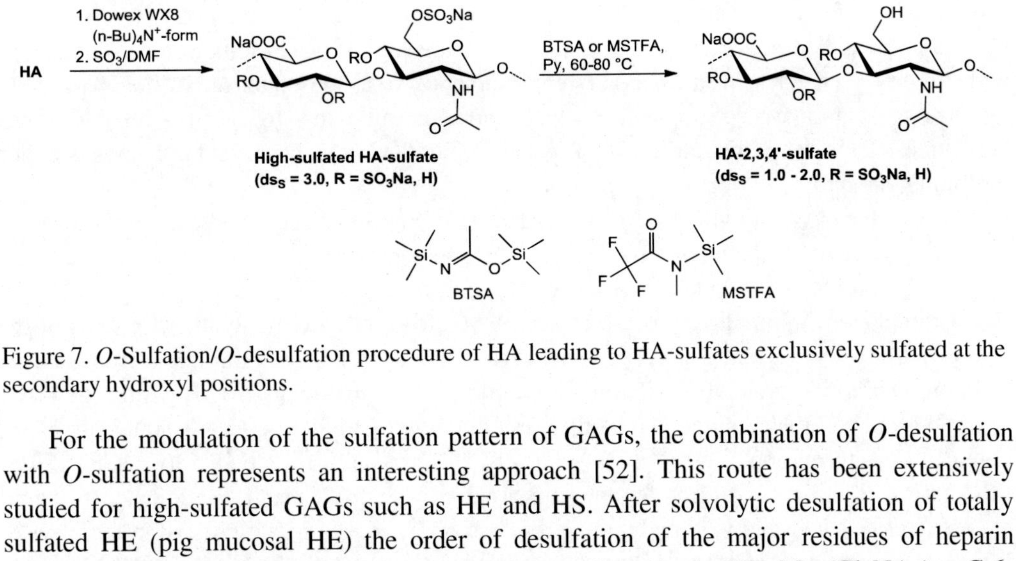

Figure 7. *O*-Sulfation/*O*-desulfation procedure of HA leading to HA-sulfates exclusively sulfated at the secondary hydroxyl positions.

For the modulation of the sulfation pattern of GAGs, the combination of *O*-desulfation with *O*-sulfation represents an interesting approach [52]. This route has been extensively studied for high-sulfated GAGs such as HE and HS. After solvolytic desulfation of totally sulfated HE (pig mucosal HE) the order of desulfation of the major residues of heparin GlcNAc and IdoA and of the minor GlcA unit was found to be NH-SO_3 (GlcNAc) > C-6-OSO_3 (GlcNAc) > C-3-OSO_3 (IdoA) ~ C-2-OSO_3 (GlcA) ~ C-3-OSO_3 (GlcNAc) > C-2-OSO_3 (IdoA) ≈ C-3-OSO_3 (GlcA), thus, affording HE with similar sulfation patterns but relatively high contents of 3-*O*-sulfated *N*-glucosamine units compared to the starting HE [75]. Under the same reaction conditions, in completely sulfated HS the C-2-SO_3 (GlcA) is desulfated preferentially in comparison to C-3-OSO_3 (GlcA) [75].

Using the same sequence of reactions (i. e. high *O*-sulfation followed by *O*-desulfation with silylamides) also non-sulfated HA and medium sulfated CS and DS have been used to obtain fully C-6' desulfated derivatives that are sulfated exclusively at the secondary C-2, C-3, and C-4' positions [55,72]. Using both, BTSA and MTSTFA, to desulfate high-sulfated

HA no significant difference concerning the effectiveness of the desulfation reaction was found. HA (pyridinium salt) with a ds = 3.1 was desulfated within 24 h at 60 °C reducing the ds down to 2.3 (Figure 7).

^{13}C-NMR investigations revealed that after 4 h at 60 °C a partial desulfation of the C-6'-OSO_3 position occurred whereas all other positions remained unchanged. After a reaction time of 24 h the C-6' position was completely desulfated but also the C-4'-OSO_3 position already showed partial desulfation. Thus, it can be concluded that the desulfation of HA takes place regioselectively at the primary hydroxy group but a complete desulfation of this position without simultaneously desulfating the secondary OH-groups can be hardly achieved [55].

4.3. Synthesis of Crosslinkable GAG Derivatives

One of the most interesting applications of GAGs in biomedical sciences is represented by hydrogels [76,77]. Those hydrogels are not only advantageous matrices in tissue regeneration due to their excellent biocompatibility, biodegradability, and increased permeability, they also mimic the native extracellular matrix with its unique biological functions including the regulation of cell proliferation and differentiation, as well as the sequestration of growth factors [78,79]. Although already native HA forms hydrogel-like structures in diluted aqueous solutions, all GAGs are highly water-soluble macromolecules with limited persistence in an aqueous environment. As a consequence, sufficiently stable hydrogels for biomedical applications can only be prepared from crosslinkable GAG derivatives. This is particularly true for the sulfated GAGs because they have smaller molecular weights.

The so far most straightforward method to crosslink GAGs is their reaction with low-molecular weight bifunctional crosslinkers such as dialdehydes (e. g. glutardialdehyde), diepoxides, or divinyl compounds such as divinyl sulfone (for a recent survey see [56]). Bifunctional crosslinking agents are often employed to react with the hydroxyl groups of the GAGs but also with their carboxyl groups because this can be used to initiate a cross-linking reaction. Due to the gradually different reactivities of the functional groups within the GAG repeating unit and their limited availability (e. g. in case of sulfated GAGs) and accessibility, complete transformation of all reactive groups by the crosslinker can be hardly achieved. For this reason, reaction conditions for the crosslinking have to be carefully adjusted to avoid cytotoxic effects caused by an incomplete reaction of the cross-linking groups. In many cases it is more promising to introduce a reactive functional group into the GAGs which is able to undergo cross-linking under specific conditions, for instance, irradiation. Such a "pre-derivatization" often enables the purification and characterization of the reactive intermediate to provide better control about the subsequent crosslinking reaction. In Table 1 an overview on selected methods for the cross-linking of GAGs is given.

A simple and well established approach to crosslinkable biopolymer derivatives represents the introduction of acrylate or methacrylate functions into GAGs. For this purpose different routes are reported.

In a common esterification reaction methacrylate esters of HA, CS and HE have been prepared using methacrylic anhydride (MAA) as acylation agent in an aqueous medium under basic conditions (see Figure 8, route A) [92,93]. Degrees of methacrylation of up to 1.0 have been obtained by this procedure. In another approach CS was treated with glycidyl methacrylate (GMA) in aqueous solution [94]. It was recently also found that after reaction times of one day transesterification dominates over common epoxide ring opening. As a consequence, glycidol is split off as leaving group and in the major product, which is formed under these conditions, the methacryloyl groups are attached directly to the primary or secondary hydroxyl groups of CS (Figure 8, route B). Using extended reaction times of 15 days, products can be isolated formed to a major extent by ring-opening of the glycidyl ring. In this case both, the ester derivative resulting from acylating of the carboxyl group of CS (Figure 8, route C) and side products with the methacryloyl glycerol residue attached to the sulfate group, can be detected [95]. Similar reaction products can be found for the treatment of HA with GMA [96].

Table 1. Selected cross-linking methods for GAGs. For abbreviations see the list of abbreviations

GAG	Crosslinker	Reaction conditions	Ref.
CS	Glutardialdehyde	Co-reactant: polyvinylalcohol, H_2O, 5% HCl, 2 h, RT	[80]
	PEG-diglycidyl ether	1 M NaOH, 24 h, RT	[81]
	1,12-diaminododecane	DMSO, DCC, dicyclohexylcarbodiimide, 12-24 h, RT	[82]
	EDC/NHS	Co-reactant: collagen, chitosan; MES (pH: 5.5), 4 h, RT	[83,84]
CS-methacrylate	γ rays	Co-reactant: HA-methacrylate, polyvinylalcohol, H_2O, RT, 5-25 kGy	[85]
CS-hydrazide	Poly(ethylene glycol) dialdehyde	H_2O, 24 h, 4 °C	[86]
CS-aldehyde		Co-reactant: gelatine, PBS, 4 °C, 24 h	[87]
CS-(Gly-Lys dipeptide)	Transglutaminase	Internal phase emulsion, PBS, Triton X-405, toluene, 50 °C, 24 h	[88]
HE	4-arm "StarPEG"-amine	EDC, sulfo-NHS	[89]
HE-thiol	Poly(ethylene glycol) diacrylate	PBS (degassed), RT	[90]
HE- cinnamate		UV irradiation (5.0 J/cm^2 at 280 nm) in the frozen state (-20 °C)	[91]

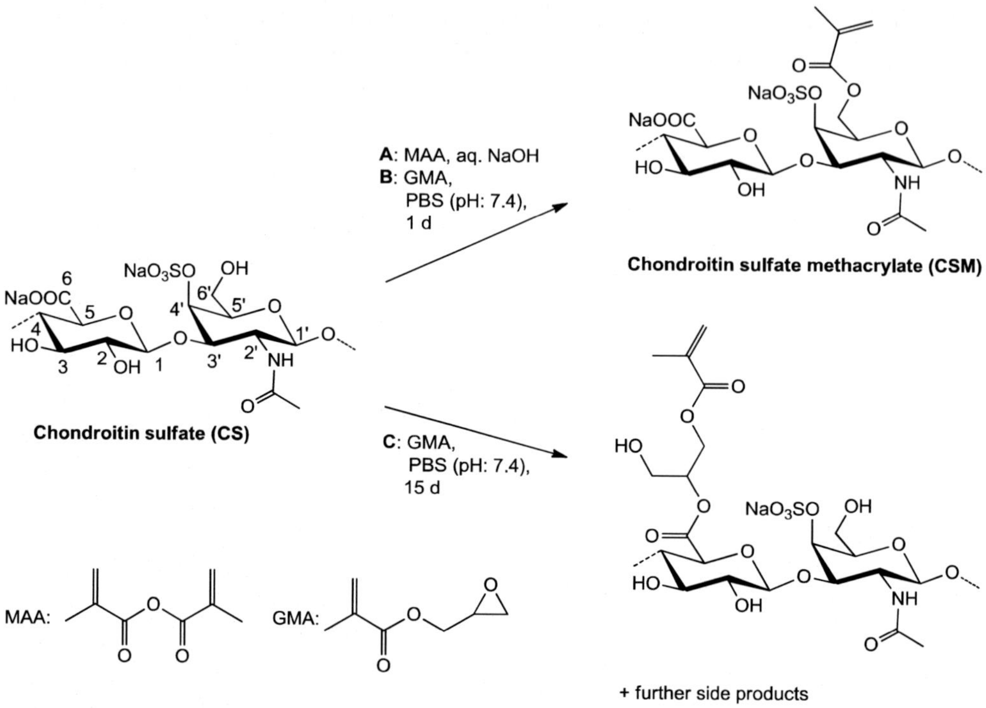

Figure 8. Synthetic routes for the methacrylation of CS using MAA (route A) and GMA (routes B and C) as methacrylation agent.

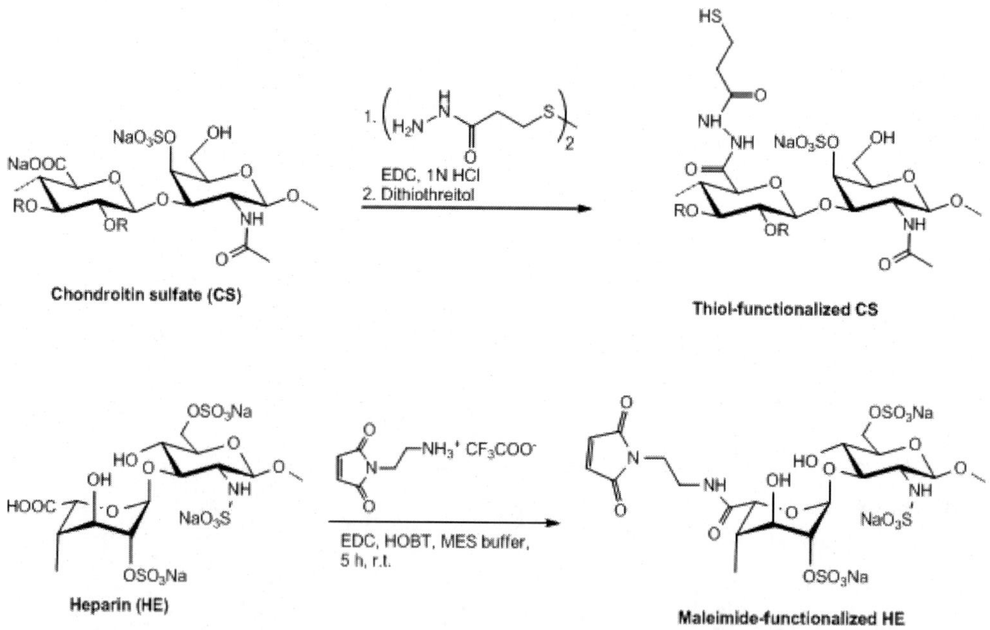

Figure 9. Synthesis of reactive thiolated CS and maleimide-functionalized HE derivatives suitable as starting materials for GAG cross-linking.

Acrylation of CS has been performed by the reaction with adipic dihydrazide and subsequent conversion of the amino groups with acrylic acid [97]. Recently, an efficient one-step acrylation of HA was described with acryl chloride under phase-transfer conditions yielding degrees of acrylation of up to 1.7 [98].

An elegant route to use the carboxyl groups of the GAGs for network formation is their carbodiimide-mediated reaction with a disulfide containing dihydrazide [99]. After reductive cleavage of the disulfide bond a reactive thiol-containing GAG intermediate is formed (Figure 9) which can be crosslinked easily either by a new disulfide formation under oxidative conditions [100] or by a reaction with diacrylates via a Michael addition [99].

In a recent study on HE-based hydrogels the carboxyl groups of HE were modified with aminoethyl maleimide as shown in Figure 9 [101]. The maleimide-functionalized HE can readily be used in crosslinking reactions with thiolated polymers derived, for instance, from poly(ethylenglycol) or other GAGs.

5. How to Characterize GAGs?

The characterization of GAGs is an important issue: using dedicated GAG preparations, can one really be sure that the polysaccharide has always the same molecular weight, the same sulfation pattern and that there are no impurities present?

Unfortunately, the detailed characterization of a given polysaccharide is extremely difficult and depends on the specialization of the involved scientist: a cell biologist or an immunologist will consider a completely different parameter set to be important than a chemist. According to our matter of expertise, we will focus on chemical aspects of (sulfated) GAG characterization and only to a minor extent on cellular or immunological aspects. The following parameters are of particular interest:

1) The molecular weight of the polysaccharide.
2) The number and the position of functional groups such as sulfate as well as the homogeneity of the sequence of the polymer repeat units.
3) The absence of impurities, for instance, solvents or precipitating agents.

Since some powerful methods (such as mass spectrometry (MS)) cannot be applied to high molecular mass GAGs, a differentiation between the intact polysaccharides and smaller oligosaccharides subsequent to (enzymatic or if necessary chemical) degradation is necessary [56]. Accordingly, selected methods of GAG degradation (i.e., the cleavage of the glycosidic linkages under retention of the residual functional groups) will also be shortly discussed. A sketch of useful analytical methods is illustrated in Figure 10.

As already indicated the focus of the next few paragraphs will be on NMR spectroscopic and mass spectrometric (MS) methods. These techniques are highly complimentary and give very detailed insights into GAG structures.

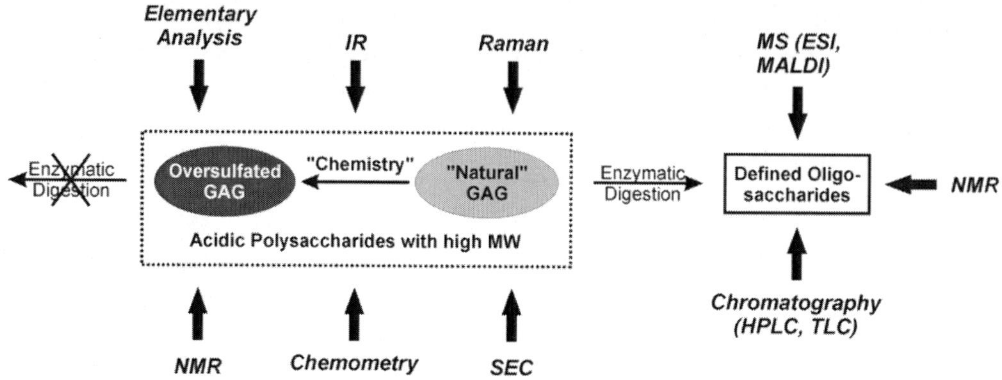

Figure 10. Survey of analytical methods to characterize GAGs. Please note that not all the indicated methods are applicable to the poly- and larger oligosaccharides. Only the techniques emphasized in the right part of this figure will be discussed here in more detail.

5.1. Determination of the Molecular Weights of Sulfated GAGs

High performance liquid chromatography (HPLC) is one of the most powerful and established tools to characterize high molecular weight GAGs and this important topic has been previously comprehensively reviewed [102]. However, the focus is normally on highly sulfated derivatives such as heparin [103,104] that are widely used in medicine [25].

The elucidation of the MW of a GAG by means of chromatography and the subsequent determination of the sulfate content by elementary analysis is a common approach to characterize (oversulfated) GAGs [56]. The combination between strong anion exchange-(SAX)-HPLC in combination with GAG detection by circular dichroism (CD) is a powerful method to separate highly acidic polysaccharides [105]. Amino phases (standard silica modified by -NH_2 residues) is a very useful stationary phase to separate native sulfated and oversulfated GAGs from human blood, whereby detection limits of the order of 20 ng could be achieved [106].

GAG separation is traditionally performed by gel permeation chromatography (GPC) - also known as "size exclusion chromatography" (SEC) [107]. A drawback of GPC is that GAG standards with defined molecular weights are required because GPC (at least if the refraction index or the UV absorption is determined) is a relative but not an absolute method of MW determination. Another problem of GPC is coming from structural aspects of GAGs [108]: the size of HA, for instance, changes in dependence on the pH and/or the salt content because screening of the charges of the carboxylate groups results in the transition from a linear into a coiled molecule. Although this is a minor problem regarding sulfated GAGs (since the sulfate represents a strong electrolyte), this is one reason why chromatographic methods are normally combined with methods of light scattering (LS) that help to evaluate the GAG structures in solution. Further reading is available in [109].

In addition to GPC, electrophoretic methods may also be used due to the charge of the native GAGs [110]. As the carboxylate and the sulfate residues of HA and CS provide very different charge densities (sulfate is a strong electrolyte, while carboxylate is a weak one), pH-dependent separation of both GAGs can be easily achieved. The most important application of electrophoresis is the determination of impurities (oversulfated CS or DS) in heparin [111]. Here, the fragmentation of heparin by nitrous acid into smaller oligosaccharides is extraordinarily helpful: this degradation does only work in the case of heparin but not in the case of other GAGs. Although mass spectrometry (MS) is unequivocally the most powerful tool of biopolymer characterization, the mass range is sometimes (in dependence on the analyte) very limited: using MS exclusively ions can be detected and the ionization process is likely to take place in the gas phase. The most widely used MS techniques of GAG characterization are electrospray ionization (ESI) MS and matrix-assisted laser desorption and ionization (MALDI) MS [112,113]. We will focus here on MALDI MS since we are most experienced in this method. For a more comprehensive description of technical aspects of MALDI MS, the excellent textbook by Franz Hillenkamp and Jasna Peter-Katalinić (now also available in its second edition) can be highly recommended [114]. The most important advantages of MALDI MS are the simple performance, the considerable tolerance of salts and other impurities as well as the high sensitivity [115,116].

Due to their high polarity (sulfate residues) and the significant molecular weight, it is extremely difficult to transfer GAG polysaccharides into the gas phase without pronounced decomposition. This can be illustrated by the following data: proteins with masses close to 1 MDa could be already detected by MS [114]. In contrast, Hao and coworkers were only able to detect dextran with an average MW of about 75 kDa [117]. However, dextran is the most easily detectable polysaccharide because there are no additional, polar functional groups: the situation gets much more difficult as soon as heparin is of interest. To these authors best knowledge only heparin oligosaccharides with masses smaller than about 6 kDa were detectable by MS so far. Therefore, MS is not the method of choice to determine the molecular weights of high mass polysaccharides [118].

5.2. Determination of the Sulfation Patterns of GAGs

Although the overall extent of sulfation can be conveniently determined by elementary analysis, this approach does not provide any information about the sulfation pattern, i.e., it remains unknown whether all polymer repeating units are sulfated to the same extent and whether the sulfate residues are homogeneously distributed along the polysaccharide chain. Since the discovery of the sulfation code [18], however, this aspect attracts increasing interest.

5.2.1. NMR Spectroscopy

Methods of magnetic resonance spectroscopy are probably the most powerful tools to characterize native and chemically modified GAGs. This particularly applies because in addition to information about the composition, also structural parameters are achievable by NMR. Additionally, NMR is also extremely useful to study GAG-protein interactions [12,119]. Although solid-state NMR is nowadays also an established research technique, we will focus here on classical high-resolution (solution) NMR because all sulfated GAGs are readily soluble. Even if the analysis of the native GAGs is basically possible by NMR, the MW of the GAGs has a significant impact on the spectral quality: the higher the MW, the more significant are the line-widths of the observed resonances. This is caused by the shortened relaxation times of the polymers [120]. This broadening effect is a particular problem for ^{1}H NMR because its range of chemical shift (about 10 ppm) is rather limited. Therefore, native GAGs are regularly characterized by ^{13}C [121] - often in combination with 2D-NMR experiments that help to improve the achievable resolution [122,123].

Native GAGs, such as HA and CS can be easily differentiated by ^{13}C-NMR [124]. As the HA polymer repeating unit comprises 14 different carbon atoms (6 of the glucuronic acid and 8 of the *N*-acetylglucosamine residue), 14 different ^{13}C resonances can be observed [125]. Unfortunately, the introduction of additional functional groups such as sulfate reduces the frequency dispersion significantly making assignments (particularly of the ring carbon atoms) much more difficult. With the increasing availability of high-field NMR spectrometers, there are now also increasing attempts to characterize GAGs by ^{15}N NMR. Although the ^{15}N nucleus is a "difficult" nucleus (very low natural abundance (0.37%) accompanied by low sensitivity), this increasingly important field has been recently reviewed [126]. The shifts of the ^{15}N amide resonances seem particularly helpful to differentiate the individual sites of sulfation (4-, 6-, or 4,6-) and resolved also the slight differences between glucos- and galactosamine [126]. However, the low natural abundance of the ^{15}N isotope in combination with its low NMR sensitivity makes even nowadays the enrichment of ^{15}N necessary although it is expensive [127].

In order to improve the achievable resolution, many structural studies (not only the ^{15}N-based studies) [128] are normally performed subsequent to the degradation of the native GAGs: subsequent to digestion with the enzyme chondroitinase ABC, for instance, the positional isomers, i.e., C-4'-S and C-6'-S can be easily differentiated by their characteristic chemical shift differences [129]. A similar approach enables the differentiation of HE, DS and CS [130]. The spectral patterns of the *N*-acetyl resonances (between about 2.00 and 2.20 ppm) seem extraordinarily helpful to discriminate the differently sulfated GAGs [33] and the advantage of this approach is that significant information can be derived from the ^{1}H NMR spectra.

5.2.2. Mass Spectrometry

As outlined above, MS is not suitable to characterize native polysaccharides and degradation into defined oligosaccharides is mandatory. Many depolymerization-causing reagents are available for "simple" carbohydrates such as dextrans that contain (beside the glycosidic linkages) no additional functional groups. However, the situation is much more complex if (oversulfated) GAGs have to be analyzed because they have more complex structures and possess with the *N*-acetyl side chain and the sulfate residues two functional groups that are sensitive to acidic as well as alkaline conditions. Fortunately, there are specific enzymes available that cleave GAG polysaccharides under generation of defined oligosaccharides. The most widely used enzymes to depolymerize native GAGs are bacterial or testicular hyaluronidases [113,131] as well as bacterial chondroitinases, in the majority of cases the ABC type from *proteus vulgaris* [132]. The latter enzyme digests chondroitin-4- (CS-A), chondroitin-6- (CS-C) as well as dermatan sulfate (CS-B) and has the additional advantage that it exhibits (beside cleavage of the glycosidic linkages) eliminase activity, leading to an introduction of a double bond into the uronic acid residue of the disaccharide [133]. The related UV absorption at 232 nm ($\varepsilon \approx 7000$ M^{-1} cm^{-1}) is helpful to quantitatively monitor the generation of the digestion products and, thus, the endpoint of the enzymatic digestion. Although bacterial hyaluronidases do also generate unsaturated products, the application of hyaluronidases is less favorable because complex oligosaccharide mixtures are obtained. It is a common feature of MS that the oligosaccharides in a mixture are not detected with the same sensitivities: smaller oligosaccharides can be more easily detected in comparison to larger ones [113,134].

Unfortunately, enzymatic digestion is only easily possible in the case of the natural GAGs, while chemically modified (e.g., oversulfated) GAGs are much more refractive to the digestion - resulting in very small oligosaccharide yields. It has been shown, for instance, by using oversulfated GAGs [135] as well as phosphorylated GAGs [136] that negatively charged functional groups lead to a nearly complete inhibition of the enzyme. Although this can be minimized by carefully adjusting the conditions of the digestion process [113], this is a very serious problem that aggravates the characterization of (bio)chemically modified GAGs. For this reason, it is not always clear whether the data determined with naturally occurring GAGs are really reliable: one could easily imagine that higher sulfated oligosaccharides are generated with lower efficiencies upon the digestion of the polysaccharide. Therefore, their concentrations might be underestimated.

Another problem is the sulfate loss that occurs in the gas phase of the mass spectrometer [137]. This is a particular problem if MALDI MS is used because (high vacuum) MALDI is less "soft" in comparison to ESI MS [138]. This sulfate loss seems to occur exclusively when protonated sulfate residues are present but not when the deprotonated or sodiated ions are generated. This is illustrated in Figure 11.

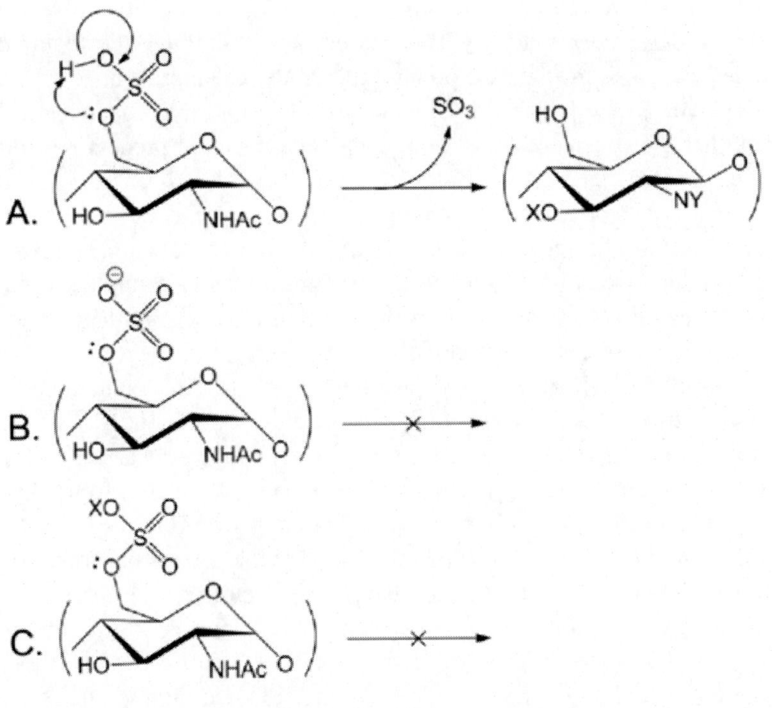

Figure 11. Illustration to show that the tendency for the loss of sulfate residues depends on the charge and the type of the ion. A, protonated sulfate groups are readily lost as SO_3. B, deprotonated sulfate groups are much more stable. Thus, the higher the density of the negative charge, the lower the extent of SO_3 loss. C, generation of the corresponding alkali metal cation (X) also reduces the extent of the sulfate loss [118].

Many efforts were already made to avoid this unwanted fragmentation, but according to the present knowledge, this process can only be minimized, for instance, by using ionic liquid crystalline matrices [139] or the cesium salts of the GAG [140], but never completely suppressed. Nevertheless, it is extremely important to use suitable reference compounds to fine-tune the mass spectrometer to minimize the sulfate loss: such compounds are represented by either Arixtra (a compound similar to heparin) or sucrose-octasulfate [140].

Nevertheless, sulfate loss is very helpful to differentiate GAG isomers: of course, the 4- and 6-isomers of the CS disaccharide, for instance, have exactly the same mass. Therefore, differentiation is possible only by MS/MS: it has been shown by using MALDI [141] as well as ESI [142] that there are helpful fragments that enable a clear differentiation of the isomers, i.e., some characteristic fragment ions are exclusively arising from a certain isomer. Although a more detailed discussion of these aspects is beyond the scope of this paper, the reader should note that the mass analyzer of the used mass spectrometer has a significant impact on the achievable information. For instance, it has been recently shown that migration of the sulfate residues may occur when an ion trap (IT) mass analyzer is used [143].

5.2.3. Mass Spectrometry in Combination with Chromatography

As the quantitative evaluation of MS/MS spectra is normally less accurate in comparison to the MS^1 spectra, separation of the individual isomeric oligosaccharides is normally the method of choice. This separation can be carried out by a variety of methods including HPLC [144] as well as thin-layer chromatography (TLC) [145]. Applications of HPLC are much more popular and have been extensively reviewed [146], although the required (complete) enzymatic digestion is still the weakest point of this analytical approach.

Even if ESI MS is characterized by a reduced sulfate loss in comparison to MALDI MS [118], ESI MS has the considerable drawback that the ionization process is strongly affected by the presence of salts.

This is why solvents or eluents with a higher salt content cannot be used - or the sample must be desalted prior to MS characterization. MALDI is by far less affected by contaminants such as salts but cannot be online coupled with LC because a solid sample is used for MALDI MS. Therefore, the direct combination between MALDI and thin-layer chromatography (TLC) is in some cases a suitable alternative.

The negative ion MALDI-TOF mass spectra of selected GAGs (subsequent to enzymatic digestion and directly recorded from a TLC plate) are shown in Figure 12.

It is obvious from Figure 12 that separation of GAG oligosaccharides can be performed by TLC and high quality mass spectra can be recorded directly from the TLC plate. Nevertheless, there is still one problem which is coming from the use of high amounts of formic acid that are required for the separation of the GAGs on a normal phase TLC plate [147]: under these conditions, esterification occurs (detectable by the mass shift of 28) and, thus, a minor moiety of the GAG is only detectable as the corresponding formyl ester. This problem could be overcome by using reversed phase TLC (instead of normal phase TLC) because under these conditions there would be no need to use highly polar solvents such as formic acid. However, to these authors best knowledge such investigations of GAG oligosaccharides have not yet been reported.

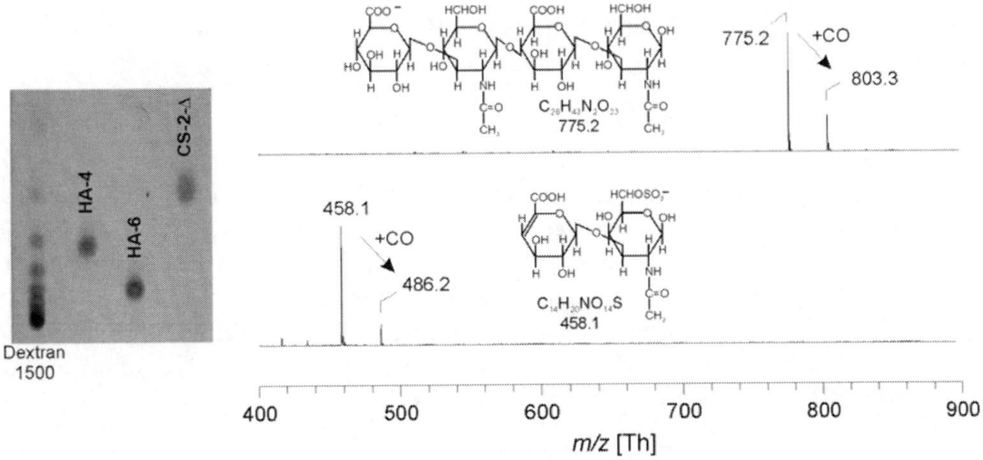

Figure 12. Negative ion MALDI-TOF mass spectra of selected oligosaccharides obtained by enzymatic digestion of HA-4 (upper spectrum) and the unsaturated disaccharide of chondroitin sulfate (lower spectrum). Mass spectra were recorded directly from the TLC plate (shown at the left). The molecular structures and the *m/z* ratios are provided along with the mass spectra. Please note that "+CO" indicates the formation of the formyl ester because the separation of the carbohydrates was performed in the presence of formic acid. Dextran 1500 as well as HA-6 are only shown for comparative purposes on the TLC plate (left) but no mass spectra of these compounds are provided. All MALDI spectra were recorded by using DHB (2,5-dihydroxybenzoic acid) as matrix. Reprinted with modification from [145], with permission.

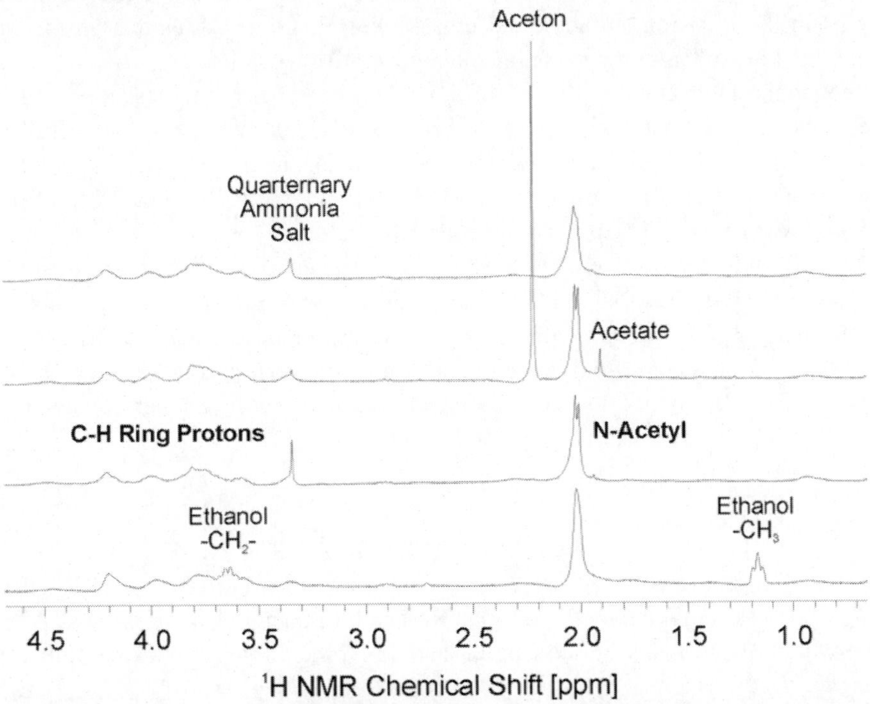

Figure 13. ^{1}H-NMR spectra of aqueous solutions of selected commercially available preparations of chondroitin sulfate. Please note that the CS resonances should fall within a small range of chemical shift: the N-acetyl at about 2.03 ppm and all other C-H resonances between about 3.2 and 4.2 ppm.

5.3. Checking the Absence of Impurities Such As Solvents or Precipitating Agents

Although many analytical techniques might be basically suitable to fulfill this purpose, NMR is in our opinion the method of choice to screen a given GAG preparation for impurities such as solvents or precipitating agents. There are two reasons: on the one hand, such compounds have a much lower MW in comparison to the native GAGs. Since NMR is a kind of a "mobility filter", low-mass compounds are detected with much higher sensitivities than the GAGs - even if they are present only in very small amounts. On the other hand, NMR detection is non specific and detects all compounds (at least if they possess comparable MWs) according to their concentration. Therefore, no prior knowledge about the type of the impurity is required. This is illustrated in Figure 13.

Conclusion

Chemically modified GAGs (particularly oversulfated ones) attract increasing scientific interest because they exhibit pharmacological effects and/or bind to physiologically relevant proteins and control in this way metabolic processes. Additionally, coating of implants with these compounds may improve the incorporation of the implant into the tissue and/or reduce immunological (inflammatory) rejections. All these effects are summarized in [4] or [148, 149]. This review was aimed to provide an overview about suitable chemical methods to modify (with the focus on sulfation and cross-linking) natural GAGs that are of particular interest in the field of tissue regeneration and regenerative medicine.

Although the chemical modification of GAGs is still a challenging task due to the need of regioselective reactions and retention of the original molecular weight, some useful synthesis routes were identified. Particular attention was paid to the introduction of sulfate groups because the elucidation of the sulfation code is nowadays a particular "hot" topic of research - in particular when applied in combination with hydrogels.

Hopefully, these authors were also able to show that significant attention has to be paid to the characterization of the modified GAGs since otherwise reproducible studies on the *in vivo* effects are not possible: modern soft ionization MS techniques and/or NMR spectroscopic approaches seem the methods of choice to perform an in-depth characterization of the GAGs of interest. However, it became hopefully clear that it is still a long way to have fully characterized GAG derivatives.

Acknowledgments

This work was supported by the German Research Council (TRR 67, projects A2, A8 and Z3). We would like to thank our colleagues and co-workers for many useful hints regarding this review.

REFERENCES

[1] Schiller, J., Fuchs, B., Arnold, K., 2006. The molecular organization of polymers of cartilage: an overview of health and disease. *Curr. Org. Chem.* 10, 1771-1789.

[2] Schiller J., Fuchs, B., Arnhold, J., Arnold, K., 2003. Contribution of reactive oxygen species to cartilage degradation in rheumatic diseases: molecular pathways, diagnosis and potential therapeutic strategies. *Curr. Med. Chem.* 10, 2123-2145.

[3] Pradhan, S., Farach-Carson, M.C., 2010. Mining the extracellular matrix for tissue engineering applications. *Regen. Med.* 5, 961-970.

[4] Schnabelrauch, M., Scharnweber, D., Schiller, J., 2013. Sulfated glycosaminoglycans as promising artificial extracellular matrix components to improve the regeneration of tissues. *Curr. Med. Chem.* 20, 2501-2523.

[5] Schulz, R.M., Bader, A., 2007. Cartilage tissue engineering and bioreactor systems for the cultivation and stimulation of chondrocytes. *Eur. Biophys. J.* 36, 539-568.

[6] Tolar, J., Blazar, B.R., Wagner, J.E., 2011. Concise review: transplantation of human hematopoietic cells for extracellular matrix protein deficiency in epidermolysis bullosa. *Stem Cells* 29, 900-966.

[7] Hoffmann, O., Zafiropoulos, G.G., 2012. Tooth-implant connection: a review. *J. Oral. Implantol.* 38, 194-200.

[8] Salbach, J., Rachner, T.D., Rauner, M., Hempel, U., Anderegg, U., Franz, S., Simon, J.C., Hofbauer, L.C., 2012. Regenerative potential of glycosaminoglycans for skin and bone. *J. Mol. Med.* 90, 625-635.

[9] DeAngelis, P.L., 2002. Evolution of glycosaminoglycans and their glycosyltransferases: implications for the extracellular matrices of animals and the capsules of pathogenic bacteria. *Anat. Rec.* 268, 317-326.

[10] Bulow, H.E., Hobert, O., 2006. The molecular diversity of glycosaminoglycans shapes animal development. *Annu. Rev. Cell Dev. Biol.* 22, 375-407.

[11] Zhang, L., 2010. Glycosaminoglycan (GAG) biosynthesis and GAG-binding proteins. *Prog. Mol. Biol. Transl. Sci.* 93, 1-17.

[12] Pichert, A., Samsonov, S.A., Theisgen, S., Thomas, L., Baumann, L., Schiller, J., Beck-Sickinger, A.G., Huster, D., Pisabarro, M.T., 2012. Characterization of the interaction of interleukin-8 with hyaluronan, chondroitin sulfate, dermatan sulfate and their sulfated derivatives by spectroscopy and molecular modeling. *Glycobiology* 22, 134-145.

[13] van der Smissen, A., Hintze, V., Scharnweber, D., Moeller, S., Schnabelrauch, M., Majok, A., Simon, J.C., Anderegg, U., 2011. Growth promoting substrates for human dermal fibroblasts provided by artificial extracellular matrices composed of collagen I and sulfated glycosaminoglycans. *Biomaterials* 32, 8938-8946.

[14] Lieleg, O., Ribbeck, K., 2011. Biological hydrogels as selective diffusion barriers. *Trends Cell. Biol.* 21, 543-551.

[15] Huster, D., 2008. Solid-state NMR studies of collagen structure and dynamics in isolated fibrils and in biological tissues. *Annu. Rep. NMR Spectrosc.* 64, 127-159.

[16] Labat-Robert, J., 2004. Cell-matrix interactions in aging: role of receptors and matricryptins. *Ageing Res. Rev.* 3, 233-247.

[17] Pan, J., Qian, Y., Zhou, X., Lu, H., Ramacciotti, E., Zhang, L., 2005. Chemically oversulfated glycosaminoglycans are potent modulators of contact system activation and different cell signaling pathways. *J. Biol. Chem.* 285, 22966-22975.

[18] Gama, C.I., Hsieh-Wilson, L.C., 2005. Chemical approaches to deciphering the glycosaminoglycan code. *Curr. Opin. Chem. Biol.* 9, 609-619.

[19] Merceron, C., Portron, S., Vignes-Colombeix, C., Rederstorff, E., Masson, M., Lesoeur, J., Sourice, S., Sinquin, C., Colliec-Jouault, S., Weiss, P., Vinatier, C., Guicheux, J., 2012. Pharmacological modulation of human mesenchymal stem cell chondrogenesis by a chemically oversulfated polysaccharide of marine origin: potential application to cartilage regenerative medicine. *Stem Cells* 30, 471-480.

[20] Schlorke, D., Thomas, L., Samsonov, S.A., Huster, D., Arnhold, J., Pichert, A., 2012. The influence of glycosaminoglycans on IL-8-mediated functions of neutrophils. *Carbohydr. Res.* 356, 196-203.

[21] Nikitovic, D., Chatzinikolaou, G., Tsiaoussis, J., Tsatsakis, A., Karamanos, N.K., Tzanakakis, G.N., 2012. Insights into targeting colon cancer cell fate at the level of proteoglycans / glycosaminoglycans. *Curr. Med. Chem.* 19, 4247-4258.

[22] Morawski, M., Brückner, G., Arendt, T., Matthews, R.T., 2012. Aggrecan: beyond cartilage and into the brain. *Int. J. Biochem. Cell Biol.* 44, 690-693.

[23] Sasisekharan, R., Raman, R., Prabhakar, V., 2006. Glycomics approach to structure-function relationships of glycosaminoglycans. *Annu. Rev. Biomed. Eng.* 8, 181-231.

[24] Sugahara, K., Mikami, T., Uyama, T., Mizuguchi, S., Nomura, K., Kitagawa, H., 2003. Recent advances in the structural biology of chondroitin sulfate and dermatan sulfate. *Curr. Opin. Struct. Biol.* 13, 612-620.

[25] Liu, H., Zhang, Z., Linhardt, R.J., 2009. Lessons learned from the contamination of heparin. *Nat. Prod. Rep.* 26, 313-321.

[26] Kogan, G., Soltés, L., Stern, R., Gemeiner, P., 2007. Hyaluronic acid: a natural biopolymer with a broad range of biomedical and industrial applications. *Biotechnol. Lett.* 29, 17-25.

[27] Fraser, J.R., Laurent, T.C., Laurent, U.B., 1997. Hyaluronan: its nature, distribution, functions and turnover. *J. Intern. Med.* 242, 27-33.

[28] Shiedlin, A., Bigelow, R., Christopher, W., Arbabi, S., Yang, L., Maier, R.V., Wainwright, N., Childs, A., Miller, R.J., 2004. Evaluation of hyaluronan from different sources: Streptococcus zooepidemicus, rooster comb, bovine vitreous, and human umbilical cord. *Biomacromolecules* 5, 2122-2127.

[29] Boeriu, C.G., Springer, J., Kooy, F.K., van den Broek, L.A.M., Eggink, G., 2013. Production methods for hyaluronan. *Int. J. Carbohydr. Chem.* ID 624967.

[30] Volpi, N., 2006. Therapeutic applications of glycosaminoglycans. *Curr. Med. Chem.* 13, 1799-1810.

[31] Mikami, T., Kitagawa, H., 2013. Biosynthesis and function of chondroitin sulfate. *Biochim. Biophys. Acta* 1830, 4719-4733.

[32] Ramacciotti, E., Clark, M., Sadeghi, N., Hoppensteadt, D., Thethi, I., Gomes, M., Fareed, J., 2011. Review: contaminants in heparin: review of the literature, molecular profiling, and clinical implications. *Clin. Appl. Thromb. Hemost.* 17, 126-135.

[33] Alban, S., Lühn, S., Schiemann, S., Beyer, T., Norwig, J., Schilling, C., Rädler, O., Wolf, B., Matz, M., Baumann, K., Holzgrabe, U., 2011. Comparison of established and

novel purity tests for the quality control of heparin by means of a set of 177 heparin samples. *Anal. Bioanal. Chem.* 399, 605-620.

[34] Schiller, J., Volpi, N., Hrabárová, E., Šoltés, L., 2011. Hyaluronic acid: a natural biopolymer. In: *Biopolymers: Biomedical and Environmental Applications*; Kalia, Ed.; Scrivener, pp. 3-34.

[35] Carney, S.L., Osborne, D.J., 1991. The separation of chondroitin sulfate disaccharides and hyaluronan oligosaccharides by capillary zone electrophoresis. *Anal. Biochem.* 195, 132-140.

[36] Pichert, A., Schlorke, D., Franz, S., Arnhold, J., 2012. Functional aspects of the interaction between interleukin-8 and sulfated gycosaminoglycans. *Biomatter* 2, 142-148.

[37] Karamanos, N.K., Syrokou, A., Vanky, P., Nurminen, M., Hjerpe, A., 1994. Determination of 24 variously sulfated galactosaminoglycan- and hyaluronan-derived disaccharides by high-performance liquid chromatography. *Anal. Biochem.* 221, 189-199.

[38] Funderburgh, J.L., 2002. Keratan sulfate biosynthesis. *IUBMB Life* 54, 187-194.

[39] Gandhi, N.S., Mancera, R.L., 2008. The structure of glycosamininoglycans and their interaction with proteins. *Chem. Biol. Drug Des.* 72, 455-482.

[40] Lamanna, W.C., Kalus, I., Padva, M., Baldwin, R.J., Merry, C.L., Dierks, T., 2007. The heparanome - the enigma of encoding and decoding heparan sulfate sulfation. *J. Biotechnol.* 129, 290-307.

[41] Sasisekharan, R., Shriver, Z., Venkataraman, G., Narayanasami. U., 2002. Roles of heparan-sulphate glycosaminoglycans in cancer. *Nat. Rev. Cancer* 2, 521-528.

[42] Bergström, S., 1935. Über die Wirkungsgruppe des Heparins. *Naturwissenschaften* 23, 706.

[43] Karrer, P., Koenig, H., Usteri, E., 1943. Zur Kenntnis blutgerinnungshemmender Polysaccharid-polyschwefelsaure-ester und ähnlicher Verbindungen. *Helv. Chim. Acta* 26, 1296-1315.

[44] Meyer, K.H., Piroue, R.P., Odier, M.E., 1952. Les derives sulfates de l'acide chondroitine-sulfurique et leur action anticoagulante. Sur les polysaccharides amines IV. *Helv. Chim. Acta* 35, 574-588.

[45] Nagasawa, K., Inoue, Y., 1971. Reaction between carbohydrates and sulfuric acid. (2). Depolymerization and sulfation of chondroitin sulfate by sulfuric acid. *Chem. Pharm. Bull.* 19, 2617-2622.

[46] Naggi, A., Torri, G., Casu, B., Pangrazzi, J., Abbadini, M., Zameita M., Donati, M.B., Lansen, J., Maffrand, J.P., 1987. "Supersulfated" heparin fragments, a new type of low-molecular weight heparin. Physico-chemical and pharmacological properties. *Biochem. Pharmacol.* 36, 1895-1900.

[47] Maroufi, R.M., Tapon-Bretaudiere, J., Mardiguian, J., Sternberg, C., Dautzenberg, M.D., Fischer, A.M., 1990. Influence of the oversulfation method and the degree of sulfation on the anticoagulant properties of dermatan sulfate derivatives. *Thromb. Res.* 59, 749-758.

[48] Nagasawa, K., Uchiyama, H., Wajima, N., 1986. Chemical sulfation of preparations of chondroitin 4 and 6-sulfate, and dermatan sulfate. Preparation of chondroitin sulfate E-like materials from chondroitin4-sulfate. *Carbohydr. Res.* 158, 183-190.

[49] Nadkarni, V.D., Toida, T., Van Gorp, C.L., Schubert, R.L., Weiler, J. M., Hansen, K.P., Caldwell, E.E.O., Linhardt, R.L., 1996. Preparation and biological activity of N-sulfonated chondroitin and dermatan sulfate derivatives. *Carbohydr. Res.* 290, 87-96

[50] Crescenzi, V., Francescangeli, A., Renier, D., Bellini, D., 2002. New cross-linked and sulfated derivatives of partially deacetylated hyaluronan: synthesis and preliminary characterization. *Biopolymers* 64, 86-94.

[51] Magnani, A., Lamponi, S., Rappuoli, R., Barbucci, R., 1998. Sulphated hyaluronic acids: a chemical and biological characterisation. *Polym. Int.* 46, 225-240.

[52] Casu, B., Naggi, A., Torri, G., 2002. Chemical derivatization as a strategy to study structure-sctivity relationships of glycosaminoglycans. *Seminars Thromb. Hemostasis* 28, 335-342.

[53] Hintze, V., Moeller, S., Schnabelrauch, M., Bierbaum, S., Viola, M., Worch, H., Scharnweber, D., 2009. Modifications of hyaluronan influence the interaction with human bone morphogenetic protein-4 (hBMP-4). *Biomacromolecules* 10, 3290-3297.

[54] Magnani, A., Albanese, A., Lamponi, S., Barbucci, R., 1996. Blood-interaction performance of differently sulphated hyaluronic acids. *Thromb. Res.* 81, 383-395.

[55] Becher, J., Möller, S., Riemer, T., Schiller, J., Hintze, V., Bierbaum, S., Scharnweber, D., Worch, H., Schnabelrauch, M., 2012. Sulfated glycosaminoglycan building blocks for the design of artificial extracellular matrices. *ACS Symp. Ser.* 1107, 315-328.

[56] Schiller, J., Becher, J., Möller, S., Nimptsch, K., Riemer, T., Schnabelrauch, M., 2010. Synthesis and characterization of chemically modified glycosaminoglycans of the extracellular matrix. *Mini-Rev. Org. Chem.* 7, 290-299.

[57] Ludwig-Baxter, K.G, Zhengchun Liu, Z., Perlin, A.S., 1991. Regioselectivity in the sulfation of dermatan sulfate and methyl 4,6-O-benzylidene-α-D-idopyranoside. *Carbohydr. Res.* 214, 245-256.

[58] Ofosu, F.A., Modi, G.J., Blajchman, M.A., Buchanan, M.R., Johnson, E.A., 1987. Increased sulphation improves the anticoagulant activities of heparan sulphate and dermatan sulphate. *Biochem. J.* 248, 889-896.

[59] Ogamo, A., Metori, A., Uchiyama, H., Nagasawa, K., 1989. Reactivity toward chemical sulfation of hydroxyl groups of heparin. *Carbohydr. Res.* 193, 165-172.

[60] Kovensky, J., Cirelli, A.F., 1997 Chemical modification of glycosaminoglycans: selective 2-sulfation of D-glucuronic acid units in heparan sulfate. *Carbohydr. Res.* 303, 119-122.

[61] Barbucci, R., Leone, G., Chiumiento, A., Di Cocco, M.E., D'Orazio, G., Raffaella Gianferri, R., Delfini, M., 2006. Low- and high-resolution nuclear magnetic resonance (NMR) characterisation of hyaluronan-based native and sulfated hydrogels. *Carbohydr. Res.* 341, 1848-1858.

[62] Möller, S., Schmidtke, M., Weiss, D., Schiller, J., Pawlik, K., Wutzler, P., Schnabelrauch, M., 2012. Synthesis and antiherpetic activity of carboxymethylated and sulfated hyaluronan derivatives. *Carbohydr. Polym.* 90, 608-615.

[63] Becher, J., Möller, S., Weiss, D., Schiller, J., Schnabelrauch, M., 2010. Synthesis of new regioselelctively sulfated hyaluronans for biomedical application. *Macromol. Symp.* 296, 446-452.

[64] Roy, A., Urner, J., 1983. On the desulphation of carbohydrate sulphates. *Carbohydr. Res.* 124, 338-343.

[65] Jaseja, M., Rej, R.N., Sauriol, F., Perlin, A.S., 1989. Novel regio- and stereoselective modifications of heparin in alkaline solution. Nuclear magnetic resonance spectroscopic evidence. *Can. J. Chemistry* 67, 1449-1456.

[66] Usov, A.I., Adamyants, K.S., Miroshnikova, L.I., Shaposhnikova, A.A., Kochetkov, N.K., 1971. Solvolytic desulphation of sulphated carbohydrates. *Carbohydr. Res.* 18, 336-338.

[67] Matsuo, M., Takano, R., Kamei-Hayashi, K., Hara, S., 1993. A novel regioselective desulfation of polysaccharide sulfates: Specific 6-O-desulfation with N,O-bis(trimethylsilyl)acetamide. *Carbohydr. Res.* 241, 209-215.

[68] Fernandez, C., Hattan, C.M., Kerns, R.J., 2006. Semi-synthetic heparin derivatives: chemical modifications of heparin beyond chain length, sulfate substitution pattern and N-sulfo/N-acetyl groups. *Carbohydr. Res.* 341, 1253-1265.

[69] Inohue, Y., Nagasawa, K., 1976. Selective N-desulfation of heparin with dimethyl sulfoxide containing water or methanol. *Carbohydr. Res.* 46, 87-95.

[70] Holme, K.R., Liang, W., Yang, Z., Lapierre, F., Shaklee, P.N., Lam, L., 1996. In: *Nonanticoagulant Actions of Glycosaminoglycans*; Harenberg, J.; Casu, B., Eds.; Plenum Press: New York, pp. 139-162.

[71] Kariya, Y., Kyogashima, M., Suzuki, K., Isomura, T., Sakamoto, T., Horie, K., Ishihara, M., Takanoi, R., Kameii, K., Harai, S., 2000. Preparation of completely 6-*O*-desulfated heparin and its ability to enhance activity of basic fibroblast growth factor. *J. Biol. Chem.* 275, 25949-25958.

[72] Toida, T., Suzuki, A., Nakajima, K., Chaidedgumjorn, A., Imanari, T., 2000. Effect of 6-O-sulfonate hexosamine residue on anticoagulant activity of fully O-sulfonated glycosaminoglycans. *Glycoconjugate J.* 17, 393-399.

[73] Kantor, T.G., Schubert, M., 1957. Method for the desulfation of chondroitin sulfate. *J. Am. Chem. Soc.* 79, 152-153.

[74] McCarthy, M.M.U., John R. Bakers, J.R., 1979. Isolation and desulphation of keratan sulphates. *Carbohydr. Res.* 69, 151-164.

[75] Naggi, A., De Cristofano, B., Bisio, A., Torri, G., Casu, B., 2001. Generation of anti-factor Xa active, 3-O-sulfated glucosamine-rich sequences by controlled desulfation of oversulfated heparins. *Carbohydr. Res.* 336, 283-290.

[76] Li, Y., Rodrigues J., Tomas, H., 2012. Injectable and biodegradable hydrogels: gelation, biodegradation and biomedical applications. *Chem. Soc. Rev.* 41, 2193-2221.

[77] Lienemann, P.S., Lutolf, M.P., Ehrbar, M., 2012. Biomimetic hydrogels for controlled biomolecule delivery to augment bone regeneration. *Adv. Drug Del. Rev.* 64, 1078-1089.

[78] Rice, J.J., Martino, M.M., De Laporte, L, Tortelli, F., Briquez, P.S., Hubbell, J.A., 2013. Engineering the Regenerative Microenvironment with Biomaterials. *Adv. Healthcare Mater.* 2, 57-71.

[79] Bierbaum, S., Hintze, V., Scharnweber, D., 2012. Functionalization of biomaterials surfaces using artificial extracellular matrices. *Biomatter* 2, 1-10.

[80] Lee, C.-T., Kung, P.-H., Lee, Y.-D., 2005. Preparation of poly(vinyl alcohol)-chondroitin sulfate hydrogel as matrices in tissue engineering. *Carbohydr. Polym.* 61, 348-354.

[81] Wang S.-C., Chen, B.-H., Wang, L.-F., Chen, J.-S., 2007. Characterization of chondroitin sulfate and its interpenetrating polymer network hydrogels for sustained-drug release. *Int. J. Pharm.* 329, 103-109.

[82] Sintov, A., Di-Capua, N., Rubinstein, A., 1995. Cross-linked chondroitin sulphate: characterization for drug delivery purposes. *Biomaterials* 16, 473-478.

[83] Pieper, J.S., A. Oosterhof, A., Dijkstra, P.J., Veerkamp, J.H., van Kuppevelt, T.H., 1999. Preparation and characterization of porous crosslinked collagenous matrices containing bioavailable chondroitin sulphate. *Biomaterials* 20, 847-858.

[84] Zhao L., Gwon, H.-J., Lim, Y.-M., Nho, Y.-C., Kim, S.Y., 2014. Hyaluronic acid/chondroitin sulfate-based hydrogel prepared by gamma irradiation technique. *Carbohydr. Polym.* 102, 598-605.

[85] Mi, F.-L., Shyu, S.-S., Peng, C.-K., Wu, Y.-B., Sung, H.-W., Wang, P.-S., Huang C.-C., 2006. Fabrication of chondroitin sulfate-chitosan composite artificial extracellular matrix for stabilization of fibroblast growth factor. *J. Biomed. Mater. Res.* 76A, 1-15.

[86] Kirker, K.R., Prestwich, G.D., 2004. Physical properties of glycosaminoglycan hydrogels. *J. Polym. Sci. B: Polym. Phys.* 42, 4344-4356.

[87] Dawlee, S., Sugandhi, A., Balakrishnan, B., Labarre, D., Jayakrishnan A., 2005. Oxidized chondroitin sulfate-cross-linked gelatin matrixes: A new class of hydrogels. *Biomacromolecules* 6, 2040-2048.

[88] De Colli, M., Massimi, M., Barbetta, A., Di Rosario, B.L., Nardecchia, S., Devirgiliis, L., Dentini, M., 2012. A biomimetic porous hydrogel of gelatin and glycosaminoglycans cross-linked with transglutaminase and its application in the culture of hepatocytes. *Biomed. Mater.* 7, 055005.

[89] Freudenberg, U., Hermann, A., Welzel, P.B., Stirl, K., Schwarz, S.C., Grimmer, M., Zieris, A., Panyanuwat, W., Zschoche, S., Meinhold, D., Storch, A., Werner, C., 2009. A star-PEG–heparin hydrogel platform to aid cell replacement therapies for neurodegenerative diseases. *Biomaterials* 30, 5049-5060.

[90] Tae, G., Kim, Y.-J., Choi, W.I., Kim, M., Stayton, P.S., Hoffman, A.S., 2007. Formation of a novel heparin-based hydrogel in the presence of heparin-binding biomolecules. *Biomacromolecules* 8, 1979-1986.

[91] Nakamura, S., Ishihara, M., Obara, K., Masuoka, K., Ishizuka, T., Kanatani, Y., Takase, B., Matsui, T., Hattori, H., Sato, T., Kariya, T., Tadaaki Maehara T., 2006. Controlled release of fibroblast growth factor-2 from an injectable 6-O-desulfated heparin hydrogel and subsequent effect on in vivo vascularisation. *J. Biomed. Mater. Res.* 78A, 364-371.

[92] Wang, L.-F., Shen, S.-S., Lu, S.-C., 2003. Synthesis and characterization of chondroitin sulfate-methacrylate hydrogels. *Carbohydr. Polym.* 52, 389-396.

[93] Bryant, S.J., Davis-Arehart, K.A., Luo, N., Shoemaker, R.K., Arthur, J.A., Anseth, K.S., 2004. Synthesis and characterization of photopolymerized multifunctional hydrogels: Water-soluble poly(vinyl alcohol) and chondroitin sulphate sulfate macromers for chondrocyte encapsulation. *Macromolecules* 37, 6726-6733.

[94] Qiang Li, Q., Williams, C.G., Sun, D.D.N., Wang, J., Leong, K., Elisseeff, J.H., 2004. Photocrosslinkable polysaccharides based on chondroitin sulfate. *J. Biomed. Mater. Res.* 68A, 28-33.

[95] Li, Q., Wang, D.A., Elisseeff, J.E., 2003. Heterogeneous-phase reaction of glycidyl methacrylate and chondroitin sulfate: mechanism of ring-opening-transesterification competition. *Macromolecules* 36, 2556-2562.

[96] Moeller, S., Weisser, J., Bischoff, S., Schnabelrauch, M., 2007. Dextran and hyaluronan methacrylate based hydrogels as matrices for soft tissue reconstruction. *Biomol. Eng.* 24, 496-504.

[97] Jo, S., Kim, S., Hwang, Y., Noh, I., 2011. Induction and biological evaluations of self cross-linking chondroitin sulfate-poly(ethylene oxide) hydrogel. *Macromol. Res.* 19, 1303-1309.

[98] Becher, J., Möller, S., Schnabelrauch, M., 2013. Phase transfer-catalyzed synthesis of highly acrylated hyaluronan. *Carbohydr. Polym.* 93, 438-441.

[99] Cai, S., Liu, Y., Shu, X.Z., Prestwich, G.D., 2005. Injectable glycosaminoglycan hydrogels for controlled release of human basic fibroblast growth factor. *Biomaterials* 26, 6054-6067.

[100] Zheng Shu, X., Liu, Y., Luo, Y., Roberts, M.C., Prestwich, G.D., 2002. Disulfide cross-linked hyaluronan hydrogels. *Biomacromolecules* 3, 1304-1311.

[101] Nie, T., Baldwin, A., Yamaguchi, N. Kiick, K.L., 2007. Production of heparin-functionalized hydrogels for the development of responsive and controlled growth factor delivery systems. *J. Controlled Rel.* 122, 287-296.

[102] Stylianou, M., Triantaphyllidou, I.E., Vynios, D.H., 2006. Advances in the analysis of chondroitin/dermatan sulfate. *Adv. Pharmacol.* 53, 141-166.

[103] Korir, A.K., Larive, C.K., 2009. Advances in the separation, sensitive detection, and characterization of heparin and heparan sulfate. *Anal. Bioanal. Chem.* 393, 155-169.

[104] Malavaki, C.J., Theocharis, A.D., Lamari, F.N., Kanakis, I., Tsegenidis, T., Tzanakakis, G.N., Karamanos, N.K., 2011. Heparan sulfate: biological significance, tools for biochemical analysis and structural characterization. *Biomed Chromatogr.* 25, 11-20.

[105] Chmielewski, A.J., Stanley, F.E., Stalcup, A.M., 2011. The use of SAX-HPLC-CD as a heparin screening strategy. *J. Chromatogr. B* 879, 2619-2623.

[106] Huang, Y., Washio, Y., Hara, M., Toyoda, H., Koshiishi, I., Toida, T., Imanari, T., 1996. Simultaneous determination of dermatan sulfate and oversulfated dermatan sulfate in plasma by high-performance liquid chromatography with postcolumn fluorescence derivatization. *Anal. Biochem.* 240, 227-234.

[107] Volpi, N., Schiller, J., Stern, R., Soltés, L., 2009. Role, metabolism, chemical modifications and applications of hyaluronan. *Curr. Med. Chem.* 16, 1718-1745.

[108] Gribbon, P., Heng, B.C., Hardingham, T.E., 2000. The analysis of intermolecular interactions in concentrated hyaluronan solutions suggest no evidence for chain-chain association. *Biochem. J.* 350, 329-335.

[109] Capila, I., Sasisekharan, R., 2004. In: Chemistry and Biology of Hyaluronan; Garg, H.G., Hales, C.A., Eds.; Elsevier: Amsterdam, pp. 21-40.

[110] Volpi, N., Maccari, F., Linhardt, R.J., 2008. Capillary electrophoresis of complex natural polysaccharides. *Electrophoresis* 29, 3095-3106.

[111] Volpi, N., Maccari, F., Suwan, J., Linhardt, R.J., 2012. Electrophoresis for the analysis of heparin purity and quality. *Electrophoresis* 33, 1531-1537.

[112] Sisu, E., Flangea, C., Serb, A., Zamfir, A.D., 2011. Modern developments in mass spectrometry of chondroitin and dermatan sulfate glycosaminoglycans. *Amino Acids* 41, 235-256.

[113] Lemmnitzer, K., Schiller, J., Becher, J., Möller, S., Schnabelrauch, M.. 2014. Improvement of the digestibility of sulfated hyaluronans by bovine testicular

hyaluronidase - A UV spectroscopic and mass spectrometric study. *BioMed Res. Int.*, 2014, 986594.

[114] Hillenkamp, F., Peter-Katalinić, J., 2007. MALDI MS - A Practical Guide to Instrumentation. Methods and Application, 1st ed.; Wiley-VCH: Weinheim.

[115] Northen, T.R., Yanes, O., Northen, M.T., Marrinucci, D., Uritboonthai, W., Apon, J., Golledge, S.L., Nordström, A., Siuzdak, G., 2007. Clathrate nanostructures for mass spectrometry. *Nature* 449, 1033-1036.

[116] Schiller, J., Arnhold, J., Benard, S., Reichl, S., Arnold, K., 1999. Cartilage degradation by hyaluronate lyase and chondroitin ABC lyase: a MALDI-TOF mass spectrometric study. *Carbohydr. Res.* 318, 116-122.

[117] Hao, C., Ma, X., Fang, S., Liu, Z., Liu, S., Song, S., Liu, J., 1998. Positive and negative-ion matrix-assisted laser desorption/ionization mass spectrometry of saccharides. *Rapid Commun. Mass Spectrom.* 12, 345-348.

[118] Zaia, J., 2013. Glycosaminoglycan glycomics using mass spectrometry. *Mol. Cell Proteomics* 12, 885-892.

[119] Rudd, T.R., Yates, E.A., Hricovíni, M., 2009. Spectroscopic and theoretical approaches for the determination of heparin saccharide structure and the study of protein-glycosaminoglycan complexes in solution. *Curr. Med. Chem.* 16, 4750-4766.

[120] Schiller, J., Huster, D., Fuchs, B., Naji, L., Kaufmann, J., Arnold, K., 2004. Evaluation of cartilage composition and degradation by high-resolution magic-angle spinning nuclear magnetic resonance. *Methods Mol. Med.* 101, 267-285.

[121] Hricovíni, M., 2004. Structural aspects of carbohydrates and the relation with their biological properties. *Curr. Med. Chem.* 11, 2565-2583.

[122] Guerrini, M., Naggi, A., Guglieri, S., Santarsiero, R., Torri, G., 2005. Complex glycosaminoglycans: profiling substitution patterns by two-dimensional nuclear magnetic resonance spectroscopy. *Anal. Biochem.* 337, 35-47.

[123] Pomin, V.H., 2014. Solution NMR conformation of glycosaminoglycans. *Prog. Biophys. Mol. Biol.* 114, 61-68.

[124] Bociek, S.M., Darke, A.H., Welti, D., Rees, D.A., 1980. The ^{13}C NMR spectra of hyaluronate and chondroitin sulphates. Further evidence on an alkali-induced conformation change. *Eur. J. Biochem.* 109, 447-456.

[125] Naji, L., Kaufmann, J., Huster, D., Schiller, J., Arnold, K., 2000. ^{13}C NMR relaxation studies on cartilage and cartilage components. *Carbohydr. Res.* 327, 439-446.

[126] Pomin, V.H., 2013. Advances in glycosaminoglycanomics by ^{15}N-NMR spectroscopy. *Anal. Bioanal. Chem.* 405, 3035-3048.

[127] Pomin, V.H., Sharp, J.S., Li, X., Wang, L., Prestegard, J.H., 2010. Characterization of glycosaminoglycans by ^{15}N NMR spectroscopy and *in vivo* isotopic labeling. Anal. Chem. 82, 4078-4088.

[128] Blundell, C.D., Almond, A., 2007. Temperature dependencies of amide ^{1}H- and ^{15}N-chemical shifts in hyaluronan oligosaccharides. *Magn. Reson. Chem.* 45, 430-433.

[129] Yamada, S., Yoshida, K., Sugiura, M., Sugahara, K., 1992. One- and two-dimensional ^{1}H-NMR characterization of two series of sulfated disaccharides prepared from chondroitin sulfate and heparan sulfate/heparin by bacterial eliminase digestion. *J. Biochem.* 112, 440-447.

[130] Beyer, T., Diehl, B., Randel, G., Humpfer, E., Schäfer, H., Spraul, M., Schollmayer, C., Holzgrabe, U., 2008. Quality assessment of unfractionated heparin using ^{1}H nuclear magnetic resonance spectroscopy. *J. Pharm. Biomed. Anal.* 48, 13-19.

[131] Prabhakar, V., Capila, I., Sasisekharan, R., 2009. The structural elucidation of glycosaminoglycans. *Methods Mol. Biol.* 534, 147-156.

[132] Linhardt, R.J., Avci, F.Y., Toida, T., Kim, Y.S., Cygler, M., 2006. CS lyases: structure, activity, and applications in analysis and the treatment of diseases. *Adv. Pharmacol.* 53, 187-215.

[133] Nimptsch, A., Schibur, S., Schnabelrauch, M., Fuchs, B., Huster, D., Schiller, J., 2009. Characterization of the quantitative relationship between signal-to-noise (S/N) ratio and sample amount on-target by MALDI-TOF MS: Determination of chondroitin sulfate subsequent to enzymatic digestion. *Anal. Chim. Acta* 635, 175-182.

[134] Busse, K., Averbeck, M., Anderegg, U., Arnold, K., Simon, J.C., Schiller, J., 2006. The signal-to-noise ratio as a measure of HA oligomer concentration: a MALDI-TOF MS study. *Carbohydr. Res.* 341, 1065-1070.

[135] Sommers, C.D., Mans, D.J., Mecker, L.C., Keire, D.A., 2011. Sensitive detection of oversulfated chondroitin sulfate in heparin sodium or crude heparin with a colorimetric microplate based assay. *Anal. Chem.* 83, 3422-3430.

[136] Leone, G., Consumi, M., Lamponi, S., Magnani, A., 2012. New hyaluroran derivative with prolonged half-life for ophthalmogical formulation. *Carbohydr. Polymers* 88, 799-808.

[137] Harvey, D.J., 2012. Analysis of carbohydrates and glycoconjugates by matrix-assisted laser desorption/ionization mass spectrometry: an update for 2007-2008. *Mass Spectrom. Rev.* 31, 183-311.

[138] Zaia, J., 2010. Mass spectrometry and glycomics. *OMICS* 14, 401-418.

[139] Laremore, T.N., Murugesan, S., Park, T.J., Avci, F.Y., Zagorevski, D.V., Linhardt, R.J., 2006. Matrix-assisted laser desorption/ionization mass spectrometric analysis of uncomplexed highly sulfated oligosaccharides using ionic liquid matrices. *Anal. Chem.* 78, 1774-1779.

[140] Laremore, T.N., Linhardt, R.J., 2007. Improved matrix-assisted laser desorption/ionization mass spectrometric detection of glycosaminoglycan disaccharides as cesium salts. *Rapid Commun. Mass Spectrom.* 21, 1315-1320.

[141] Schulz, R., Höhle, S., Zernia, G., Zscharnack, M., Schiller, J., Bader, A., Arnold, K., Huster, D., 2006. Analysis of extracellular matrix production in artificial cartilage constructs by histology, immunocytochemistry, mass spectrometry, and NMR spectroscopy. *J. Nanosci. Nanotechnol.* 6, 2368-2381.

[142] Zaia, J., Li, X.Q., Chan, S.Y., Costello, C.E., 2003. Tandem mass spectrometric strategies for determination of sulfation positions and uronic acid epimerization in chondroitin sulfate oligosaccharides. *J. Am. Soc. Mass Spectrom.* 14, 1270-1281.

[143] Kenny, D.T., Issa, S.M., Karlsson, N.G., 2011. Sulfate migration in oligosaccharides induced by negative ion mode ion trap collision-induced dissociation. *Rapid Commun. Mass Spectrom.* 25, 2611-2618.

[144] Yamada, H., Miyauchi, S., Hotta, H., Morita, M., Yoshihara, Y., Kikuchi, T., Fujikawa, K., 1999. Levels of chondroitin sulfate isomers in synovial fluid of patients with hip osteoarthritis. *J. Orthop. Sci.* 4, 250-254.

[145] Nimptsch, K., Süss, R., Riemer, T., Nimptsch, A., Schnabelrauch, M., Schiller, J., 2010. Differently complex oligosaccharides can be easily identified by matrix-assisted laser desorption and ionization time-of-flight mass spectrometry directly from a standard thin-layer chromatography plate. *J. Chromatogr. A* 1217, 3711-3715.

[146] Zaia, J., 2009. On-line separations combined with MS for analysis of glycosaminoglycans. *Mass Spectrom. Rev.* 28, 254-272.

[147] Zhang, Z., Xie, J., Zhang, F., Linhardt, R.J., 2007. Thin-layer chromatography for the analysis of glycosaminoglycan oligosaccharides. *Anal. Biochem.* 371, 118-120.

[148] Salbach, J., Rachner, T.D., Rauner, M., Hempel, U., Anderegg, U., Franz, S., Simon, J.C., Hofbauer, L.C., 2012. Regenerative potential of glycosaminoglycans for skin and bone. *J. Mol. Med. (Berl.)* 90, 625-635.

[149] Salbach, J., Kliemt, S., Rauner, M., Rachner, T.D., Goettsch, C., Kalkhof, S., von Bergen, M., Möller, S., Schnabelrauch, M., Hintze, V., Scharnweber, D., Hofbauer, L.C., 2012. The effect of the degree of sulfation of glycosaminoglycans on osteoclast function and signaling pathways. *Biomaterials* 33, 8418-8429.

In: Sulfated Polysaccharides
Editors: M. Gama, H. Nader, H. de Oliveira Rocha
ISBN: 978-1-63482-975-5

Chapter 4

UNUSUAL HEPARIN-LIKE COMPOUNDS FROM MARINE INVERTEBRATES

Suely Ferreira Chavante[1,*], Adriana da Silva Brito[2], Rômulo dos Santos Cavalcante[1] and Giulianna Paiva Viana de Andrade[1]

[1]Laboratório de Glicoconjugados Bioativos (GLICOBIO), Departamento de Bioquímica, Universidade Federal do Rio Grande do Norte (UFRN), Natal/RN, Brasil
[2]Faculdade de Ciências da Saúde do Trairi (FACISA), Universidade Federal do Rio Grande do Norte (UFRN), Santa Cruz/RN, Brasil

ABSTRACT

The wide diversity of marine organisms offers a rich source of structurally different sulfated polysaccharides. Among them, unusual heparin-like compounds from a variety of marine invertebrate species have been described in literature, mainly to molluscs, crustaceans and tunicates. These compounds share structural similarities with heparin mammalian, but with some peculiarities, which vary according to the species. Significant differences in terms of the ratios of iduronic acid to glucuronic acid are notable, when compared to mammalian heparin. However, a relatively high sulfation degree for all these compounds was identified. Dissulfated and trissulfated disaccharides represent the main structures of heparin-like compounds in some mollusc and shrimp species that mimic the anticoagulant activity of mammalian heparin. Non-anticoagulant activities as anti-inflammatory and anti-angiogenic properties have also been described to a non-hemorrhagic peculiar heparin-like compound from shrimp. The discovery of these compounds inside special "mast-like" cells on different tissues of molluscs and basophil-like cells in the hemolymph of tunicates, associated to their structural features, suggests the maintenance of these molecules through evolution. Although many structural details and the biological role of these polymers still remain to be established, the ability to

* E-mail: chavante@ufrnet.br.

interfere in processes such as coagulation, inflammation and cancer related provides strong evidence of the promising therapeutic potential of these unusual polysaccharides.

1. Introduction

During the evolutionary process, marine organisms certainly developed biochemical mechanisms of adaptation to extreme environmental conditions. Because of these adaptations, several biomolecules, such as polysaccharides and especially glycosaminoglycans (GAG), show peculiar structures species-specific with great chemical diversity, absent in their counterpart of terrestrial animals [1, 2]. Thus, these marine compounds become potential candidates for studies of new drugs, such as heparin. This GAG obtained from mammalian tissues is widely used in the treatment and prevention of thromboembolic diseases [3], although other non-anticoagulant activities are well known [4-9]. However, hemorrhagic complications [10] and occurrence of contamination related to heparin [11-13] have stimulated the search of compounds that mimic its biological properties, but with reduced collateral effect.

In this context, over the last decades, heparin-like compounds (HLC) and heparan sulfate-like compounds (HSLC) have been studied in marine invertebrates.

A wide and updated overview of the distribution of these compounds in invertebrates is illustrated in the cladogram below (Figure 1). HSLC has a ubiquitous distribution in all organisms with tissue organization, from cnidarians to urochordates. Studies on the great relevance of HSLC and of hybrid compounds bearing structural features of heparin and heparan sulfate have been described in several species [14-20]. On the other hand, the occurrence of HLC has been reported on a few marine invertebrates species, consistent with its limited distribution along the evolutionary tree, where HLC appeared for the first time in molluscs [21]. Despite a lower distribution among species of marine invertebrates, the HLC are well documented in the literature with a large collection of data, especially for molluscs, crustaceans, and tunicates.

Therefore, this chapter will present a brief review about the occurrence of heparin-like compounds, with emphasis on some structural aspects and tissue distribution associated with pharmacological activities.

2. Heparin-Like Compounds from Molluscs

Marine molluscs represent a group of organisms with a wide diversity of species. Such invertebrates have been extensively studied since the beginning of the 20th century because they are a rich source of byproducts with pharmaceutical activities. Some of these activities are attributed to the occurrence of sulfated polysaccharide, including heparin-like compounds.

The first report on these compounds in mollusc was described by Thomas in 1951 [21], and it identified a material containing metachromatic property and anticoagulant activity in the bivalve *Spisula (Mactra) solidissima*. In the following years, Frommhagen and coworkers [22] obtained compounds with high anticoagulant activity from extracts of *S. solidissima* and *Cyprina* (*Artica*) *islandica* calling them mactina A and B, respectively.

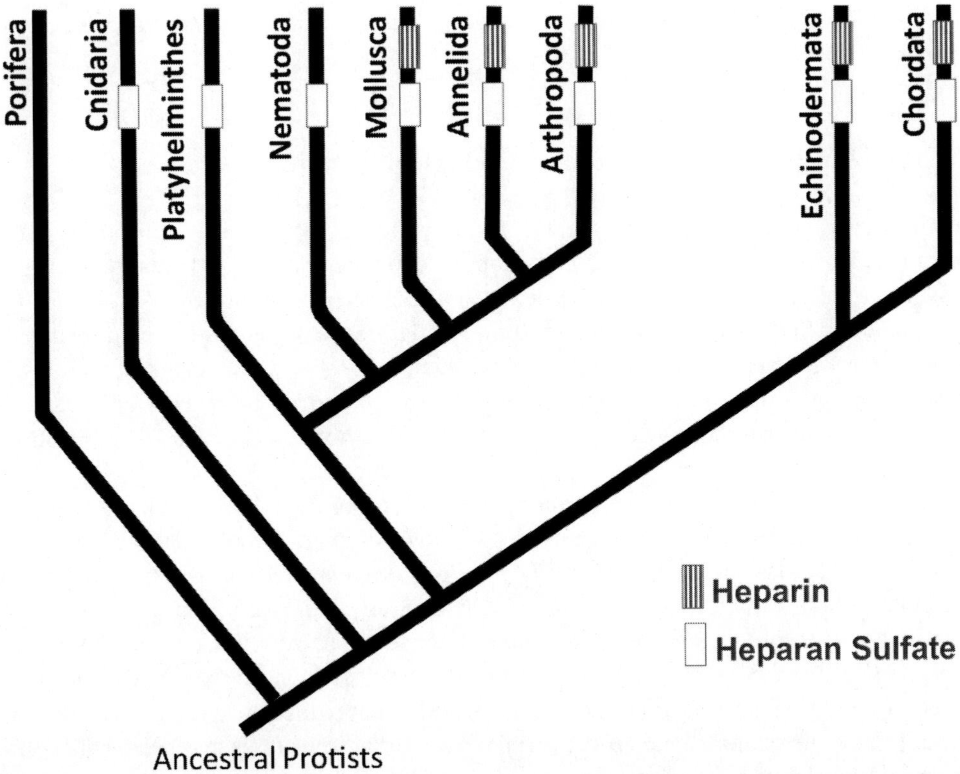

Figure 1. Distribution of heparin-like compounds and heparan sulfate-like compounds in invertebrate animals.

Preliminary analyses of these compounds revealed the presence of polysaccharide containing glucuronic acid, glucosamine and sulfate, which possessed anticoagulant activity [23-25]. Since then, other studies have reported the presence of HLC in various molluscs. The first and more detailed study about structure of HLC present in molluscs came from the extensive work of Dietrich and their collaborators. From the screening studies on the occurrence of HLC in a variety of species, these academic researchers led to the suggestion that heparin had maintained their main structural features through evolution [26]. Among a variety of species of molluscs analyzed, *Anomalocardia brasiliana, Tivela mactroides* and *Donax striatus* had caught special attention to researchers due the discovery of high content of HLC in these clams [26-30]. Furthermore, another significant finding by this group was the presence of "mast-like" cells containing heparin in different tissues of *A. brasiliana* [31].

In the mid 1980s, an unusual heparin was also isolated from the clam *Mercenaria mercenaria* by Jordan and Marcum [32]. Later, Ulrich and Boon performed histological studies of various tissues of this mollusc and showed the presence of HLC inside granules in the special cells [33]. A few years later, another group used a similar analytical approach to describe the occurrence and the characterization of HLC from *Tapes phylippinarum* [34] and *Callista chione* [35].

Although the presence of HLC in other Molluscs such as *Anodonta sp* [36], *Mesodersma donacium, Lolligo brasiliensis, Octopus* sp [14], *Turritella attenuata* [37] and *Meretrix meretrix* [38] is already documented in literature, this chapter is addressed with special emphasis on those ones which structural studies are more detailed and will be described bellow.

2.1. Structural Features and Pharmacological Activities

Classical colorimetrical methods, electrophoretic approach, molecular weight determination, chemical analysis, enzymatic depolymerization and nuclear magnetic resonance (NMR) spectroscopy were undertaken to characterize the structure of unusual HLC from molluscs. Such analysis had shown some similarities between mollusc and mammalian heparins, but with some structural peculiarities which define the compounds from molluscs as unusual heparins. Since these structural features were based on the results of a variety of methodological approaches, which were important to refer to the compounds from mollusc as unusual heparin, some of these analyzes are described here.

A relatively great structural variation of HLC according to mollusc species analyzed was observed [26, 30]. The preliminary analysis had shown a range of molecular weights varying from 11 to 32 KDa. The HLC from *T. Phylippinarum* and *M. mercenaria* exhibited a molecular weight closer to bovine heparin. In contrast, the HLC from *C. chione* had a molecular weight lower than that of mammalian heparin (Table 1). Except for HLC from *C. chione,* all of the other compounds from molluscs were distinguished from bovine heparin by presenting a higher anticoagulant activity [26, 28]. These differences did not seem to be relevant, since the mammalian heparin population with these features had been isolated from commercial preparations. Structural peculiarities of HLC from different molluscs have been discovered through enzymatic depolimerization with *Flavobacterium heparinum* enzymes.

These studies revealed the variability of distinct regions according to the degree of susceptibility to *F. heparinum* heparinase and heparitinase II. Despite of all the HLC from mollusk containing the same type of disaccharides as mammalian heparins, the quantitative differences are remarkable (Table 2).

Table 1. Molecular weight and *in vitro* anticoagulant activities of heparin-like compounds from marine molluscs and bovine intestinal heparin

Origin	MW (KDa)	In vitro (UI/mg)		References
		USP	APTT	
A. brasiliana	32	320	345	[27]
T. mactroides	25	220	150	[26]
D. striatus	20	180	-	[26]
M. mercenaria	18	348*	-	[32]
T. hylippinarum	14	-	347	[34]
C. chione	11	-	97	[35]
Bovine Intestine	15	140	135	[30]

*Indicates that anticoagulant activity was assayed by anti-Xa assay; MW, molecular weight; USP, United State Pharmacopeia; aPTT, activated prothrombine thromboplastin time.

Table 2. Percentage of the main saccharide units constituent from molluscs and bovine heparins

Origin	ΔIdoA,2S-GlcNS,6S	ΔIdoA,2S-GlcNS	ΔGlcA-GlcNS,6S	ΔTetra
A. brasiliana	21.0	5.3	34.8	18.2
D. striatus	16.2	4.4	34.8	15.1
T. macrtroides	16.3	3.0	45.0	16.2
Bovine Lung	40.3	2.5	1.5	40.8
Bovine Intestine	22.7	2.0	26.7	24.0

ΔIdoA,2S: 2-O-sulfated iduronic acid; ΔGlcA: glucuronic acid; GlcNS,6S: N,6-disulfated glucosamine; GlcNS: N-sulfated glucosamine; ΔTetra: tetrasaccharide.

A large amount of disaccharide, containing glucuronic acid (GlcA) and N,6-disulfated glucosamine (GlcNS,6S) (ΔGlcA-GlcNS,6S), formed by the action of heparitinase II upon HLC from *T. mactroides, A. brasiliana and D. striatus*, was identified. Considering the specificity of this enzyme, it was concluded that HLC from these molluscs contained higher amounts of glucuronic acid residues in their structures when compared to mammalian heparins [39]. The structural characterization of HLC from these molluscs was confirmed by NMR spectroscopic [26].

Another systematic study on the structure of HLC from *A. brasiliana* confirmed the great variation in terms of the ratios of total iduronic acid (IdoA) to total glucuronic acid (GlcA) when compared to heparin preparations. The analysis of uronic acid content by acid hydrolysis of HLC from this mollusc revealed that it contains large amounts of glucuronic acid (64%), contrasting with lung heparin that contains a higher proportion of iduronic acid (>90%). These ratios are lower for bovine mucosa heparin and they reflect on its relative proportions of disaccharides formed by lyases from *F. heparinum* (Table II). Considering this structural feature, HLC from *A. brasiliana* is structurally closer to bovine intestinal heparin than to lung bovine heparin [30]. It is also notable the increased proportions of disaccharide ΔIdoA,2S-GlcNS in the structure of the HLC from molluscs when compared to mammalian heparins. According to these findings, Pejler and their collaborators suggested that the 6-O-sulfation of these sequences could be hindered during biosynthetic step [28].

An additional structural analysis of HLC from *A. brasiliana* revealed the presence of fractions containing up to three antithrombin (AT) binding sites per chain [28]. In mammalian heparins, the high content of 3-O-sulfated glucosamine residues (GlcNS,3S,6S), followed by GlcA residues, has been regarded as marker of the pentasaccharidic binding site for AT and has been associated with the high anticoagulant activity [40]. The HLC from *A. brasiliana* showed 25-30% of the total disaccharide units containing 3-O-sulfated saccharide sequences (GlcNS-GlcA-GlcNS,3S and GlcNS-GlcA-GlcNS,3S,6S). The content of these sequences increased in chains according with their affinity for antithrombin and with the number of antithrombin-binding region per chain. An unusual sequence containing 3-O-sulfated glucosamine was also identified as GlcNS-IdoA-GlcNS,3S,6S and represented 3-4% of the total disaccharide units.

In another structural evidences, Jordan and Marcum [32] had also isolated a HLC with high affinity to AT from *Mercenaria mercenaria*. In such compound, it was identified the occurrence of 3-O-sulfated glucosamine residues. Oligosaccharides containing 3-O-sulfated glucosamine residues were also identified in the HLC from *T. Phylippinarum* by Cesaretti and their collaborators [34].

HLC from Mediterranean bivalve *Callista chione* was also characterized by the highest proportion of disulfated disaccharide, mainly bearing nonsulfated iduronic acid and glucuronic acids [35]. In contrast, the HLC from *T. Phylippinarum* has shown a high content of trisulfated disaccharide [34].

2.2. Distribution on Tissues

Regarding the localization of HLC in mollusc tissues, two different studies have revealed its presence inside granules confined to specialized cells of these invertebrates. In the beginning of this century, Ulrich and Boon showed that the HLC from clam *M. mercenaria* is located in cell populations near epithelia. In the intestine and palps, HLC may bind to proteases and assist them in the process of digestion. Occurrence of such compounds was also identified near organs close to the environment, as pallium, siphon and ctenidia, which suggests a role in defense mechanisms [33]. Histological and histochemical analyzes in different tissues from *A. brasiliana* also revealed the presence of HLC in granules inside the cytoplasm of "mast-like" cells, found exclusively in the mantle, ctenidium, intestine, labial palp and foot tissues. By USP assay, the anticoagulant activity assessed in these tissues ranged from 20 UI/mg, in the foot, until 210 UI/mg in the labial palp [31]. These findings gave support to the suggestion that HLC from molluscs are involved in their defense mechanism.

3. Heparin-Like Compounds from Crustaceans

Crustaceans represent a large group of animals classified within the phylum Arthropoda, where the crabs, lobsters, crayfish, shrimp and barnacles are inserted. The occurrence of heparin-like compounds in these organisms was first related by Cassaro and Dietrich in 1977 [14], through a systematic study of polysaccharide distribution in several invertebrates' species, among which some crustaceans appear, such as lobster *Scyllarides brasiliensis*. Years later, in 1981, Ehrlich, Patel and Stivala [41] also reported the occurrence of these compounds in the red crab *Geryon guinquedens*. Despite of these findings, for nearly two decades no significant study on the structural characterization of HLC from crustaceans was published. The interest in research on such compounds only became intensified when Dietrich and coworkers, in 1999 [42], characterized for the first time a HLC from the heads of shrimp *Penaeus brasiliensis*.

In the following year, Medeiros and coworkers [17] expanded these studies reporting the presence of heparin-like compounds at others crustacean species, such as *Dedrocephalus brasiliensis, Ucides cordatus* and *Chaceon fenneri*. Recently, more detailed studies addressing to structural characterization and pharmacological activities of HLC were described to other crustaceans species such as *Litopenaeus vannamei* shrimp [43-45] and *Goniopsis cruentata* crab [46], and they will be detailed on the following.

3.1. Structural Features and Pharmacological Activities

An important contribution in research about heparin-like compounds from crustaceans was achieved after the discovery of a HLC with low molecular weight obtained in the heads of *Penaeus brasiliensis* shrimp. NMR spectroscopy and enzymatic analysis revealed that shrimp HLC is rich in disulfated disaccharides (ΔUA-GlcNS,6S and ΔUA,2S-GlcNS) and has a higher degree of glucuronic and non-sulfated iduronic acid residues, when compared to mammalian heparin. Interestingly, HLC shrimp showed potent "in vitro" anti-Xa and "in vivo" antithrombotic activities [42]. Since then, over the last decade, the studies initiated by Dietrich and coworkers were extended by the researches of HLC in the heads of the main shrimp specie cultivated in the northeastern region of Brazil, the *Litopenaeus vannamei.* This study, not only expands the knowledge about the sulfated polysaccharides from marine invertebrates, but also becomes economically and environmentally attractive, since the reuse of the shrimp head waste to obtain of HLC could contribute to solve issues related to improper discard of this organic residue.

Detailed structural studies about *L. vannamei* HLC, as well as its pharmacological activities, have been reported recently. These studies showed the occurrence of two different heparin-like compounds with several structural peculiarities. The first one, has an average molecular weight of 36 kDa and it is rich in glucuronic acid and disulfated disaccharide units (ΔUA-GlcNS,6S and ΔUA,2S-GlcNS) [43]. For the first time, the detailed characterization of this compound identified through NMR spectroscopy and enzymatic analysis revealed an unusual trisaccharide sequence containing high degree of 3-O-sulfated glucosamine groups (~15%) [45]. Interestingly, shrimp HLC exhibited a negligible anticoagulant activity (~5UI/mg) when compared to mammalian heparin, despite to bind and induce conformational change in AT, similar to heparin and synthetic pentasaccharide [45, 47]. Moreover, this compound exhibits anti-inflammattory and anti-angiogenic properties [43, 48]. The second HLC isolated from the *L. vannamei* shrimp heads is an atypical molecule which shares some structural features with heparin - such as the high degree of N- and 6-O-sulfation and minor N-acetylation - and with heparan sulfate - regarding the high content of glucuronic acid. Additionally, this molecule has an average molecular weight of 15 kDa, similar to porcine heparin (~16 kDa), and lower than bovine heparan sulfates (~25 kDa). Different to the first HLC obtained from shrimp, this heparin/heparan hybrid displayed a high anti-Xa activity and was able to stabilize the AT [44]. These findings suggest that the anticoagulant activity of the shrimp compounds may, among other factors, be associated to its ability to bind and stabilize AT [45, 47].

Recently, Andrade and coworkers [46] described the most complete study about the structure of heparin-like compounds obtained from crabs. For the first time, these authors isolated a natural 2-O-sulfated glucuronic acid-rich HLC with negligible anticoagulant activity from the *Goniopsis cruentata.* Crab heparin-like compound also showed a high degree of disulfated and N-sulfated disaccharides (ΔUA, 2S-GlcNS, ΔUA-GlcNS,6S and ΔUA-GlcNS) and absence of non-sulfated iduronic acid. Such peculiarities in crab HLC suggest a particular mode of biosynthesis in these compounds. In mammalian, after the epimerization step, a majority of iduronic acid is 2-O-sulfated [48]. In crustaceans, curiously, this 2-O-sulfation seems to occur in all iduronic acid residues and, surprisingly, before the step of epimerization to crab HLC. Heparin-like compounds from shrimp and crab display a poor bleeding potential [43, 44, 46]. Therefore, the compounds described here may direct further studies about new therapeutical agents without side effects.

4. HEPARIN-LIKE COMPOUNDS FROM TUNICATE (UROCHORDATA)

Tunicates are filter-feeding marine invertebrates that belong to the subphylum Urochordata. These organisms are mainly characterized by the presence of an external tunic that protects the animal's body. The most prominent tunicates are the sea squirts (ascidians). Different sulfated polysaccharides were described in this class of tunicate [49-52]. An important representative of ascidians is the benthic invertebrate *Styela plicata* – specie with the ability to adapt to widely fluctuating environments, e.g., regarding temperature and salinity [53-55], and tolerance to highly polluted water [56]. Interestingly, accumulating evidence has shown that *Styela plicata* also represents an important source of heparin-like compounds. Some structural features and pharmacological potential of these compounds are discussed below.

4.1. Occurrence on Tissues, Structural Features and Pharmacological Activities

In tunicates, compounds structurally related to heparin were first identified in the viscera of *S. plicata* by Pavão and coworkers in 1994 [52], and since then they have been widely studied and characterized. The results of such studies have shown that ascidian HLC and mammalian heparin are composed by the same disaccharide units, but in different proportions. Similar to that observed in molluscs and crustaceans, HLC from the body of *S. plicata* is formed mainly by disulfated disaccharide ΔUA2S-GlcNS (47.5%), followed by trisulfated disaccharide ΔUA2S-GlcNS6S (38.3%). Disaccharides containing 3-O-sulfation on glucosamine residues are also present, but in smaller proportion [57].

Such disaccharide composition appears to vary according to tissue location. Data from enzymatic depolymerization studies with heparin and heparan lyases and RMN spectroscopy revealed that HLC isolated from intestine, heart and pharynx are mainly composed by trisulfated disaccharide and small amounts of disulfated disaccharide. On the other hand, the disulfated disaccharide is prevalent unit in the HLC from cloak [58, 59]. In all studied tissues, HLC occurs in intracellular granules. In the eggs, these molecules are composed mainly by trisulfated disaccharide IdoA2S-GlcNS,6S, followed by a lower content (~30%) of disulfated disaccharide IdoA-GlcNS,6S [58] and they are found inside of cell monolayers named test cells, which surround the eggs of the sea squirt. In the adult animal, HLC is located in cytoplasmic granules of epithelial cells of organs that are in close contact with the external environment [57, 58] and in intracellular granules of circulating "basophil-like" cells [60].

When compared to mammalian heparin, HLC from *S. plicata* body has a high average molecular mass (32 kDa) and reduced anticoagulant activity as estimated by activated Partial Thromboplastin Time (aPTT assay) [57]. However, the polysaccharide from ascidian is able to activate heparin cofactor II by the same extent of the heparin counterpart [61]. The reduced anticoagulant activity can justify the low effect of ascidian HLC on venous thrombosis inhibition. On the other hand, these compounds are able to inhibit the arterial thrombosis to the same extent as mammalian heparin. Studies on the pharmacological potential of the HLC from ascidian are not restricted to its effects on hemostasis. Belmiro and coworkers [62] studied the effect of a heparin-like compound from *S. plicata* in a colitis model and discovered that it was more effective than mammalian heparin in reducing the cell infiltration to the injury site. Interestingly, the treatment with the ascidian compound does not alter the clotting time of the treated animals. Recently it was demonstrated that ascidian HLC also diminish the inflammatory response in a diversion colitis model [63]. In addition to its anti-inflammatory potential, heparin-like compound from *S. plicata*, similar to crustaceans HLC, has a low hemorrhagic effect [57]. This is an interesting property since hemorrhagic complications are an important disadvantage of the clinical use of heparin.

5. Biological Role of Heparin–Like Compounds in Marine Invertebrates

As previously discussed, the presence of different HLC in marine invertebrates has been identified from molluscs to urochordates. It was also reported that some of these compounds have binding regions "apparently" specific to antithrombin (AT) and thus they are able of inactivating "in vitro" some coagulation protease cascade from mammals. This fact is intriguing and raises the discussion about the biological role of HLC in marine invertebrates, since such organisms are devoid of coagulation system similar to those of mammalians. In vertebrates, heparin is located inside mast cells of tissues that are in direct contact with the external environment as skin, lung and intestine [29, 58, 64-67]. The studies of tissue distribution of HLC in molluscs and tunicates revealed that such compounds show a distribution similar to heparin in vertebrates. Therefore, it has been proposed that, similarly to vertebrate's heparin, the HLC from marine invertebrates also plays a role in defensive mechanism, protecting the organisms against foreign bodies, such as bacteria and viruses [67]. The discovery of "mast-like" and "basophil-like" cells in molluscs and tunicates, respectively [31, 33, 60], was particularly significant to reinforce this idea. Despite these evidences, the real function of heparin-like compounds in marine invertebrates is still mysterious and further studies may help to elucidate this question.

Conclusion

Among marine invertebrates, heparin-like compounds from molluscs, crustaceans and tunicates have been the most extensively studied. The detailed structural characterization of such compounds demonstrated that the heparin structure had been maintained throughout evolution and the unusual features described here lead us to think about the occurrence of a particular mode of HLC biosynthesis. The discovery of HLC with a high content of 3-O-sulfated glucosamine in some species of molluscs and crustaceans was particularly significant. Initially the occurrence of this rare residue was associated with the capacity to activate AT. However, the discovery of a HLC containing high levels of 3-O-sulfation, but with negligible anti-Xa activity, in *L. vannamei* shrimp, brought to light the discussion about the importance of this residue for AT activation and reinforces the idea that the structure-activity relationship between heparin and protein relies on the complex relationships between substitution patterns, conformation, dynamics and stabilization of protein-heparin complex [45, 47]. Regarding to pharmaceutical potential, the discovery of HLC with reduced anticoagulant activity and low hemorrhagic effect in crustaceans and tunicates opened new perspectives to study mechanisms in which anticoagulation is a disadvantage, such as anti-inflammatory and anti-angiogenic aplications. Although there are many evidences on the pharmacological activities of HLC from marine invertebrates, questions about its physiological role in these organisms still need to be answered. The studies of tissue distribution of HLC in molluscs and tunicates, and the discovery of "mast-like" and "basophil-like" cells contained heparin-like compounds was a great step to help elucidanting this question.

REFERENCES

[1] Haefner, B., 2003. Drugs from the deep: marine natural products as drug candidates, *Drug Discovery Today*, 8, 536-544.

[2] Senni, K., Pereira, J., Gueniche, F., Delbarre-Ladrat, C., Sinquin, C., Ratiskol, J., Godeau, G., Fischer, A. M., Helley, D., Colliec-Jouault, S., 2011. Marine polysaccharides: a source of bioactive molecules for cell therapy and tissue engineering, *Mar. Drugs*, 9, 1664-1681.

[3] Jorpes, J. E., 1959. Heparin: a mucopolysaccharide and an active antithrombotic drug, *Circulation*, 19, 87-91.

[4] Folkman, J., Taylor, S., Spillberg, C., 1983. The role of heparin in angiogenesis, *Ciba Found Symp.*, 100, 132-149.

[5] Rabenstein, D. L., 2002. Heparin and heparan sulfate: structure and function, *Natural Product Reports*, 19, 312-331.

[6] Wang, J. G., Mu, J. S., Zhu, H. S., Geng, J. G., 2002. N-desulfated non-anticoagulant heparin inhibits leukocyte adhesion and transmigration in vitro and attenuates acute peritonitis and ischemia and reperfusion injury in vivo, *Inflamm. Res.*, 51, 435-443.

[7] Casu, B., Vlodavsky, I., Sanderson, R. D., 2008. Non-anticoagulant heparins and inhibition of cancer, *Pathophysiol. Haemost. Thromb*, 36, 195-203.

[8] Guerrini, M. H., M.; Torri, G., 2007. Interaction of heparins with fibroblast growth factors: conformational aspects, *Curr. Pharm. Des.*, 2045–2056.

[9] Young, E., 2008. The anti-inflammatory effects of heparin and related compounds, *Thromb Res.*, 122, 743-752.

[10] Nader, H. B., Tersariol, I. L., Dietrich, C. P., 1989. Antihemostatic activity of heparin disaccharides and oligosaccharides obtained by chemical and enzymatic fragmentation: reversal of the hemorrhagic activity by ATP and myosin, *Thromb Res.*, 54, 207-214.

[11] Blossom, D. B., Kallen, A. J., Patel, P. R., Elward, A., Robinson, L., Gao, G., Langer, R., Perkins, K. M., Jaeger, J. L., Kurkjian, K. M., Jones, M., Schillie, S. F., Shehab, N., Ketterer, D., Venkataraman, G., Kishimoto, T. K., Shriver, Z., McMahon, A. W., Austen, K. F., Kozlowski, S., Srinivasan, A., Turabelidze, G., Gould, C. V., Arduino,

M. J. and Sasisekharan R., 2008. Outbreak of adverse reactions associated with contaminated heparin, *The New England journal of medicine*, 359, 2674-2684.

[12] Kishimoto, T. K., Viswanathan, K., Ganguly, T., Elankumaran, S., Smith, S., Pelzer, K., Lansing, J. C., Sriranganathan, N., Zhao, G., Galcheva-Gargova, Z., Al-Hakim, A., Bailey, G. S., Fraser, B., Roy, S., Rogers-Cotrone, T., Buhse, L., Whary, M., Fox, J., Nasr, M., Dal Pan, G. J., Shriver, Z., Langer, R. S., Venkataraman, G., Austen, K. F., Woodcock, J., Sasisekharan, R., 2008. Contaminated heparin associated with adverse clinical events and activation of the contact system, *N. Engl. J. Med.*, 358, 2457-2467.

[13] Rudd, T. R., Gaudesi, D., Lima, M. A., Skidmore, M. A., Mulloy, B., Torri, G., Nader, H. B., Guerrini, M., Yates, E. A., 2011. High-sensitivity visualisation of contaminants in heparin samples by spectral filtering of 1H NMR spectra, *Analyst*, 136, 1390-1398.

[14] Cassaro, C. M., Dietrich, C. P., 1977. Distribution of sulfated mucopolysaccharides in invertebrates, *J. Biol. Chem.*, 252, 2254-2261.

[15] Hovingh, P., Linker, A., 1982. An unusual heparan sulfate isolated from lobsters (Homarus americanus), *J. Biol. Chem.*, 257, 9840-9844.

[16] Nader, H. B., Ferreira, T. M., Paiva, J. F., Medeiros, M. G., Jeronimo, S. M., Paiva, V. M., Dietrich, C. P., 1984. Isolation and structural studies of heparan sulfates and chondroitin sulfates from three species of molluscs, *J. Biol. Chem.*, 259, 1431-1435.

[17] Medeiros, G. F., Mendes, A., Castro, R. A., Bau, E. C., Nader, H. B., Dietrich, C. P., 2000. Distribution of sulfated glycosaminoglycans in the animal kingdom: widespread occurrence of heparin-like compounds in invertebrates, *Biochim. Biophys. Acta.*, 1475, 287-294.

[18] Chavante, S. F., Santos, E. A., Oliveira, F. W., Guerrini, M., Torri, G., Casu, B., Dietrich, C. P., Nader, H. B., 2000. A novel heparan sulphate with high degree of N-sulphation and high heparin cofactor-II activity from the brine shrimp Artemia franciscana, *Int. J. Biol. Macromol.*, 27, 49-57.

[19] Gomes, A. M., Kozlowski, E. O., Pomin, V. H., de Barros, C. M., Zaganeli, J. L., Pavao, M. S., 2010. Unique extracellular matrix heparan sulfate from the bivalve Nodipecten nodosus (Linnaeus, 1758) safely inhibits arterial thrombosis after photochemically induced endothelial lesion, *J. Biol. Chem.*, 285, 7312-7323.

[20] Chen, J., Yamada, S., Hama, Y., Shetty, A. K., Kobayashi, T., Oda, H., Seiki, K., Kim, E., Kimura, T., Takahashi, N., Hidari, K. I., Suzuki, T., Suzuki, Y., Sugahara, K., 2011. Unique heparan sulfate from shrimp heads exhibits a strong inhibitory effect on infections by dengue virus and Japanese encephalitis virus, *Biochem. Biophys. Res. Commun.*, 412, 136-142.

[21] Thomas, L. J., 1951. A blood anticoagulant from surf clams, *Biol. Bull.*, 101, 230.

[22] Frommhagen, L. H., Fahrenbach, M. J., Brockman, J. A., Stockstad, E. L. R.,1953. Heparin-like aticoagulants from molusca, *Proc. Soc. Exp. Biol. Med.*, 82(2), 280-283.

[23] Thomas, L. J., 1954. The localization of heparin-like blood anticoagulant substances in the tissues of Spisula solidíssima, *Biol. Bull.*, 106(1), 129-138.

[24] Burson, S. L., Fahrenbach, M. J., Frommhagen, L. H., Riccardi, B. A., Brown, R. A., Brockman, J. A., Lewry, H. V., Stockstad E. L. R., 1956. Isolation and purification of mactins, heparin-like anticoagulants from molusca, *J. Am. Chem. Soc.*, 78, 5874-5878.

[25] Cifonelli, J. A., Mathews, M. B., 1972. A comparision of the chemical structures of mactins with mammalian heparins, *Conn. Tissue Res.*, 1 (2), 121-129.

[26] Dietrich, C. P., Nader, H. B., de Paiva, J. F., Santos, E. A., Holme, K. R., Perlin, A. S., 1989. Heparin in molluscs: chemical, enzymatic degradation and 13C and 1H n.m.r. spectroscopical evidence for the maintenance of the structure through evolution, *Int. J. Biol. Macromol.*, 11, 361-366.

[27] Dietrich, C. P., de Paiva, J. F., Moraes, C. T., Takahashi, H. K., Porcionatto, M. A., Nader, H. B., 1985. Isolation and characterization of a heparin with high anticoagulant activity from Anomalocardia brasiliana, *Biochim. Biophys. Acta.*, 843, 1-7.

[28] Pejler, G., Danielsson, A., Bjork, I., Lindahl, U., Nader, H. B., Dietrich, C. P., 1987. Structure and antithrombin-binding properties of heparin isolated from the clams Anomalocardia brasiliana and Tivela mactroides, *J. Biol. Chem.*, 262, 11413-11421.

[29] Nader, H. B., Chavante, S. F., dos-Santos, E. A., Oliveira, T. W., de-Paiva, J. F., Jeronimo, S. M., Medeiros, G. F., de-Abreu, L. R., Leite, E. L., de-Sousa-Filho, J. F., Castro, R. A., Toma, L., Tersariol, I. L., Porcionatto, M. A., Dietrich, C. P., 1999. Heparan sulfates and heparins: similar compounds performing the same functions in vertebrates and invertebrates?, *Braz. J. Med. Biol. Res.*, 32, 529-538.

[30] Nader, H. B., Lopes, C. C., Rocha, H. A., Santos, E. A., Dietrich, C. P., 2004. Heparins and heparinoids: occurrence, structure and mechanism of antithrombotic and hemorrhagic activities, *Curr. Pharm. Des.*, 10, 951-966.

[31] Santos, E. A., Rocha, L. R., Pereira, N. M., Andrade, G. P., Nader, H. B., Dietrich, C. P., 2002. Mast cells are present in epithelial layers of different tissues of the mollusc Anomalocardia brasiliana. In situ characterization of heparin and a correlation of heparin and histamine concentration, *Histochem. J.*, 34, 553-558.

[32] Jordan, R. E., Marcum, J. A., 1986. Anticoagulantly active heparin from clam (Mercenaria mercenaria), *Arch. Biochem. Biophys.*, 248, 690-695.

[33] Ulrich, P. N., Boon, J. K., 2001. The histological localization of heparin in the northern quahog clam, Mercenaria mercenaria, *J. Invertebr. Pathol.*, 78, 155-159.

[34] Cesaretti, M., Luppi, E., Maccari, F., Volpi, N., 2004. Isolation and characterization of a heparin with high anticoagulant activity from the clam Tapes phylippinarum: evidence for the presence of a high content of antithrombin III binding site, *Glycobiology*, 14, 1275-1284.

[35] Luppi, E., Cesaretti, M., Volpi, N., 2005. Purification and characterization of heparin from the Italian clam Callista chione, *Biomacromolecules*, 6, 1672-1678.

[36] Rahemtulla, F., Lovtrup, S., 1975. The comparative biochemistry of invertebrate mucopolysaccharides. IV. Bivalvia. Phylogenetic implications, *Comp. Biochem. Physiol. B.*, 50, 631-635.

[37] Arumugam, M., Shanmugam, A., 2004. Extraction of heparin and heparin-like substance from marine mesogastropod mollusc Turritella attenuata (Lamarck, 1779), *Indian J. Exp. Biol.*, 42, 529-532.

[38] Saravanan, R., Shanmugam, A., 2010. Isolation and characterization of low molecular weight glycosaminoglycans from marine mollusc Amussium pleuronectus (linne) using chromatography, *Appl. Biochem. Biotechnol.*, 160, 791-799.

[39] Paiva, J. F., Santos, E. A., Jeske, W., Fareed, J., Nader, H. B., Dietrich, C. P., 1995. A comparative study on the mechanism of the anticoagulant action of mollusc and mammalian heparins, *Comp. Biochem. Physiol. A. Physiol.*, 111, 495-499.

[40] Casu, B., Naggi, A., Torri, G., 2010. Heparin-derived heparan sulfate mimics to modulate heparan sulfate-protein interaction in inflammation and cancer, *Matrix Biol.*, 29, 442-452.

[41] Ehrlich, J., Patel, B., Stivala, S. S., 1981. Comparative studies of mucopolysaccharides from marine animals. II. Illex illecebrosvs (Lesueur) and various crustaceans, *J. exp. mar. Bio. Ecol.*, 95-101.

[42] Dietrich, C. P., Paiva, J. F., Castro, R. A., Chavante, S. F., Jeske, W., Fareed, J., Gorin, P. A., Mendes, A., Nader, H. B., 1999. Structural features and anticoagulant activities of a novel natural low molecular weight heparin from the shrimp Penaeus brasiliensis, *Biochim. Biophys. Acta.*, 1428, 273-283.

[43] Brito, A. S., Arimateia, D. S., Souza, L. R., Lima, M. A., Santos, V. O., Medeiros, V. P., Ferreira, P. A., Silva, R. A., Ferreira, C. V., Justo, G. Z., Leite, E. L., Andrade, G. P., Oliveira, F. W., Nader, H. B., Chavante, S. F., 2008. Anti-inflammatory properties of a heparin-like glycosaminoglycan with reduced anti-coagulant activity isolated from a marine shrimp, *Bioorg. Med. Chem.*, 16, 9588-9595.

[44] Brito, A. S., Cavalcante, R. S., Palhares, L. C., Hughes, A. J., Andrade, G. P., Yates, E. A., Nader, H. B., Lima, M. A., Chavante, S. F., 2014. A non-hemorrhagic hybrid heparin/heparan sulfate with anticoagulant potential, *Carbohydr. Polym.*, 99, 372-378.

[45] Chavante, S. F., Brito, A. S., Lima, M., Yates, E., Nader, H., Guerrini, M., Torri, G., Bisio, A., 2014. A heparin-like glycosaminoglycan from shrimp containing high levels of 3-O-sulfated D-glucosamine groups in an unusual trisaccharide sequence, *Carbohydr. Res.*, 390, 59-66.

[46] Andrade, G. P., Lima, M. A., de Souza Junior, A. A., Fareed, J., Hoppensteadt, D. A., Santos, E. A., Chavante, S. F., Oliveira, F. W., Rocha, H. A., Nader, H. B., 2013. A heparin-like compound isolated from a marine crab rich in glucuronic acid 2-O-sulfate presents low anticoagulant activity, *Carbohydr. Polym.*, 94, 647-654.

[47] Lima, M. A., Hughes, A. J., Veraldi, N., Rudd, T. R., Hussain, R., Brito, A. S., Chavante, S. F., Tersariol, I. I., Siligardi, G., Nader, H. B., Yates, E. A., 2013. Antithrombin stabilisation by sulfated carbohydrates correlates with anticoagulant activity, *Med. Chem. Comm.*, 4, 870-873.

[48] Kreuger, J., Kjellen, L., 2012. Heparan sulfate biosynthesis: regulation and variability, *J. Histochem. Cytochem.*, 60, 898-907.

[49] Pavao, M. S., Albano, R. M., Lawson, A. M., Mourao, P. A., 1989. Structural heterogeneity among unique sulfated L-galactans from different species of ascidians (tunicates), *J. Biol. Chem.*, 264, 9972-9979.

[50] Pavao, M. S., Mourao, P. A., Mulloy, B., 1990. Structure of a unique sulfated alpha-L-galactofucan from the tunicate Clavelina, *Carbohydr. Res.*, 208, 153-161.

[51] Albano, R. M., Pavao, M. S., Mourao, P. A., Mulloy, B., 1990. Structural studies of a sulfated L-galactan fromStyela plicata (Tunicate): analysis of the Smith-degraded polysaccharide, *Carbohydr. Res.*, 163-174.

[52] Pavao, M. S., Rodrigues, M. A., Mourao, P. A., 1994. Acidic polysaccharides of the ascidian Styela plicata. Biosynthetic studies on the sulfated L-galactans of the tunic, and preliminary characterization of a dermatan sulfate-like polymer in body tissues, *Biochim. Biophys. Acta.*, 1199, 229-237.

[53] Sims, L. L., 1984. Osmoregulatory Capabilities of 3 Macrosympatric Stolidobranch Ascidians, Styela clava Herdman, Styela plicata (Lesueur), and Styela montereyensis (Dall), *J. Exp. Mar. Biol. Ecol.,* 117–129.

[54] Thiyagarajan, V., Qian, P.-Y., 2003. Effect of temperature, salinity and delayed attachment on development of the solitary ascidian Styela plicata (Lesueur), *Journal of Experimental Marine Biology and Ecology*, 290, 133-146.

[55] Pineda, M. C., Lopez-Legentil, S., Turon, X., 2011. The whereabouts of an ancient wanderer: global phylogeography of the solitary ascidian Styela plicata, *PLoS One*, 6, e25495.

[56] Naranjo, S. A., Carballo, J. L., García-gómez, J. C., 1996. Effects of environmental stress on ascidian populations in Algeciras Bay (southern Spain). Possible marine bioindicators?, *Mar. Ecol. Prog. Ser.,* 119–131.

[57] Santos, J. C., Mesquita, J. M., Belmiro, C. L., da Silveira, C. B., Viskov, C., Mourier, P. A., Pavao, M. S., 2007. Isolation and characterization of a heparin with low antithrombin activity from the body of Styela plicata (Chordata-Tunicata). Distinct effects on venous and arterial models of thrombosis, *Thromb Res.*, 121, 213-223.

[58] Cavalcante, M. C., Allodi, S., Valente, A. P., Straus, A. H., Takahashi, H. K., Mourao, P. A., Pavao, M. S., 2000. Occurrence of heparin in the invertebrate styela plicata (Tunicata) is restricted to cell layers facing the outside environment. An ancient role in defense?, *J. Biol. Chem.*, 275, 36189-36186.

[59] Gandra, M., Cavalcante, M., Pavao, M., 2000. Anticoagulant sulfated glycosaminoglycans in the tissues of the primitive chordate Styela plicata (Tunicata), *Glycobiology*, 10, 1333-1340.

[60] de Barros, C. M., Andrade, L. R., Allodi, S., Viskov, C., Mourier, P. A., Cavalcante, M. C., Straus, A. H., Takahashi, H. K., Pomin, V. H., Carvalho, V. F., Martins, M. A., Pavao, M. S., 2007. The Hemolymph of the ascidian Styela plicata (Chordata-Tunicata) contains heparin inside basophil-like cells and a unique sulfated galactoglucan in the plasma, *J. Biol. Chem.*, 282, 1615-1626.

[61] Pavao, M. S., 2002. Structure and anticoagulant properties of sulfated glycosaminoglycans from primitive Chordates, *An. Acad. Bras. Cienc.*, 74, 105-112.

[62] Belmiro, C. L., Castelo-Branco, M. T., Melim, L. M., Schanaider, A., Elia, C., Madi, K., Pavao, M. S., de Souza, H. S., 2009. Unfractionated heparin and new heparin analogues from ascidians (chordate-tunicate) ameliorate colitis in rats, *J. Biol. Chem.*, 284, 11267-11278.

[63] Alvarenga, V., Pacheco, R. G., Esposito, C. C., Buongusto, F., Castelo-Branco, M. T. L., Madi, K., Belmiro, C. R., Pavão, M. S. G., de Souza, H. S. P., Schanaider, A., 2014. Ascidian (chordate-tunicate) and mammalian heparin enemas attenuate experimental diversion colitis, *Surgery*, 155, 217-227.

[64] Nader, H. B., Takahashi, H. K., Straus, A. H., Dietrich, C. P., 1980. Selective distribution of the heparin in mammals: conspicuous presence of heparin in lymphoid tissues, *Biochim. Biophys. Acta.*, 627, 40-48.

[65] Straus, A. H., Nader, H. B., Dietrich, C. P., 1982. Absence of heparin or heparin-like compounds in mast-cell-free tissues and animals, *Biochim. Biophys. Acta.*, 717, 478-485.

[66] Straus, A. H., Sant'anna, O. A., Nader, H. B., Dietrich, C. P., 1984. An inverse relationship between heparin content and antibody response in genetically selected mice. Sex effect and evidence of a polygenic control for skin heparin concentration, *Biochem. J.*, 220, 625-630.

[67] Nader H. B., Dietrich, C. P., 1989. Natural occurrence and possible biological role of heparin, Edward Arnold London.

In: Sulfated Polysaccharides
Editors: M. Gama, H. Nader, H. de Oliveira Rocha
ISBN: 978-1-63482-975-5

Chapter 5

HEPARAN/HEPARIN POLYSACCHARIDE SYNTHESIS: VISITING THE CHEMOENZYMATIC SYNTHESIS

Hugo Alexandre Oliveira Rocha*
Laboratório de Biotecnologia de Polímeros Naturais (BIOPOL),
Departamento de Bioquímica,
Universidade Federal do Rio Grande do Norte (UFRN),
Natal/RN, Brazil

ABSTRACT

Heparan/heparin complex polysaccharides consist of disaccharide units containing iduronic acid (IdoAU) or glucuronic acid (GlcAU), and glucosamine (GlcN). Several of these sugar residues are sulfated. In addition, heparin has more iduronic acid residues and sulfate groups than heparin does. Their biosynthetic pathways are very similar. More than 70 years have passed since heparin was first used as a drug, and it is still being used as the main drug in several clinical anticoagulant treatments. Heparin has typically been extracted from animal sources, leading to concerns about contamination of the final heparin product. Moreover, religious concerns exist about the sources of heparin; Jews and Muslims do not accept porcine heparin, whereas Hindus do not accept bovine heparin. To develop new heparins, as well as understand the mode of action of these compounds, heparin/heparan polysaccharides and oligosaccharides with well-defined structures need to be synthesized.

Despite the success of fondaparinux, a fully synthetic anticoagulant pentasaccharide, chemical synthesis has proven extremely difficult. Over the past decade, several enzymes for heparan/heparin oligosaccharide synthesis have been cloned and expressed in bacteria, leading to the development of a new chemoenzymatic approach to heparan/heparin synthesis.

Here we review the progress in the development and use of the chemoenzymatic synthesis method.

---

* E-mail: hugo-alexandre@uol.com.br.

1. INTRODUCTION

Due to the abundance of hydroxyl and sulfate groups and, in many cases, the presence of carboxyl groups, sulfated polysaccharides (SPs) exhibit a hydrophilic, cation-binding physiochemical nature. In addition, many different oligosaccharide sequences that differ in sulfate and monosaccharide content and prevalence are present in the sulfated polysaccharides, giving them the potential to bind to a wide range of protein ligands and effectors in a selective manner. This structural variety allows for sulfated polysaccharides to play essential roles in a number of biological processes, as reviewed in several chapters of this book. This structural variety of SPs could also make them part of a saccharide-based information code, as has been suggested for sulfated glycosaminoglycans [1, 2, 3]. However, in contrast to proteins, peptides, DNA or RNA, complex polysaccharides, including SPs, have been underutilized for a long time, mainly because there are no simple standardized methods for oligosaccharide/polysaccharide synthesis starting from monomeric building blocks.

The number of researchers in this field is now steadily increasing owing to a growing desire to understand this "polysaccharide code" and identify new carbohydrate-binding proteins, which can lead to the development of new drugs using SPs as a structural scaffold.

The greatest example of these efforts was the discovery of a rare pentasaccharide fragment that is sulfated at O-3 in the middle GlcN residue of heparin (Hep), which is responsible for Hep/antithrombin binding [4, 5]. Based on these findings, a fully synthetic pentasaccharide, called fondaparinux (trade name: Arixtra), was produced and eventually approved by the US Food and Drug Administration (FDA) for the treatment of deep-vein thrombosis.

SPs biosynthesis *in vivo* results in a variety of chain sizes and subunit modification patterns; thus, there will always be some variability in SPs obtained from natural sources. Furthermore, seasonal variations, environmental factors, feedstock's, as well as the inherent differences between different living beings all add to the variability of naturally occurring SPs. Unfortunately, large-scale physicochemical separation technologies cannot completely resolve intact SPs; thus, most commercial preparations are quite often heterogeneous with respect to size and/or levels of sulfation.

Therefore, despite of several efforts, it is very difficult to obtain a sufficient amount of pure and chemically defined SPs from natural sources for the determination of detailed structure–activity relationships. As an alternative to this problem, several studies have suggested approaches for the production of synthetic sulfated polysaccharides [6, 7]. The major advances have been made in the synthesis of heparan/heparin (HS/Hep) oligosaccharides/ polysaccharides.

2. *IN VITRO* SYNTHESIS HEPARAN/HEPARIN

To establish more detailed structure–activity relationships and correlate specific sequences and sulfation patterns with protein binding and biological activity, it is essential to be able to synthesize HS/Hep oligosaccharides/ polysaccharides in a controlled manner. Glycomic arrays could also incorporate HS/Hep oligosaccharides, provided their structures are known. Such arrays could be used, for example, to define pharmaceutical grade heparins that contain active structures, including those with the most favorable on-target activities and minimal side effects. HS/Hep oligosaccharides/ polysaccharides can be synthesized using two different approaches: chemical synthesis and chemoenzymatic synthesis.

2.1. Chemical Synthesis

HS/Hep oligosaccharides consist of a disaccharide repeating unit of either iduronic acid (IdoAU) or glucuronic acid (GlcAU), and glucosamine (GlcN) residues, each capable of carrying sulfate groups. The numbers of sulfate groups and uronic acids and their locations are related to the functions of the polysaccharides.

The creation of specific glycosidic bonds presents a formidable synthetic challenge for several reasons. Protecting groups, potential participating groups, promoters or catalysts, reaction conditions, donor leaving groups and acceptors must be carefully designed because they are crucial for generating the correct regiochemistry and stereochemistry of the created glycosidic bond [8]. In the case of HS/Hep oligosaccharides, the presence of sulfate groups is an additional factor that increases the complexity of their synthesis.

Hence, multiple factors must be considered for a successful synthetic design of HS/Hep oligosaccharides. These include (a) an efficient gram-scale preparation of L-idose or L-iduronic acid; (b) the choice of uronic acid residue or the corresponding sugar as building blocks; (c) formation of the 1,2-cis linkage from the GlcN donor; (d) use of complex protecting/deprotecting procedures to add the sulfated groups; and (e) methods used for elongation of the backbone sequence [9].

As L-idose and IdoAU are not easily obtained from natural sources, they must be synthesized. Therefore, since the early 1980s, several methods have been proposed for the synthesis of L-ido building blocks [10]. Many approaches starting with glucose or glucuronic acid involve C-5 epimerization and/or C-6 oxidation. However, C-5 epimerization occasionally must be done under extreme conditions such as strongly basic conditions, and the range of usable protective groups is limited. C-6 oxidation can also be problematic in terms of yield and reproducibility [11, 12]. Other routes to IdoAU employing L-arabinose [13] or D-xylose [14] were suggested. Nevertheless, these routes raise issues related to yield and reproducibility. Therefore, the synthesis of iduronic acid building blocks remains a major challenge in the synthesis of the desired HS/Hep oligosaccharides, since current methods require many (8–15) steps for preparation of a single monosaccharide building block (IdoAU).

The resulting monosaccharide moieties are then used to synthesize HS/Hep disaccharide building blocks possessing glucosamine-glucuronic acid (GlcN-GlcAU) or iduronic acid (GlcN-IdoAU) sequences. The disaccharide building blocks are further modified by sugar-chain elongation, selective deprotection, sulfation, and finally deprotection. Although many approaches have been suggested for the synthesis of the HS/Hep oligosaccharide backbone, all of them contain a large number of steps. In addition, chemical synthesis of HS/Hep oligosaccharides with more than six monosaccharides is particularly difficult using the currently available chemical methods [15].

Overall, HS/Hep oligosaccharide synthesis is not amenable through current organic synthesis methods, which are quite laborious and often require substantial optimization of each step, owing to the requirement for complex protecting/deprotecting procedures. In addition, the necessary preparation of various precursor compounds makes the process very expensive. As a result, HS oligosaccharide synthesis can be completed by expert chemists in only a small number of highly specialized labs.

Over the last 10 years, a chemoenzymatic alternative to strict chemical synthesis of HS/Hep oligosaccharide/polysaccharides has emerged that employs several enzymes involved in the synthesis of glycosaminoglycans.

2.2. Chemoenzymatic Synthesis

2.2.1. Semi-Synthetic Synthesis

In recent years, there has been a growing interest in the use of enzymes in the synthesis of oligosaccharides/polysaccharides as an alternative to chemical synthesis. Enzymatic glycosylation occur stereo- and regioselectively under mild conditions, without protecting group manipulation [16]. In addition, even very sterically demanding couplings can be performed selectively. Enzymes catalyzing HS/Hep oligosaccharide/polysaccharide synthesis are usually glycosyltransferases (GTases), sulfotransferases (STases) and epimerases, some of which exist in multiple isoforms [17].

Intracellularly, the biosynthesis of Hep/HS is performed by multiple enzymes located in the Golgi apparatus (Figure 1). The exostosin (EXT) family of genes encodes glycosyltransferases involved in heparan sulfate biosynthesis. EXT1 and EXT2 form a complex that polymerizes HS and Hep chains into a basic backbone consisting of β1–4-linked D-glucuronic acid and α-1–4-linked N-acetyl-D-glucosamine (GlcNAc) [18]. In addition to this basic backbone synthesis, a series of enzymatic processing reactions occurs, including removal of the N-acetyl group and subsequent N-sulfation by N-deacetylase/N-sulfotransferase (NDST); conversion of glucuronic acids to IdoAU by the action of C5-epimerase; addition of sulfate groups by O-sulfotransferases (2-OST for sulfating O-2, of IdoAU, 6-OST for sulfating O-6 of GlcN, and 3-OST for sulfating O-3 of GlcN).

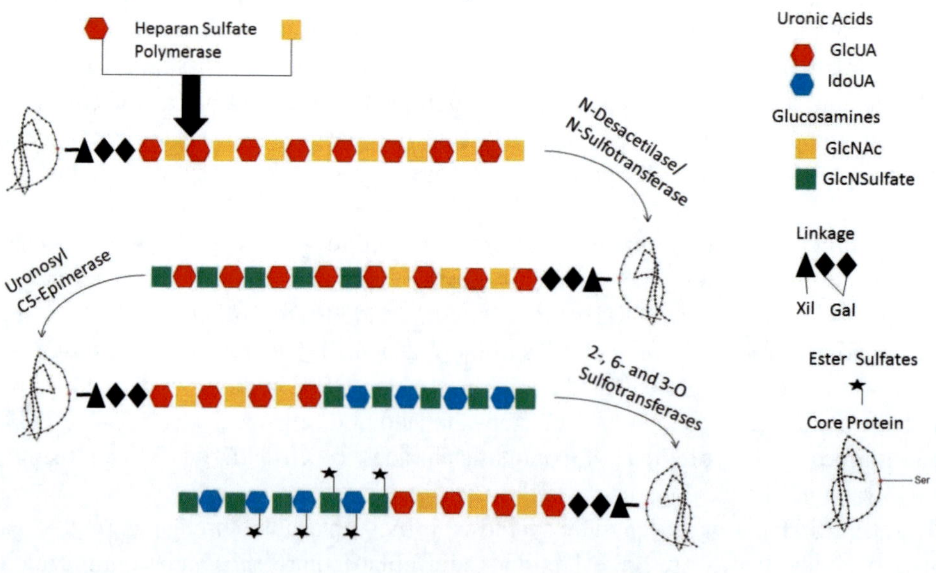

Figure 1. Scheme of the biosynthetic assembly of Hs/Hep proteoglycans.

Of all these enzymes, NDST is the only one that has two activities: it removes N-acetyl groups from GlcNAc residues (N-deacetylase activity), and transfers a sulfate group to D-glucosamine residues (N-sulfotransferase activity) [19].

Kuberan and colleagues [20] were the first to demonstrate the use of enzymes for HS/Hep oligosaccharide synthesis. They cloned and expressed the sulfotransferases and C-5 epimerase in the baculovirus system, and purified heparosan from an *Escherichia coli* K5 strain. Heparosan produced by *E. coli* K5 has the same [GlcUAα- (1-4)GlcNAc-β(1-4)]n structure as the basic backbone [20] generated in heparin biosynthesis. Then, using a semi-synthetic approach, the heparosan (backbone polymer) was submitted to N-deacetylation by either hydrazinolysis at 100°C or alkaline treatment.

The molecule obtained (heparosan amino group-free) was then chemically N-sulfated. Finally, it was modified with the cloned/expressed enzymes (Figure 2) to obtain HS containing antithrombin binding sites with anticoagulant activity. However, only 1 μg of product was generated by this approach.

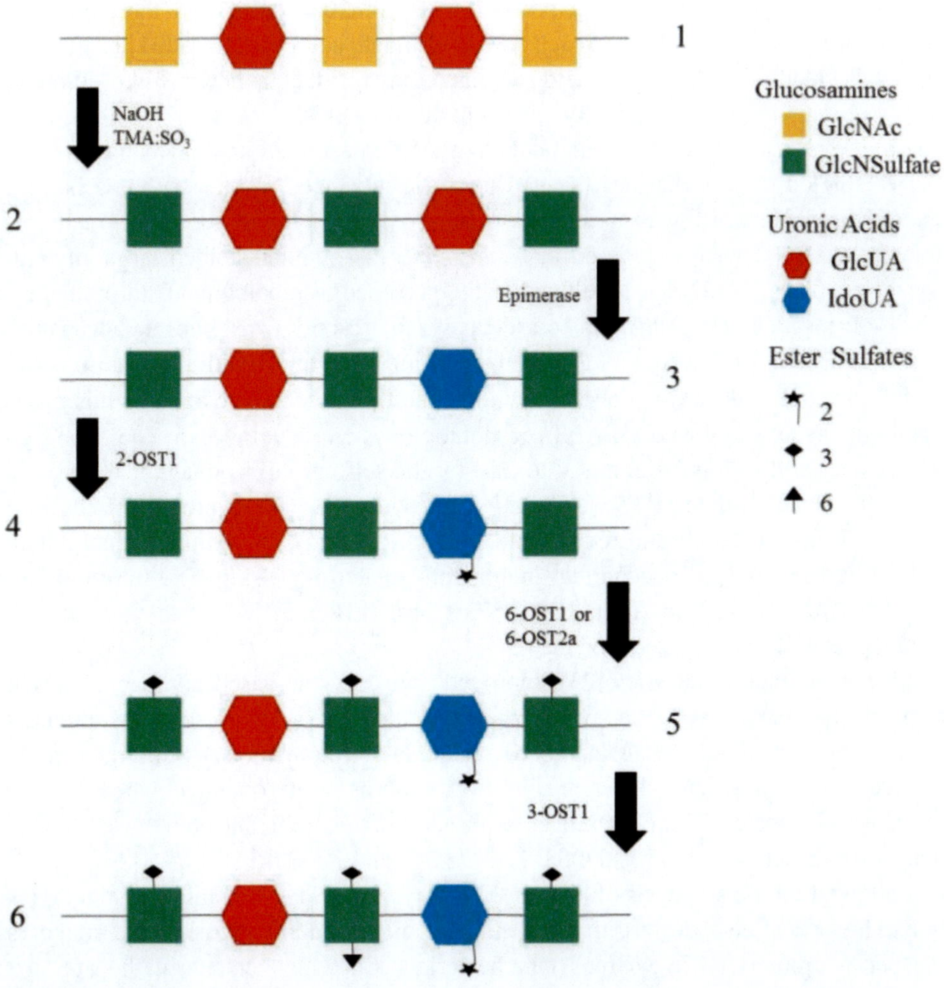

Figure 2. Scheme of the chemoenzymatic synthesis of heparan sulfated starting from heparosan. Adapted from Kuberan and colleagues [20] with permission from the American Society for Biochemistry and Molecular Biology.

In the same year, Lindahl and colleagues [21] reported an alternative chemoenzymatic approach for the synthesis of anticoagulant heparin from heparosan. After chemical N-deacetylation followed by N-sulfation of the heparosan (~ 90% N-sulfated sample), C5 epimerase was used to generate IdoAU residues, followed by chemical persulfation and selective desulfation. This last step was used to remove IdoA 3-O-sulfate groups, since the presence of this sulfate residue in heparin chains inhibits their anticoagulant (antifactor Xa) activity. This approach yielded a low molecular weight heparin (Mr ≈ 8000 Da), in which ~70% of the IdoUA was sulfated (IdoA 2-O-sulfated 60% and IdoA 3-O-sulfated 10%). In addition, almost 50% of GlcN units were 3-O-sulfated, which is more than that observed in natural heparin. However, this new anticoagulant compound had a proportion of 3-O-sulfated GlcUA residues that was higher than those reported in natural heparin. Significantly, however, this approach yielded about 5 g of heparin-like compound.

The two above-mentioned approaches do have some weaknesses. Although Kuberan and colleagues [20] used several enzymes to catalyze Hep/HS oligosaccharide/polysaccharide synthesis, only approximately 1 μg of heparin-like compound was generated.

In addition, the enzymes were expressed in mammalian cells, and therefore are not available in large quantities. The approach used by Lindahl and colleagues [21] resulted in the addition of 3-O-sulfo-D-glucuronic acid, an unnatural residue in heparin-like compounds, suggesting a limitation in the selectivity of the chemical steps.

Considering these results, Chen and colleagues [22] described an enzymatic approach to synthesize HS/Hep polysaccharides that led to major advances. First, desulfated N-sulfated heparin (CDSNS Heparin) was used as a basic backbone. In addition, the recombinant sulfotransferases used were expressed in *Escherichia coli*, which yielded large amounts of these enzymes. Thus, CDSNS was sulfated in the presence of recombinant sulfotransferases. However, during the HS sulfotransferase-catalyzed reactions, 3′-phosphoadenosine 5′-phosphate (PAP) was produced, and this compound inhibited the O-sulfotransferase activities with IC50 values of ~100 μM under the reaction conditions used. To overcome this problem, Chen and colleagues employed a PAPS regeneration system developed by the Wong group [23], which uses aryl sulfotransferase IV to transfer the sulfo group from an inexpensive sulfo donor, p-nitrophenol sulfate (PNPS), to PAP to produce PAPS (Figure 3). This approach prevented PAP accumulation and reduced the cost of synthesis. In addition, it permitted the synthesis of heparin-like polysaccharides in amounts greater than 1 mg. The sulfotransferases could also be immobilized on AminoLink® Plus beads (Pierce), where they expressed the same activity as non-immobilized enzyme [24].

In 2007, Chen and colleagues [25] improved their enzyme-based approach. They used heparosan as the basic backbone, which was N-deacetylated by alkaline treatment. This compound was then N-sulfated using recombinant N-sulfotransferase. Subsequently, eight heparin-like compounds with different sulfation patterns were obtained when N-sulfated heparosan was incubated in the presence or absence of recombinant epimerase along with different sulfotransferases (2-OST, 6-OST-1, 6-OST-3, and 3-OST-1).

This allowed for the synthesis of heparin-like polysaccharides containing distinct types of sulfation. This article also showed that the HS/Hep biosynthetic enzymes were coexpressed with chaperon proteins (GroES/GroEL) in bacteria [26], which enhanced the appropriate folding of HS/Hep biosynthetic enzymes and improved their specific activity and solubility. Another significant finding involved determining whether 2-O-sulfo-IdoUA was important for the anticoagulant activity of heparin.

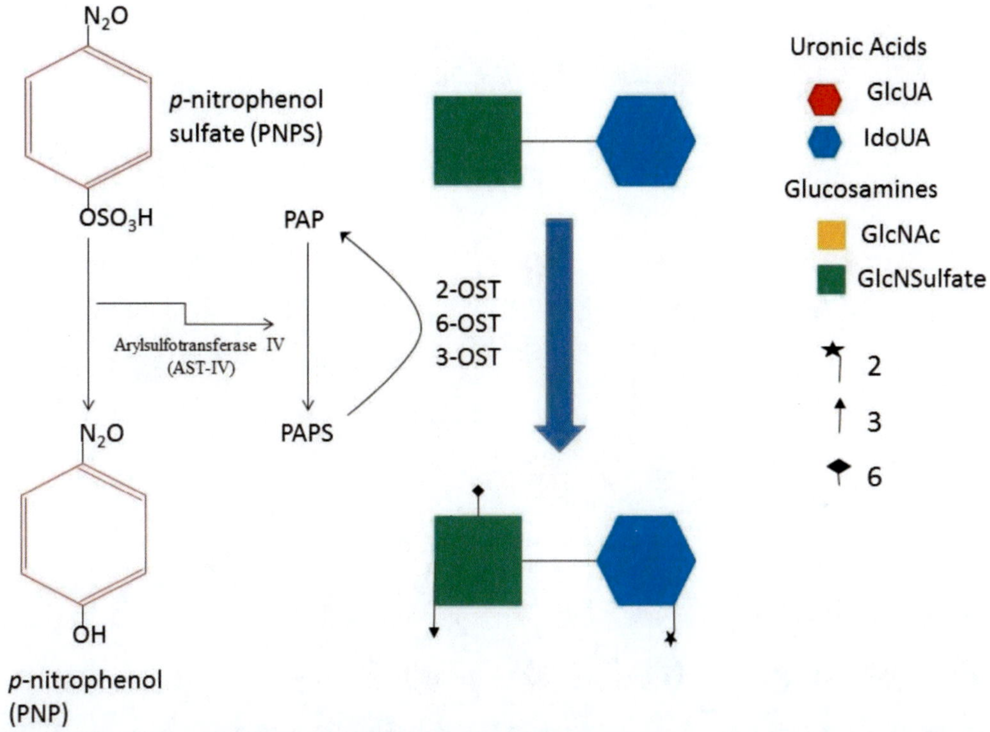

Figure 3. Scheme of PAPs regeneration system.

Prior to that report, the presence of 2-O-sulfo-IdoUA was considered crucial for binding of the chemically synthesized pentasaccharide to antithrombin [27]. However, results from an antithrombin bind assay prompted Chen and colleagues [25] to suggest that this residue was not important for the AT binding of decasaccharides, dodecasaccharides, larger oligosaccharides or polysaccharides, whereas in small oligosaccharides (from tetra to octasaccharide), the residue plays a major role in antithrombin binding [28]. A summary of the enzyme-based approach used by this group is shown in Figure 4.

These studies showed that the synthesized polysaccharides possessed structural heterogeneity, since at least two different oligosaccharides were obtained after hydrolysis. This could have resulted from at least two scenarios:

a) The action of the enzymes used was not complete
b) The starting material itself was heterogeneous

Therefore, a crucial step for the synthesis of HS/heparin using this semi-synthetic chemoenzymatic approach is to control the precise production of the heparosan backbone polymer. The backbone needs to be of a defined molecular weight (monodisperse or nearly so) and contain a homogeneous disaccharide composition. Moreover, for the complete process to be economical, the yield of heparosan needs to be high.

During the 1980s, several methods for preparing heparosan were described, including those by Vann and colleagues [29]. However, all the methods had a very low yield. In 2008, Viskov and colleagues [30] reported a high heparosan yield (10.2 g/L) from *E. coli* K5 fermentation culture supernatant.

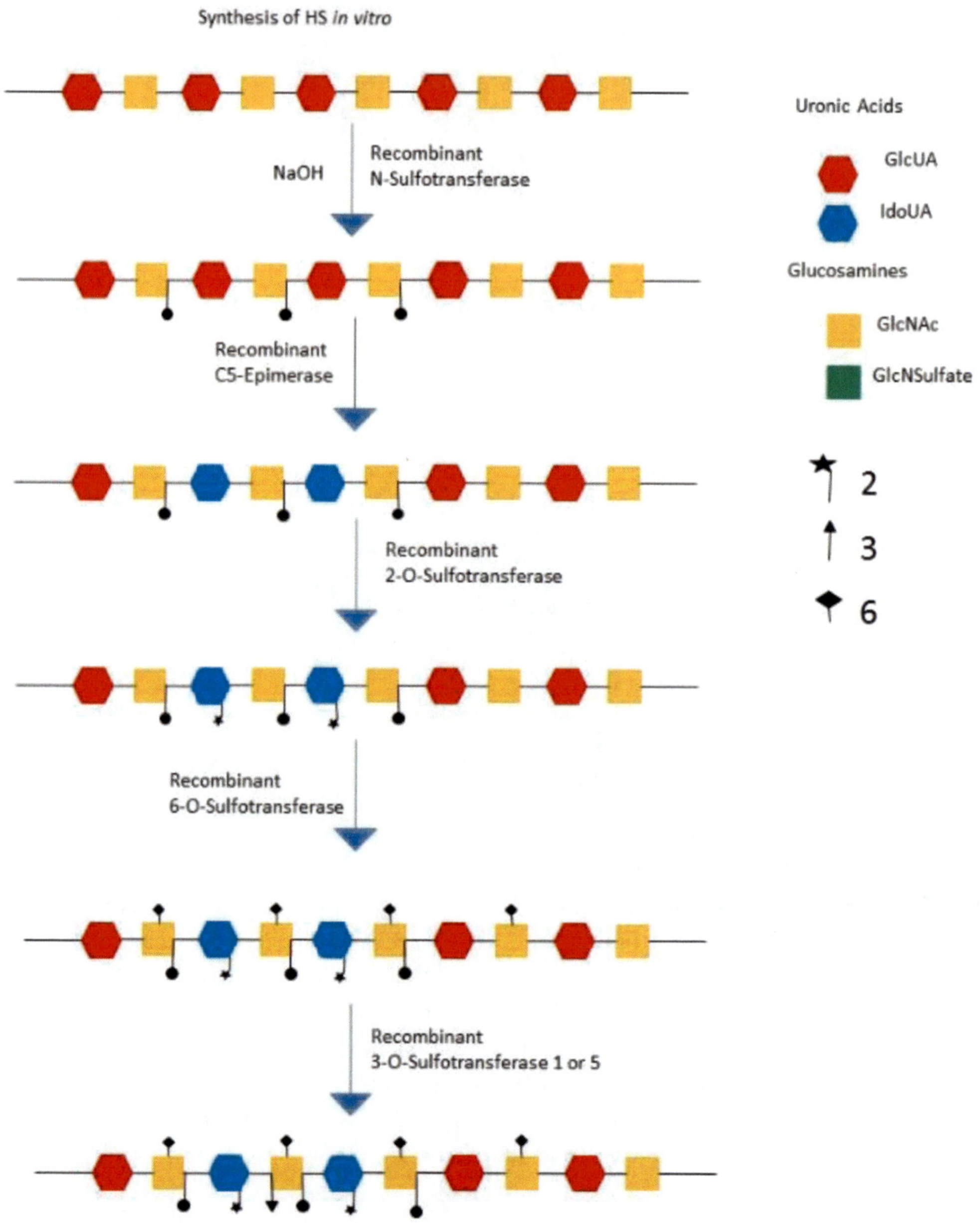

Figure 4. Scheme of HS synthesis *in vitro* using heparosan as backbone for the action of several enzymes catalyzing Hep/HS oligosaccharide/polysaccharide synthesis.

Thereafter, Wang and colleagues [31] described an improvement in the medium used to produce heparosan from cultures of *E. coli* K5. These authors suggested a complex LB medium with glucose using exponentially fed-batch culture as the best condition to obtain heparosan; this condition yielded 15 g/L heparosan.

One problem of this approach, however, was the presence of *E. coli* K5 lyase, the action of which resulted in the addition of an unnatural non-reducing monosaccharide residue, unsaturated uronic acid, at the end of the chain. Since this residue is not present in normal heparin, an additional step will be required to remove this monosaccharide.

To use enzyme-based synthesis and modifications, the enzymes need to be readily available. Hep/HS biosynthetic enzymes are mostly of mammalian origin, presenting challenges for making these recombinant proteins accessible in large quantities. However, several of these enzymes have recently been cloned and expressed in bacteria [32]. The major exceptions are EXT-1 and EXT-2; however, these enzymes can be replaced by bacterial GTases like KfiA, KfiC, PmHS1, and PmHS2. *E. coli* KfiC and KfiA enzymes are glycosyltransferases that are able to synthesize heparosan. Initially KfiC was described to possess both GlcAU transferase and GlcNAc transferase activities [33]. Later, this same group reported that *E. coli* KfiA was actually the GlcNAc-transferase required for heparosan polymerization [34]. Heparosan can also be polymerized by PmHS1 and PmHS2 glycosyltransferases from the gram-negative bacterium *Pasteurella multocida* [35]. In fact, PmHS1, but not PmHS2, can synthesize a high molecular weight heparosan (100– 800 kDa) in vitro. However, only PmHS2 is able to use unnatural UDP-sugar analogs as substrates *in vitro* [36]. Another exception to straightforward cloning and expression of mammalian enzyme in *E. coli* is NDST. Although the enzyme is known to possess two activities, only the N-sulfotransferase (NST) activity was expressed in the protein isolated from *E. coli* [37].

2.2.2. Enzymatic Synthesis of HS and Heparin Oligosaccharides

Many of the functions exhibited by HS/Hep polysaccharides are dependent on only small portions (oligosaccharides) of the full-length molecules. Therefore, efforts are being made to develop enzymatic processes that yield small bioactive oligosaccharides.

Copelend and colleagues [38] showed that 3-O-sulfated octasaccharides inhibit the entry of herpes simplex virus-1. To generate these oligosaccharides, heparin was subjected to treatment with heparin lyase, and the resulting low molecular weight oligosaccharides (average molecular weight < 5000 Da) were fractionated on a Bio Gel P-10 (fine). Each purified oligosaccharide was then sulfated in the presence of 3-OST-3 enzyme.

Heparosan itself can also be used to obtain oligosaccharides, including pentasaccharides with a high capacity to bind antithrombin. Dulaney [9] showed that N-sulfated heparosan can be depolymerized by heparitinase, yielding a range of oligosaccharides, including hexasaccharides. After purification, the oligosaccharides could then be enzymatically modified via epimerization, sulfation, and removal of unsaturated uronic acid.

Following another path, others have demonstrated the synthesis of oligosaccharides starting from mono- or disaccharides. Liu and colleagues [39] synthesized a library of sulfated oligosaccharides with different sizes and sulfation patterns by using heparosan as a source of a large amount of disaccharides and an unnatural UDP-monosaccharide (trifluoroacetylglucosamine, GlcNTFA) as a sugar donor. Heparosan was degraded by nitrous acid, which yielded the disaccharide GlcAU-AnMannose. This disaccharide was then incubated with GlcNTFA in the presence of KfiA, (the authors did not use UDP-GlcN because it is not a substrate for KfiA). After the complete consumption or removal of UDP-GlcNTFA, the trisaccharide formed was incubated with PmHS2 in the presence of UDP-GlcAU. Importantly, since PmHS2 can also transfer GlcNTFA to trisaccharides, this step was needed to be carried out in the absence of GlcNTFA. The reaction cycle could be repeated to obtain oligosaccharides composed of eight, eleven, or twelve monosaccharides. After mild alkaline hydrolysis, GlcNTFA lost the TFA group, allowing N-sulfation using NST and PAPs. Using recombinant epimerase together with different sulfotransferases, other oligosaccharides were obtained. A major contribution of this article was its description on the use of unnatural sugar donors that allowed for the controlled placement of GlcN and GlcNAc.

In 2012, Xu and colleagues [40] suggested a new approach that combines chemical and chemoenzymatic syntheses. Fondaparinux is a synthetic compound that specifically binds to antithrombin, potentiating neutralization of factor Xa, and thus inhibiting thrombin formation [41]. The current chemical approach requires ~50 steps to synthesize fondaparinux. Xu and colleagues [40] suggested an approach that uses only 15 steps to synthesize oligosaccharides with anticoagulant activities. A hexasaccharide backbone was first chemically synthesized [40, 42]. Then, N-sulfation, epimerization of GlcUA to IdoUA, and O-sulfation were carried out [40, 43]. Thus, the authors obtained a nonasaccharide with high anticoagulant actvity. In another study, this same group, usind this approach, has shown the synthesis of several low molecular heparins [44]. The data suggeted that further development of the this chemoenzymatic approach is critical for next-generation of heparin-like anticoagulants.

3. Concluding Remarks and Future Perspectives

Several data showed chemoenzymatic synthesis as an alternative to chemical approach in order to synthesize heparin and anticoagulant oligosaccharides. In addition, chemoenzymatic approaches offer shorter synthetic routes. However, the yield of these approaches is still low, and it was not so far possible to synthesize unnatural structural motifs since the several enzymes used chemoenzymatic synthesis are substrate specific. To overcome this problem, a combination of chemical and chemoenzymatic methods is required to further optimize the HS/Hep synthesis. For example, using chemical approach to produce high amount of oligosaccharide backbone for the action of HS/Hep enzymes. In addition, unnatural motifs could be added using chemical methods. On the other hand, polysaccharide/oligosaccharide library obtained with this approach could be employed in arrays in order to identify relationships between structural and biological activities of HS and Hep. Besides, in order to improve chemoenzymatic synthesis other unnatural monosaccharides using as a donor leaving groups must be found, as well as, new HS/Hep enzymes, such as heparosan-N-sulfate-glucuronate 5-epimerase from *Achatina fulica*, this enzyme was recently discovered and epimerizes GlcUA residues in heparosan, but not in N-sulfated-heparosan [45].

References

[1] Ly, M., Leach, F. E., Laremore, T. N., Toida, T., Amster, I. J., Linhardt, R. J. 2011. The proteoglycan bikunin has a defined sequence. *Nat. Chem. Biol.,* 7, 827–833.

[2] Kreuger, J., Spillmann, D., Li, J. P., Lindahl, U. 2006. Interactions between heparin sulfate and proteins: The concept of specificity. *J. Cell Biol.,* 174, 323–327.

[3] Bülow, H. E., Hobert, O. 2006. The molecular diversity of glycosaminoglycans shapes animal development. *Annu. Rev. Cell Dev. Biol.*, 22, 375–407.

[4] Casu, B., Naggi, A., Torri, G. 2015. Re-visiting the structure of heparin. *Carbohydr. Res.*, 403, 60–68.

[5] Dementiev, A., Petitou, M., Herbert, J. M., Gettings, P. G. 2004. The ternary complex of antithrombin–anhydrothrombin–heparin reveals the basis of inhibitor specificity. *Nat. Struct. Mol. Biol.,* 11, 863–867.

[6] Zhou, J., Yang, L., Hu W. 2014. Stereoselective synthesis of a sulfated tetrasaccharide corresponding to a rare sequence in the galactofucan isolated from *Sargassum polycystum. J. Org. Chem.,* 79, 4718-4726.

[7] Tu, Z, Hsieh, H. W., Tsai, C. M., Hsu, C. W., Wang, S. G., Wu, K. J., Lin, K. I., Lin, CH 2013. Synthesis and characterization of sulfated Gal-β-1,3/4-GlcNAc disaccharides through consecutive protection/ glycosylation steps. *Chem. Asian J.,* 8, 1536-1550.

[8] Hanson, S., Best, M., Bryan, M. C., Wong, C. H. 2004. Chemoenzymatic synthesis of oligosaccharides and glycoproteins. *Trends Biochem. Sci.,* 29, 656-663.

[9] Dulaney, S. B, Huang, X. 2012. Strategies in Synthesis of heparin/heparan sulfate oligosaccharides: 2000–Present. *Adv. Carbohydr. Chem. Biochem.,* 67, 95–136.

[10] Poletti, L., Lay, L. 2003. Chemical contributions to understanding heparin activity: Synthesis of related sulfated oligosaccharides. *European J. Org. Chem.*, 2999-3024.

[11] Hansen, S. U., Baráth, M., Salameh, B. A., Pritchard, R. G., Stimpson, W. T., Gardiner, J. M., Jayson, G. C. 2009. Scalable synthesis of L-iduronic acid derivatives via stereocontrolled cyanohydrin reaction for synthesis of heparin-related disaccharides. *Org. Lett.*, 11, 4528-4531.

[12] Tatai, J., Osztrovszky, G., Kajtár-Peredy, M., Fügedi, P. 2008. An efficient synthesis of L-idose and L-iduronic acid thioglycosides and their use for the synthesis of heparin oligosaccharides. *Carbohydr. Res.,* 343, 596-606.

[13] Timmer, M. S., Adibekian, A., Seeberger, P. H. 2005. Short de novo synthesis of fully functionalized uronic acid monosaccharides. *Angew Chem. Int. Ed. Engl.*, 44, 7605-7607.

[14] Adibekian, A., Bindschädler, P., Timmer, M. S., Noti, C., Schützenmeister, N., Seeberger, P. H. 2007. De novo synthesis of uronic acid building blocks for assembly of heparin oligosaccharides. *Chemistry,* 13, 4510-4522.

[15] de Paz, J. L., Noti, C., Seeberger, P. H. 2006. Microarrays of synthetic heparin oligosaccharides. *J. Am. Chem. Soc.*, 128, 2766–2767.

[16] Sears, P., Wong, C.-H. 2001. Toward automated synthesis of oligosaccharides and glycoproteins. *Science,* 291, 2344–2350.

[17] Esko, J. D., Selleck, S. B. 2002. Order out of chaos: Assembly of ligand binding sites in heparan sulfate. *Annu. Rev. Biochem.,* 71, 435–471.

[18] Busse, M., Feta, A., Presto, J., Wilén, M.; Grønning, M., Kjellén, L., Kusche-Gullberg, M. 2007. Contribution of EXT1, EXT2, and EXTL3 to heparan sulfate chain elongation. *J. Biol. Chem.*, 282, 32802-32810.

[19] Nader, H. B., Lopes, C. C., Rocha, H. A., Santos, E. A., Dietrich, C. P. 2004. Heparins and heparinoids: occurrence, structure and mechanism of antithrombotic and hemorrhagic activities. *Curr. Pharm. Des.,* 10, 951-966.

[20] Kuberan, B., Beeler, D. L., Lech, M., Wu, Z. L., Rosenberg, R. D. 2003. Chemoenzymatic Synthesis of Classical and Non-classical Anticoagulant Heparan Sulfate Polysaccharides. *J. Biol. Chem.*, 278, 52613-52621.

[21] Lindahl, U., Li, J. P., Kusche-Gullberg, M.; Salmivirta, M.; Alaranta, S., Veroma, T.; Emeis, J.; Roberts, I.; Taylor, C., Oreste, P., Zoppetti, G.; Naggi, A.; Torri, G.; Casu, B.

2005. Generation of "Neoheparin" from *E. coli* K5 Capsular Polysaccharide. *J. Med. Chem.,* 48, 349-352.

[22] Chen, J., Avci, F. Y., Muñoz, E. M., McDowell, L. M., Chen, M., Pedersen, L. C., Zhang, L., Linhardt, R. J., Liu, J. 2005. Enzymatic Redesigning of Biologically Active Heparan Sulfate. *J. Biol. Chem.*, 280, 42817-42825.

[23] Burkart, M. D., Izumi, M., Chapman, E., Lin, C. H., Wong, C. H. 2000. Regeneration of PAPS for the enzymatic synthesis of sulfated oligosaccharides. *J. Org. Chem.*, 65, 5565–5574.

[24] Munoz, E., Xu, D., Avci, F., Kemp, M., Liu, J., Linhardt, R. J. 2006. Enzymatic synthesis of heparin related polysaccharides on sensor chips: rapid screening of heparin-protein interactions. *Biochem. Biophys. Res. Commun.*, 339, 597–602.

[25] Chen, J., Jones, C. L., Liu, J. 2007. Using an Enzymatic Combinatorial Approach to Identify Anticoagulant Heparan Sulfate Structures. *Chem. Biol.,* 14, 986–993.

[26] Chen, M., Bridges, A., Liu, J. 2006. Determination of the substrate specificities of N-acetyl-d-glucosaminyltransferase. *Biochemistry,* 45, 12358-12365.

[27] Das, S. K., Mallet, J. M., Esnault, J., Driguez, P. A., Duchaussoy, P., Sizun, P., Herault, J. P., Herbert, J. M., Petitou, M., Sinaÿ, P. 2001. Synthesis of conformationally locked L-iduronic acid derivatives: direct evidence for a critical role of the skew-boat 2S0 conformer in the activation of antithrombin by heparin. *Chemistry,* 7, 4821–4834.

[28] Linhardt, R. J., Kim, J. H. 2007. Combinatorial Enzymatic Synthesis of Heparan Sulfate. *Chem. Biol.*, 14, 972-973.

[29] Vann, W. F., Schmidt, M. A., Jann, B., Jann, K. 1981. The Structure of the Capsular Polysaccharide (K5 Antigenn) of Urinary-Tract-Infective Escherichia coli 010:K5:H4. *Eur. J. Biochem.,* 116, 359–364.

[30] Viskov, C., Lux, F., Gervier, R., Colas, G. 2008. Method for producing K5 polysaccharide. US Patent. 2008/0032349.

[31] Wang, Z., Ly, M., Zhang, F., Zhong, W., Suen, A., Hickey, A. M., Dordick, J. S., Linhardt, R. J. 2010. *E. coli* K5 fermentation and the Preparation of Heparosan, a Bioengineered Heparin Precursor. *Biotechnol. Bioeng.,* 107, 964–973.

[32] Peterson, S., Frick, A., Liu, J. 2009. Design of biologically active heparan sulfate and heparina using an enzyme-based approach. *Nat. Prod. Rep.*, 26, 610–627.

[33] Griffiths, G., Cook, N. J., Gottfridson, E., Lind, T., Lidholt, K., Roberts, I. S. 1998. Characterization of the Glycosyltransferase Enzyme from the *Escherichia coli* K5 Capsule Gene Cluster and Identification and Characterization of the Glucuronyl Active Site. *J. Biol. Chem.*, 273, 11752–11757.

[34] Hodson, N., Griffiths, G., Cook, N., Pourhossein, M., Gottfridson, E., Lind, T., Lidholt, K.; Roberts, I. S. 2000. Identification that KfiA, a protein essential for the biosynthesis of the *Escherichia coli* K5 capsular polysaccharide, is an alpha -UDP-GlcNAc glycosyltransferase. The formation of a membrane-associated K5 biosynthetic complex requires KfiA, KfiB, and KfiC. *J. Biol. Chem.*, 275, 27311–27315.

[35] DeAngelis, P. L., White, C. L. 2002. Identification and Molecular Cloning of a Heparosan Synthase from Pasteurella multocida Type D. *J. Biol. Chem.,* 277, 7209–7213.

[36] Sismey-Ragatz, A. E., Green, D. E., Otto, N. J., Rejzek, M., Field, R. A., DeAngelis, P. L. 2007. Chemoenzymatic Synthesis with Distinct *Pasteurella* Heparosan Synthases. *J. Biol. Chem.,* 282, 28321–28327.

[37] Liu, J., Linhardt, R. J. 2014. Chemoenzymatic synthesis of heparan sulfate and heparin. *Nat. Prod. Rep.,* 31, 1676–1685.

[38] Copeland, R., Balasubramaniam, A., Tiwari, V., Zhang, F., Bridges, A., Linhardt, R. J., Shukla, D., Liu, J. 2008. Using a 3-O-sulfated heparin octasaccharide to inhibit the entry of herpes simplex virus type 1. *Biochemistry,* 47, 5774-5783.

[39] Liu, R., Xu Y., Chen, M., Weïwer, M., Zhou, X, Bridges, A. S., DeAngelis, P. L. Zhang, Q., Linhart, R. J, Liu, J. 2010. Chemoenzymatic design of heparan sulfate oligosaccharides. *J. Biol. Chem.*, 285, 34240–34249.

[40] Xu, Y., Wang, Z., Liu, R., Bridges, A. S., Huang, X., Liu, J. 2012. Directing the biological activities of heparan sulfate oligosaccharides using a chemoenzymatic approach. *Glycobiology,* 22, 96–106.

[41] Lormeau, J. C., Herault, J. P. 1995. The effect of the synthetic pentasaccharide sr 90107/org 31540 on thrombin generation *ex vivo* is uniquely due to ATIII mediated neutralization of factor Xa. Thromb Haemost., 74, 1474–1477.

[42] Huang, X., Huang, L., Wang, H., Ye, X.-S. 2004. Iterative one-pot synthesis of oligosaccharides. *Angew Chem. Int. Ed. Engl.,* 43, 5221–5224.

[43] Xu, Y., Masuko, S., Takieddin, M., Xu, H., Liu, R., Jing, J., Mousa, S. A., Linhardt, R. J., Liu, J. 2011. Chemoenzymatic Synthesis of Homogeneous Ultralow Molecular Weight Heparins. *Science,* 334, 498–501.

[44] Xu, Y., Cai, C., Chandarajoti, K., Hsieh, P. H., Li, L., Pham, T. Q., Sparkenbaugh, E. M., Sheng, J., Key, N. S., Pawlinski, R., Harris, E. N., Linhardt, R. J., Liu, J. 2014. Homogeneous low-molecular-weight heparins with reversible anticoagulant activity. *Nat. Chem. Biol.,* 10, 248-252.

[45] Mochizuki, H., Yamagishi, K., Suzuki, K., Kim, Y. S., Kimata, K. 2015 Heparosan-glucuronate 5-epimerase: Molecular cloning and characterization of a novel enzyme. *Glycobiology*, 1–10.

In: Sulfated Polysaccharides
Editors: M. Gama, H. Nader, H. de Oliveira Rocha

ISBN: 978-1-63482-975-5

Chapter 6

SULFATED POLYSACCHARIDES FROM GREEN SEAWEED

Anabela Alves[*]
Centre of Biological Engineering, University of Minho,
Campus de Gualtar, Braga, Portugal

ABSTRACT

Marine aquatic organisms are diverse and unique. Algae in particular represent a distinctive group known for its ubiquitous use either as food, for fuel production and/or materials exploitation. Recently, there has been a revival of interest in marine algae as sources of (bioactive) molecules, in particular sulfated polysaccharides (SPs). Within this context, green algae are attracting considerable attention and inspiring scientists in natural product research. Apart from industrial uses, green algae-origin SPs are emerging as important classes of biotechnological natural-based products, with distinct chemistries and functional activities without parallelism on other organisms. This uniqueness results in a low rate of redundancy in both structural type and mechanism of action, which defines the renewed enthusiasm in green algae as novel sources of SPs.

Within this exploratory chapter, the reader will be introduced to the circle of SPs synthesized by green algae. This journey includes not only a description of these polysaccharides and an overview on their chemistries, but also a synopsis on the methodologies used to obtain them. Finally, a transition between fundamental knowledge towards a practical and translational approach will be attempted by surveying the properties and limits of (potential) applicability of these SPs.

By combining the originality of these molecules with the progress in natural products research, a strategic innovation hotspot is being created, which is expected to impact the industrial and the biotechnological sectors, both with known societal relevance.

[*] Centre of Biological Engineering, University of Minho, Campus de Gualtar, 4710-057 Braga, Portugal. Email: anabela.pinto@ceb.uminho.pt.

1. Introduction

Algae in general are diverse, complex and intriguing organisms. Furthermore, related studies are a continuously evolving field, in constant change, specifically classification and phylogeny [1-3].

Just like algae in general, green algae are also diverse, complex and intriguing. These algae include simple unicellular and colonial organisms as well as complex multicellular and differentiated macroscopic organisms [4, 5]. They are ubiquitous in a wide range of habitats, namely marine, freshwater and terrestrial, but can also develop symbiotic associations [4].

Just like their traditional name suggests, these algae are green in color and rich in diverse pigments, including chlorophylls a and b, α-, β- and γ-carotene and lutein and prasinoxanthin; phycobilins are absent in green algae [4, 5]. Green algae can be photoautotrophic or heterotrophic and starch, as α-1,4-glucan, is the reserve polysaccharide stored by these algae [4]. Reproduction and life cycle can be as diverse as the classes that represent the Chlorophyta division [4].

The not so glamorous side of green algae is associated with their ability to rapidly grow in nutrient rich environments, with known deleterious effects to local fauna and flora, but also with remarkable socio-economical impact [6, 7]. In a different perspective, these algal blooms can be regarded as a new and rich source of algal biomass that can be used for diverse commercial applications, instead of its usual destination as compost or garbage [7]. There are already some reports on the literature investigating the use of this proliferative biomass as raw material for the extraction of valuable compounds, namely polysaccharides [8, 9]. It can also be investigated as raw material for bio-energy production [10]. In this sense, exploitation of green algae can be sustained both by wild harvesting as well as by farming and controlled culturing, in order to obtain valuable novel products in a large scale [11, 12].

When compared with other algae, green algae remain rather unexplored. In 1993, Arne Jensen discussed on the present and future needs for algae and their products [11]. Although more than two decades have passed since that paper was published, his reflections remain valid, particularly regarding the limited importance of algae to Man considering the available biomass worldwide. Progress may be slow in this area, but research focused on green algae and its products has been gradual and steadily increasing, as reflected by the number of published papers and patents over the years. In Arne Jensen's own words, "*Future uses of marine algae will be decisively influenced by the effort put into and the results coming out of seaweed research*" [11].

2. Green Sulfated Polysaccharides: Simple Chemistries, Complex Polysaccharides?

Green algae cell walls and extracellular coverings are diverse and complex. If some are polysaccharide-based others are covered with scales, some are cellulosic others synthesize SPs [13-15]. This richness, naturally associated with a variety of functional properties, positions these organisms as a valuable source of materials, studied for emerging technologies or to be used in applications that can range from the food industry to biotechnological, pharmaceutical or medicinal uses, among others.

Despite this putative unlimited value, commercial exploitation of algal polysaccharides, in general, is still mostly channeled to industries that use them for their gelling, viscosifying or emulsifying properties.

Green algae SPs present themselves with rich and diverse structures, with intriguing biological functions. In fact, the function of these polysaccharides is still subject of intense speculation. They may be involved in ion retention and exchange, protection of algae from desiccation or an adaptation to saline environments [16, 17].

There are several known SPs extracted from green algae. Each polysaccharide is regarded as a novel compound with potentially innovative properties and pharmacological activities.

These SPs usually present themselves as a complex mixture of both neutral and acidic sugar moieties. Carlberg and co-workers have proposed that the majority of these neutral carbohydrates could be regarded as precursors of the more complex acidic polysaccharides [18].

Table 1. Brief summary of some of the sulfated polysaccharides identified in green algae over the years, divided according to their uronic acids content

Alga	Group	Possible compositions	Literature
Ulva fasciata, U. lactuca, U. reticulata, U. rigida, U. pertusa, U. armoricana, U. rotundata, U. clathrata, U. conglobata, U. meridionalis	Uronic acid-rich	Sulfated rhamnan Sulfated glucuronorhamnan Sulfated mannan	[8, 21, 24, 26, 28, 30-33, 35-55]
Enteromorpha linza, E. compressa, E. prolifera, E. intestinalis			
Gayralia oxysperma			
Monostroma nitidum			
Codium divaricatum, C. adhaerence, C. latum, C. fragile, C. vermilara, C. cylindricum, C. dwarkense, C. isthmocladum, C. yezoense, C. pugniformis	Trace amounts of uronic acids	Sulfated arabinan Sulfated galactan Sulfated arabinogalactan Sulfated rhamnan Sulfated mannan	[18, 23, 34, 35, 51, 56-84]
Caulerpa racemosa, C. akamurai, C. brachypus, C. lentillifera			
Bryopsis plumosa			
Capsosiphon fulvescens			
Chaetomorpha anteninna			
Spongomorpha indica			
Cladophora socialis			
Monostroma nitidum, M. latissimum			
Enteromorpha intestinalis, E. clathrata			
Codiolum pusillum			
Urospora wormskioldii			

Table 1 compiles data from diverse reports present in the literature regarding SPs obtained and identified in green algae. These polysaccharides are generally described as polydisperse heteropolysaccharides, although homopolysaccharides are also described, similar in their general plan, but diverse in their sugar moieties, branching and/or sulfation [17, 18]. Considering their composition, these SPs may be arranged in groups, as proposed by Percival and co-workers in 1979, mostly defined by their uronic acid content [14, 17, 19]. Uronic acid-rich polysaccharides are characterized by their richness on uronic acids (either glucuronic or iduronic acids), with varying amounts of other sugar moieties, like rhamnose, xylose and galactose; uronic acid-limited polysaccharides are mostly composed of galactose and/or arabinose, with smaller amounts of other sugars, like xylose [14, 17, 19]. However, sugar composition may vary due to the influence of several factors, namely algae species, seasonality or ecophysiology [8, 20-31]. On the other hand, extraction, purification procedures and characterization methods used for sugar composition determination may also account for some reported variability [21, 25, 27, 28, 32, 33]. In this sense, grouping of these polysaccharides should be regarded carefully and as a simplification of a more complex scenario. As an example, Hayakawa and co-workers [34] describe arabinose as the main sugar moiety present in *Codium fragile* SP; on another publication, Wang and colleagues [35] describe the SP extracted from *Codium fragile* as being composed of uronic acids and galactose, with minute amounts of arabinose.

3. Can I Get a Little Bit of That Sulfated Polysaccharide, Please?

There are a number of products that can be obtained from natural organisms, including general extracts, general compounds and bioactive compounds [1]. Methodologies employed to obtain these natural products is largely dependent on their intended use, targeted purity or expected performance. In the particular case of green algae, references in the literature are vast.

Green algae SPs are soluble in water. Hence, it is easy to realize that water should be an ideal solvent for the extraction of these polysaccharides. However, there are a number of problems associated with water-based extraction methodologies. Presence of salts, bacterial and fungal growth or concentration of the aqueous extract are just a few examples of what makes polysaccharide extraction a baffling task [85]. On the other hand, the target SP is usually just a part of the total complex mixture composing the water extract and the isolation of pure compounds is generally regarded as a challenging mission [85, 86]. There are of course tricks to circumvent these obstacles and allow the extraction and purification of SPs suitable for the most demanding applications [85-87]. These include, but are not limited to, algae treatments prior to extraction, extraction solvent variations (including polarity, temperature, pH or ionic strength), enzymatic treatments or precipitation techniques [86, 87]. Extracted crude polysaccharides may be further purified through ion exchange, gel permeation and affinity chromatography, with eluent variations [87].

Ideally, there should always be a balance between the methodologies used to obtain the SP of interest and the preservation of the desired property and/or bioactivity of said polysaccharide. In this sense, it is easy to understand the lack of standardized methodologies to obtain these natural products.

4. Complex to Simple: Enzymatic Degradation of Sulfated Polysaccharides

The interest in lower molecular weight derivatives from SPs obtained from green algae has fueled research on ways to control degradation of these polysaccharides. Low molecular weight derivatives may be obtained through chemical, physical or enzymatic methodologies, being the later suitable if one aims for a coordinated degradation with control over molecular weight and protection of functional groups [19, 88].

Research on enzymatic degradation of green algae polysaccharides usually focuses on the screening of green algae degrading bacteria, although enzymatic activity has been found also in fungi or mollusks (Table 2).

Table 2. Enzymes with known or potential hydrolase or lyase activity over green algae sulfated polysaccharides

Enzyme	Origin	Literature
β-Glucuronyl hydrolase	*Nonlabens ulvanivorans*, land snail or abalone entrails	[32, 89, 90]
α-Amylase	*Alteromonas macleodii* strain B7	[88]
Ulvan lyase	Gram-negative bacteria	[91-94]
Glucuronan lyase	Filamentous fungi	[95]
Pool of diverse enzymes (25% cellulose complex, 25% xylanase, 25% β-glucosidase and 25% β-glucanase and xylanase mixture)	-	[32]
Enteromorpha prolifera polysaccharides degradase	*Alteromonas* sp. A321	[96]

5. Can You Imagine How Versatile Green Algae Polysaccharides Can Be?

Green algae are particularly rich in nutrients, used by Man as food since ancient times [1]. Good examples are *Codium edule*, *Ulva fasciata*, *Caulerpa lentillifera* and *Monostroma nitidum* [4]. Green algae are also familiar to (folk) medicine. This pervasive presence in human life makes algae one of the chief marine organisms explored and cultured by Man [1, 4].

Green algae can be used as a whole or can be valorized for its extracts. Good examples of commercially exploited extracts obtained from green algae are astaxanthin from *Haenatococcus pluvialis* and β-carotenes obtained from *Dunaliella salina* [4]. However, the interest of the scientific community and industry is moving towards their richness in complex SPs. In fact, green algae are starting to define themselves as key sources of SPs of non-animal origin. The value and potential application of these polysaccharides is obvious; however this is a fairly young research area with a consequently short literature timeline. Diverse reports have been published on green algae SPs, demonstrating their ability to produce molecules distinct from those found in other algal species or terrestrial and/or animal organisms, with undeniable complexity and unconstrained multiplicity of properties. The use of these polysaccharides is largely related with their physicochemical properties, typically with associated bioactivities [1].

Green algae compounds are inherently bioactive and, generally speaking, both bioactivity and mechanisms of action depend on their chemical structure, molecular weight and chain conformations. This multi-factorial contribution that underlies reported inherent functional properties is well documented. A special emphasis is given to the influence of sulfation on polysaccharide functionality; however, this does not lead to direct correlations between different activities attributed to one given molecule [97-100]. If in some cases the removal of sulfate groups results in loss of functional activity [34], on other cases, desulfation does not appear to affect the final activity of the polysaccharide [63]. One should also account the influence of other factors, including backbone composition, polysaccharide structure and configuration, to state a few [34, 46, 59, 73, 99-102]. On the other hand, difficulties in unveiling the chemical structure of algal polysaccharides, alongside with molecular weight determination and chain conformation characterization, hinder the understanding of the relationship between polysaccharide structure and biological activity [63, 101, 102]. Complexity of the extracted SPs fuels the challenge; advanced structure-function correlations are simpler to attain for less heterogeneous and well-defined structures [17, 41, 100, 103].

To try to answer the above question of how versatile these SPs can be, in a tentative simple manner, their biological activities and physicochemical properties will be defined and used as a division tag.

In this section, key biological activities and properties reported for green algae extracts will be highlighted, namely for SPs or oligosaccharides. Many times research on a particular property or activity is associated with a specific niche of application. Nevertheless, the reader should always remember that it might not be limited to that role and it can be studied for applicability in other areas of action - knowing your polysaccharides and apply that knowledge by thinking outside the box.

5.1. Biological Activities or the 'Anti'-action

5.1.1. Blood Anticoagulant Activity

Researchers have been actively mining for compounds with anticoagulant activity, particularly as valid and valuable alternatives to the widespread use of heparins, which are many times associated with adverse side effects like bleeding or thrombocytopenia. In this context, it has been reported that green algae SPs demonstrate blood anticoagulant activity. In fact, this must be the most investigated biological activity related to these polysaccharides.

SPs obtained from several species of Codium revealed anticoagulant properties: *Codium fragile*, *C. vermilara, C. isthmocladum, C. divaricatum, C. adhaerence*, *C. latum, C. dwarkense, C. cylindricum* and *C. pugniformis* [34, 59, 62, 72, 73, 78, 80, 97, 104]. *Caulerpa cupressoides, C. prolifera, C. sertularioides, C. akamurai* and *C. brachypus, Ulva lactuca, U. conglobata, Enteromorpha linza, E. clathrata, Monostroma nitidum* and *M. latissimum* also possess active polysaccharides [23, 24, 30, 34, 54, 64, 70, 71, 75, 82, 83, 97, 105].

Different SPs and its derivatives, extracted from green algae, reveal diverse anticoagulant activities and patterns when compared with heparin, normally used as control in these studies.

The presence of sulfate groups is usually associated with thrombin inhibition and polysaccharides with higher degrees of sulfation tend to reveal stronger profiles of activity [24, 34, 59, 64, 80, 87, 99, 104]. Additionally, the position of these sulfate groups within the polysaccharidic chain influences functional activity [24, 87, 99, 105]. Sugar moieties also play a role; in fact, diverse authors associate anticoagulant activity with increased contents of arabinose [24, 34, 59, 62, 78, 99]. On the other hand, different researchers postulate on the influence of molecular size on anticoagulant properties of the extracted polysaccharides [70, 82, 87].

5.1.2. Free-radical Scavengers and Antioxidants

Antioxidants are involved in several common pathologies and degenerative conditions, as oxidative stress plays an important role in the disruption of normal cellular functions [106]. In recent years, SPs extracted from green algae have revealed their active role as free-radical scavengers and antioxidants in the inhibition of oxidative damage [106].

Researchers have claimed the effect of molecular weight, sulfation and uronic acid content on polysaccharide's activity [46, 47, 53, 55, 107-111]. On the other hand, a clear correlation between antioxidant and antitumor activities could not found [46, 47].

SPs from *Ulva fasciata, U. lactuca, U. pertusa, Enteromorpha linza, E. prolifera, Caulerpa cupressoides, C. prolifera, C. sertularioides* and *Codium isthmocladum* have revealed scavenging and antioxidant activities [46, 47, 53-55, 97, 108, 109, 111].

5.1.3. Antineoplasty

Green SPs have been studied for their role in the prevention or inhibition of tumor growth or malignant cells spreading with positive outcomes.

Antiproliferative activities towards diverse cancer cells *in vitro* and/or *in vivo* models have been reported for *Caulerpa cupressoides, C. prolifera, C. racemosa* and *C. sertularioides, Ulva fasciata, U. lactuca, U. pertusa, Enteromorpha intestinalis, Monostroma nitidum, Codium isthmocladum, C. fragile* and *Capsosiphon fulvescens*, with varying potencies [40, 46, 47, 49, 54, 65-67, 79, 81, 97, 112]. There appears to be a correlation between the antiproliferative activity and molecular weight, sulfate content and uronic acid or sugar composition of the polysaccharide [46, 47, 65, 66]. On the other hand, Jiao and co-workers discusse on an association between the immunomodulatory effect of the polysaccharides and their antitumor activity, particularly considering the limited cytotoxic effect of some of these SPs towards tumor cells [66].

5.1.4. Anti-inflammatory

The knowledge about the inflammatory process is still fragmentary and although anti-inflammatory activity is associated with SPs extracted from green algae, their mechanism of action is yet to be elucidated.

Both antinociceptive and anti-inflammatory actions have been attributed to SPs extracted from *Caulerpa cupressoides* and *C. racemosa* [113, 114].

5.1.5. Immunomodulation

Immunomodulatory substances with the potential to modulate immune response, *in vitro* and/or *in vivo*, have been obtained from green algae, particularly associated with SPs. These can subsequently be proposed as immunomodulators in patients with compromised immune systems, including cancer patients, children and elderly people.

Immunostimulatory activity has been reported on SPs obtained from *Codium fragile, Caulerpa lentillifera, Enteromorpha prolifera, E. linza, E. intestinalis, Ulva rigida, U. pertusa* and *Monostroma nitidum*, inducing proliferation, production of inflammatory mediators such as nitric oxide and pro- and anti-inflammatory cytokines, and phagocytosis [40-42, 49, 52, 55, 66, 69, 79, 115-118]. Furthermore, this immunomodulatory activity was confirmed on SPs obtained from *Enteromorpha prolifera* in mice, *in vivo* [41, 118]. The presence of sulfate groups appears to be essential in the modulation of phagocyte activity; furthermore, sugar composition and molecular weight also influence the end biological activity of the polysaccharide [42, 52, 66, 117].

5.1.6. Plant Elicitor Activity and Priming

There is an increasing need for compounds with effective antibacterial or antifungal activities to control diverse infections and associated pathologies. These activities have been researched particularly for the control of plant diseases by prompting plant defense and/or resistance, as an alternative to environmentally detrimental chemicals.

Green algae SPs are being studied for their role in plant defense and resistance mechanisms, inducing protection against infection rather than possessing a true direct activity over infectious organisms [119]. Side by side with these studies, these SPs are also being studied for their priming effect in plants, i. e., triggering a state of induced resistance with rapid and effective responses to biotic or abiotic stress [43].

Table 3 presents some reported studies of diverse green algae extracts with activities against different organisms.

Table 3. Examples of reported elicitor or priming activity of sulfated polysaccharides extracted from green algae

Origin of the sulfated polysaccharide	Plant	Pathogen	Literature
Ulva lactuca and *Caulerpa sertularioides*	Tomato plants	*Alternaria solani* *Fusarium oxysporum*	[120, 121]
A mixture of several Ulva species (mostly *U. armoricana*)	*Medicago truncatula*	*Colletotrichum trifolii*	[122]
Ulva armoricana	Common bean (*Phaseolus vulgaris*), grapevine plants and cucumber (*Cucumis sativus*)	*Erysiphe polygoni, E. necator* and *Sphareotheca fuliginea*	[8]
Ulva fasciata	Wheat and barley plants	*Blumeria graminis*	[43]

5.1.7. Antiviral

Sulfated polysaccharides are known for their antiviral properties and the number of reports on screening assays for antiviral activities on green algae extracts is continuously increasing. This has led to the identification of SPs with effective inhibitory actions towards different animal viruses, including important human pathogenic agents.

Jiao and co-workers demonstrated the antiviral activity of polysaccharides extracted from diverse algae and postulated on the role of uronic acid content to warrant this bioactivity [39]. On the other hand, the antiviral properties of sulfated rhamnans and sulfated galactans, with residual amounts of uronic acids, were studied and their potent activity against Herpes Simplex Virus type 2 was demonstrated [74, 123]. These results demonstrate the complexity of structure-function relations and emphasize the importance of elucidative studies to unveil the true nature of SPs activities.

Table 4. Examples of reported antiviral activity of sulfated polysaccharides extracted from green algae

Origin of the sulfated polysaccharide	Reported antiviral activity	Literature
Caulerpa racemosa	Herpes Simplex Virus type 1 (HSV-1) and type 2 (HSV-2) Cox B3 Virus	[63, 124]
Ulva lactuca	Influenza A/PR/8/34 (H1N1) virus Herpes Simplex Virus type 1 (HSV-1) Japanese encephalitis virus (JEV)	[39, 54, 125]
Monostroma nitidum	Herpes Simplex Virus type 2 (HSV-2)	[123]
Ulva fasciata	Human metapneumovirus (HMPV) NL/1/00	[25]
Gayralia oxysperma	Herpes Simplex Virus type 1 (HSV-1) and type 2 (HSV-2)	[36]
Codium fragile	Herpes Simplex Virus type 2 (HSV-2)	[74, 116]

5.1.8. Toxicity

Green algae SPs are being studied for different applications, many of which take advantage of their physicochemical properties and biological activities. Within this scenario, toxicological reports become ever more crucial to support the applicability of these polysaccharides. However, despite the numerous research works on SPs extracted from green algae only a few focus on toxicity studies.

Both *in vitro* and *in vivo* tests with Ulva sp. polysaccharides appear to indicate their non-cytotoxic behavior, in a dose dependent fashion [126, 127]. Qi and co-workers also demonstrate the effect of these SPs over a range of parameters including hematology, serum biochemistry parameters and liver, and between females and males mice [127].

5.2. Physicochemical Properties: Hydration or the Moisture Absorption/Retention Ability

A feature usually associated with SPs is their ability to absorb and retain water. Wang and co-workers report on this particular characteristic studied for SPs extracted from *Bryopsis plumose*, *Codium fragile* and *Enteromorpha linza* in comparison to hyaluronic acid [35]. These researchers have related these moisture-absorption and moisture-retention abilities to sulfate groups content and molecular weight [35].

6. THE COMMERCIAL MAGNETISM: A POLYSACCHARIDE SEDUCTION TALE

Green algae SPs are being studied for decades now. However, their research evolution and consequently their commercial exploitation can be described as rather slow. Diverse hypotheses have been raised to try to account for this phenomenon, particularly when compared with other algal (sulfated) polysaccharides that invaded the market and are widespread in human life and economy. But the truth of the matter is that there hasn't been a significant change: green algae SPs are still strongly entangled with academic research and the bridge that should connect it to the market is yet to be built.

Despite the fact that this research-market relationship is still on its early years, green algae SPs are paving, gradually but securely, their way and proving their applicability in a wide range of applications that can be incorporated into different market segments. Sustained technical innovation and market claim will support and maintain progress and expansion of commercially exploitable species and their products.

SPs obtained from green algae are meritable to be used for industrial applications for a number of reasons. First, these green algae SPs are novel products to be introduced into market. Second, these polysaccharides are obtained from renewable resources with the associated advantage that its exploitation can be related with environmental management of wastes generated by the excessive growth of proliferative green algae in nutrient-rich environments. Third, non-animal origin ensures safety regarding animal associated risk factors. Fourth, potential environmental and biological compatibility associated with these polysaccharides. Fifth, green algae SPs biodegradation profile; while there have been studies on isolated enzymes able to degrade these SPs they appear to resist degradation by enzymes naturally present in human body. Sixth, sulfation degree of these polysaccharides with associated functionalities. Seventh, the numerous applications and markets for green algae SPs, including cosmetics, nutrition, agriculture, pharmaceutics, medicine, etc.

These facts open up the possibilities to invest on the production of SPs and associated derivatives as well as novel applications. They rank green algae polysaccharides in an outstanding position as compared with other sulfated materials. Furthermore, they support the growing research of these polysaccharides in areas as diverse as tissue regeneration, repair and protection, preventive and therapeutic properties, macronutrients and functional food ingredients, food supplements, animal feed supplement, and/or agriculture.

Natural products, in general, play an invaluable role in diverse academic and industrial sectors. Thus, investigation and exploitation of SPs from green algae, a different source of commercially valuable products, has proved to be a promising, fast growing and profitable area. Their potential has been recognized not only by academia but also on an industrial and governmental level. However, there is a gap between discovery and commercialization, cluttered with fragmented knowledge. This should be bridged by the development of bioprocess technologies for the production of these SPs, ensuring biomass supply, product development and product recovery. This overwhelming and challenging task can only be achieved through a dynamic strategy, by the integration of collaborative expertise, including biology, chemistry and engineering.

Of course, the exploitation of algae and its constituents must comply with consumer safety regulations. However, this is not as easy and straightforward as it should be –

legislation on algae and algae utilization varies from country to country and it is constantly changing [128]. A company interested in marketing an algal compound, based on health claims for instance, has to deal with diverse legislative bodies and face elevated costs [128]. These aspects hamper industrial development of green algae SPs and may partially justify their limited commercial exploitation.

7. See You Soon, Green Algae Sulfated Polysaccharide

Throughout these pages, a general overview on green algae SPs was given, both a broader image on their pros (and cons) and the technological advances already made. There is an increasing interest and demand for these polysaccharides, essentially justified for their uniqueness. However, there is still a lot to be done, from biomass to polysaccharide production and processing with global legislative concerns. Nevertheless, research and investment in these SPs has increased in the last decades and expectations are that in the nearest future green algae SPs-based industry will position itself as a strong competitor on different market levels.

References

[1] Cardozo, K. H. M., Guaratini, T., Barros, M. P., Falcao, V. R., Tonon, A. P., Lopes, N. P., Campos, S., Torres, M. A., Souza, A. O., Colepicolo, P., Pinto, E., 2007. Metabolites from algae with economical impact. *Comp Biochem Phys C*. 146, 60-78.

[2] Fucikova, K., Leliaert, F., Cooper, E. D., Skaloud, P., D'hondt, S., De Clerck, O., Gurgel, F., Lewis, L. A., Lewis, P. O., Lopez-Bautista, J., Delwiche, C. F., Verbruggen, H., 2014. New phylogenetic hypotheses for the core Chlorophyta based on chloroplast sequence data. *Frontiers in Ecology and Evolution*. 2.

[3] Hayden, H. S., Blomster, J., Maggs, C. A., Silva, P. C., Stanhope, M. J., Waaland, J. R., 2003. Linnaeus was right all along: Ulva and Enteromorpha are not distinct genera. *Eur J Phycol*. 38, 277-294.

[4] Barsanti, L., Gualtieri, P., *Algae: Anatomy, Biochemistry and Biotechnology*. 2006, USA: Taylor & Francis Group.

[5] Leliaert, F., Smith, D. R., Moreau, H., Herron, M. D., Verbruggen, H., Delwiche, C. F., De Clerck, O., 2012. Phylogeny and Molecular Evolution of the Green Algae. *Crit Rev Plant Sci*. 31, 1-46.

[6] Zhao, J., Jiang, P., Liu, Z. Y., Wei, W., Lin, H. Z., Li, F. C., Wang, J. F., Qin, S., 2013. The yellow sea green tides were dominated by one species, Ulva (Enteromorpha) prolifera, from 2007 to 2011. *Chinese Sci Bull*. 58, 2298-2302.

[7] Charlier, R. H., Morand, P., Finkl, C. W., Thys, A., 2007. Green Tides on the Brittany Coasts. *Environmental Research, Engineering and Management*. 3, 52-59.

[8] Jaulneau, V., Lafitte, C., Corio-Costet, M. F., Stadnik, M. J., Salamagne, S., Briand, X., Esquerre-Tugaye, M. T., Dumas, B., 2011. An Ulva armoricana extract protects plants against three powdery mildew pathogens. *Eur J Plant Pathol*. 131, 393-401.

[9] Stromme, M., Mihranyan, A., Ek, R., 2002. What to do with all these algae? *Mater Lett.* 57, 569-572.

[10] Chirapart, A., Praiboon, J., Puangsombat, P., Pattanapon, C., Nunraksa, N., 2014. Chemical composition and ethanol production potential of Thai seaweed species. *J Appl Phycol.* 26, 979-986.

[11] Jensen, A., 1993. Present and Future-Needs for Algae and Algal Products. *Hydrobiologia.* 261, 15-23.

[12] Lorbeer, A. J., Tham, R., Zhang, W., 2013. Potential products from the highly diverse and endemic macroalgae of Southern Australia and pathways for their sustainable production. *J Appl Phycol.* 25, 717-732.

[13] Cronshaw, J., Myers, A., Preston, R. D., 1958. A Chemical and Physical Investigation of the Cell Walls of Some Marine Algae. *Biochim Biophys Acta.* 27, 89-103.

[14] Domozych, D. S., Ciancia, M., Fangel, J. U., Mikkelsen, M. D., Ulvskov, P., Willats, W. G. T., 2012. The cell walls of green algae: a journey through evolution and diversity. *Front Plant Sci.* 3.

[15] Eslick, E. M., Beilby, M. J., Moon, A. R., 2014. A study of the native cell wall structures of the marine alga Ventricaria ventricosa (Siphonocladales, Chlorophyceae) using atomic force microscopy. *Microscopy-Jpn.* 63, 131-140.

[16] Aquino, R. S., Grativol, C., Mourao, P. A. S., 2011. Rising from the Sea: Correlations between Sulfated Polysaccharides and Salinity in Plants. *Plos One.* 28, 6 (4), 1.

[17] Percival, E., 1979. Polysaccharides of Green, Red and Brown Seaweeds: Their Basic Structure, Biosynthesis and Function. *Brit Phycol J.* 14, 103-117.

[18] Carlberg, G. E., Percival, E., 1977. Carbohydrates of Green Seaweeds Belonging to Genus Urospora .2. Carbohydrates of Green Seaweeds Urospora-Wormskioldii and Codiolum-Pusillum. *Carbohyd Res.* 57, 223-234.

[19] Jiao, G. L., Yu, G. L., Zhang, J. Z., Ewart, H. S., 2011. Chemical Structures and Bioactivities of Sulfated Polysaccharides from Marine Algae. *Mar Drugs.* 9, 196-223.

[20] Devaki, T., Sathivel, A., BalajiRaghavendran, H. R., 2009. Stabilization of mitochondrial and microsomal function by polysaccharide of Ulva lactuca on D-Galactosamine induced hepatitis in rats. *Chem-Biol Interact.* 177, 83-88.

[21] Hernandez-Garibay, E., Zertuche-Gonzalez, J. A., Pacheco-Ruiz, I., 2011. Isolation and chemical characterization of algal polysaccharides from the green seaweed Ulva clathrata (Roth) C. Agardh. *J Appl Phycol.* 23, 537-542.

[22] Lahaye, M., Axelos, M. A. V., 1993. Gelling Properties of Water-Soluble Polysaccharides from Proliferating Marine Green Seaweeds (Ulva Spp). *Carbohyd Polym.* 22, 261-265.

[23] Maeda, M., Uehara, T., Harada, N., Sekiguchi, M., Hiraoka, A., 1991. Heparinoid-Active Sulfated Polysaccharides from Monostroma-Nitidum and Their Distribution in the Chlorophyta. *Phytochemistry.* 30, 3611-3614.

[24] Mao, W. J., Zang, X. X., Li, Y., Zhang, H. J., 2006. Sulfated polysaccharides from marine green algae Ulva conglobata and their anticoagulant activity. *J Appl Phycol.* 18, 9-14.

[25] Mendes, G. D., Soares, A. R., Martins, F. O., de Albuquerque, M. C. M., Costa, S. S., Yoneshigue-Valentin, Y., Gestinari, L. M. D., Santos, N., Romanos, M. T. V., 2010. Antiviral Activity of the Green Marine Alga Ulva Fasciata on the Replication of Human Metapneumovirus. *Rev Inst Med Trop Sp.* 52, 3-10.

[26] Robic, A., Rondeau-Mouro, C., Sassi, J. F., Lerat, Y., Lahaye, M., 2009. Structure and interactions of ulvan in the cell wall of the marine green algae Ulva rotundata (Ulvales, Chlorophyceae). *Carbohyd Polym*. 77, 206-216.

[27] Robic, A., Sassi, J. F., Dion, P., Lerat, Y., Lahaye, M., 2009. Seasonal Variability of Physicochemical and Rheological Properties of Ulvan in Two Ulva Species (Chlorophyta) from the Brittany Coast. *J Phycol*. 45, 962-973.

[28] Robic, A., Sassi, J. F., Lahaye, M., 2008. Impact of stabilization treatments of the green seaweed Ulva rotundata (Chlorophyta) on the extraction yield, the physico-chemical and rheological properties of ulvan. *Carbohyd Polym*. 74, 344-352.

[29] Yamamoto, M., 1980. Physicochemical Studies on Sulfated Polysaccharides Extracted from Seaweeds at Various Temperatures. *Agr Biol Chem Tokyo*. 44, 589-593.

[30] Qi, X. H., Mao, W. J., Chen, Y., Chen, Y. L., Zhao, C. Q., Li, N., Wang, C. Y., 2013. Chemical characteristics and anticoagulant activities of two sulfated polysaccharides from Enteromorpha linza (Chlorophyta). *J Ocean U China*. 12, 175-182.

[31] Abdel-Fattah, A. F., Edrees, M., 1972. A study on the polysaccharide content of Ulva lactucal L. *Plant Food Hum Nutr*. 22, 15-22.

[32] Costa, C., Alves, A., Pinto, P. R., Sousa, R. A., da Silva, E. A. B., Reis, R. L., Rodrigues, A. E., 2012. Characterization of ulvan extracts to assess the effect of different steps in the extraction procedure. *Carbohyd Polym*. 88, 537-546.

[33] Yaich, H., Garna, H., Besbes, S., Paquot, M., Blecker, C., Attia, H., 2013. Effect of extraction conditions on the yield and purity of ulvan extracted from Ulva lactuca. *Food Hydrocolloid*. 31, 375-382.

[34] Hayakawa, Y., Hayashi, T., Lee, J. B., Srisomporn, P., Maeda, M., Ozawa, T., Sakuragawa, N., 2000. Inhibition of thrombin by sulfated polysaccharides isolated from green algae. *Bba-Protein Struct M*. 1543, 86-94.

[35] Wang, J., Jin, W. H., Hou, Y., Niu, X. Z., Zhang, H., Zhang, Q. B., 2013. Chemical composition and moisture-absorption/retention ability of polysaccharides extracted from five algae. *Int J Biol Macromol*. 57, 26-29.

[36] Cassolato, J. E. F., Noseda, M. D., Pujolb, C. A., Pellizzaric, F. M., Damonte, E. B., Duarte, M. E. R., 2008. Chemical structure and antiviral activity of the sulfated heterorhamnan isolated from the green seaweed Gayralia oxysperma. *Carbohyd Res*. 343, 3085-3095.

[37] Dereviers, B., Leproux, A., 1993. Characterization of Polysaccharides from Enteromorpha-Intestinalis (L) Link, Chlorophyta. *Carbohyd Polym*. 22, 253-259.

[38] Jaulneau, V., Lafitte, C., Jacquet, C., Fournier, S., Salamagne, S., Briand, X., Esquerre-Tugaye, M. T., Dumas, B., 2010. Ulvan, a Sulfated Polysaccharide from Green Algae, Activates Plant Immunity through the Jasmonic Acid Signaling Pathway. *J Biomed Biotechnol*.525291. doi: 10.1155/2010/525291

[39] Jiao, G. L., Yu, G. L., Wang, W., Zhao, X. L., Zhang, J. Z., Ewart, S. H., 2012. Properties of polysaccharides in several seaweeds from Atlantic Canada and their potential anti-influenza viral activities. *J Ocean U China*. 11, 205-212.

[40] Karnjanapratum, S., You, S., 2011. Molecular characteristics of sulfated polysaccharides from Monostroma nitidum and their *in vitro* anticancer and immunomodulatory activities. *Int J Biol Macromol*. 48, 311-318.

[41] Kim, J. K., Cho, M. L., Karnjanapratum, S., Shin, I. S., You, S. G., 2011. *In vitro* and *in vivo* immunomodulatory activity of sulfated polysaccharides from Enteromorpha prolifera. *Int J Biol Macromol.* 49, 1051-1058.

[42] Leiro, J. M., Castro, R., Arranz, J. A., Lamas, J., 2007. Immunomodulating activities of acidic sulphated polysaccharides obtained from the seaweed Ulva rigida C. Agardh. *Int Immunopharmacol.* 7, 879-888.

[43] Paulert, R., Ebbinghaus, D., Urlass, C., Moerschbacher, B. M., 2010. Priming of the oxidative burst in rice and wheat cell cultures by ulvan, a polysaccharide from green macroalgae, and enhanced resistance against powdery mildew in wheat and barley plants. *Plant Pathol.* 59, 634-642.

[44] Pena-Rodriguez, A., Mawhinney, T. P., Ricque-Marie, D., Cruz-Suarez, L. E., 2011. Chemical composition of cultivated seaweed Ulva clathrata (Roth) C. Agardh. *Food Chem.* 129, 491-498.

[45] Ray, B., 2006. Polysaccharides from Enteromorpha compressa: Isolation, purification and structural features. *Carbohyd Polym.* 66, 408-416.

[46] Shao, P., Chen, X. X., Sun, P. L., 2013. *In vitro* antioxidant and antitumor activities of different sulfated polysaccharides isolated from three algae. *Int J Biol Macromol.* 62, 155-161.

[47] Shao, P., Pei, Y. P., Fang, Z. X., Sun, P. L., 2014. Effects of partial desulfation on antioxidant and inhibition of DLD cancer cell of Ulva fasciata polysaccharide. *Int J Biol Macromol.* 65, 307-313.

[48] Siddhanta, A. K., Goswami, A. M., Ramavat, B. K., Mody, K. H., Mairh, O. P., 2001. Water soluble polysaccharides of marine algal species of Ulva (Ulvales, Chlorophyta) of Indian waters. *Indian J Mar Sci.* 30, 166-172.

[49] Tabarsa, M., Han, J. H., Kim, C. Y., You, S. G., 2012. Molecular Characteristics and Immunomodulatory Activities of Water-Soluble Sulfated Polysaccharides from Ulva pertusa. *J Med Food.* 15, 135-144.

[50] Tabarsa, M., Lee, S. J., You, S., 2012. Structural analysis of immunostimulating sulfated polysaccharides from Ulva pertusa. *Carbohyd Res.* 361, 141-147.

[51] Tsubaki, S., Hiraoka, M., Hadano, S., Nishimura, H., Kashimura, K., Mitani, T., 2014. Functional group dependent dielectric properties of sulfated hydrocolloids extracted from green macroalgal biomass. *Carbohyd Polym.* 107, 192-197.

[52] Castro, R., Piazzon, M. C., Zarra, I., Leiro, J., Noya, M., Lamas, J., 2006. Stimulation of turbot phagocytes by Ulva rigida C. Agardh polysaccharides. *Aquaculture.* 254, 9-20.

[53] Li, B., Liu, S., Xing, R., Li, K. C., Li, R. F., Qin, Y. K., Wang, X. Q., Wei, Z. H., Li, P. C., 2013. Degradation of sulfated polysaccharides from Enteromorpha prolifera and their antioxidant activities. *Carbohyd Polym.* 92, 1991-1996.

[54] El-Baky, H. H. A., Baz, F. K. E., Baroty, G. S. E., 2009. Potential Biological Properties of Sulphated Polysaccharides Extracted from the Macroalgae *Ulva lactuca* L. *Academic Journal of Cancer Research.* 2, 01-11.

[55] Zhang, Z. S., Wang, X. M., Zhao, M. X., Yu, S. C., Qi, H. M., 2013. The immunological and antioxidant activities of polysaccharides extracted from Enteromorpha linza. *Int J Biol Macromol.* 57, 45-49.

[56] Bilan, M. I., Vinogradova, E. V., Shashkov, A. S., Usoy, A. I., 2007. Structure of a highly pyruvylated galactan sulfate from the Pacific green alga Codium yezoense (Bryopsidales, Chlorophyta). *Carbohyd Res*. 342, 586-596.

[57] Chattopadhyay, K., Adhikari, U., Lerouge, P., Ray, B., 2007. Polysaccharides from Caulerpa racemosa: Purification and structural features. *Carbohyd Polym*. 68, 407-415.

[58] Ciancia, M., Alberghina, J., Arata, P. X., Benavides, H., Leliaert, F., Verbruggen, H., Estevez, J. M., 2012. Characterization of Cell Wall Polysaccharides of the Coencocytic Green Seaweed Bryopsis Plumosa (Bryopsidaceae, Chlorophyta) from the Argentine Coast. *J Phycol*. 48, 326-335.

[59] Ciancia, M., Quintana, I., Vizcarguenaga, M. I., Kasulin, L., de Dios, A., Estevez, J. M., Cerezo, A. S., 2007. Polysaccharides from the green seaweeds Codium fragile and C vermilara with controversial effects on hemostasis. *Int J Biol Macromol*. 41, 641-649.

[60] Farias, E. H. C., Pomin, V. H., Valente, A. P., Nader, H. B., Rocha, H. A. O., Mourao, P. A. S., 2008. A preponderantly 4-sulfated, 3-linked galactan from the green alga Codium isthmocladum. *Glycobiology*. 18, 250-259.

[61] Fernandez, P. V., Estevez, J. M., Cerezo, A. S., Ciancia, M., 2012. Sulfated beta-D-mannan from green seaweed Codium vermilara. *Carbohyd Polym*. 87, 916-919.

[62] Fernandez, P. V., Quintana, I., Cerezo, A. S., Caramelo, J. J., Pol-Fachin, L., Verli, H., Estevez, J. M., Ciancia, M., 2013. Anticoagulant Activity of a Unique Sulfated Pyranosic (1 -> 3)-beta-L-Arabinan through Direct Interaction with Thrombin. *J Biol Chem*. 288, 223-233.

[63] Ghosh, P., Adhikari, U., Ghosal, P. K., Pujol, C. A., Carlucci, M. J., Damonte, E. B., Ray, B., 2004. *In vitro* anti-herpetic activity of sulfated polysaccharide fractions from Caulerpa racemosa. *Phytochemistry*. 65, 3151-3157.

[64] Harada, N., Maeda, M., 1998. Chemical structure of antithrombin-active rhamnan sulfate from Monostrom nitidum. *Biosci Biotech Bioch*. 62, 1647-1652.

[65] Ji, H. W., Shao, H. Y., Zhang, C. H., Hong, P. Z., Xiong, H. P., 2008. Separation of the polysaccharides in Caulerpa racemosa and their chemical composition and antitumor activity. *J Appl Polym Sci*. 110, 1435-1440.

[66] Jiao, L. L., Jiang, P., Zhang, L. P., Wu, M. J., 2010. Antitumor and Immunomodulating Activity of Polysaccharides from Enteromorpha intestinalis. *Biotechnol Bioproc E*. 15, 421-428.

[67] Kwon, M. J., Nam, T. J., 2007. A polysaccharide of the marine alga Capsosiphon fulvescens induces apoptosis in AGS gastric cancer cells via an IGF-IR-mediated PI3K/Akt pathway. *Cell Biol Int*. 31, 768-775.

[68] Love, J., Percival, E., 1964. Polysaccharides of Green Seaweed Codium Fragile .2. Water-soluble Sulphated Polysaccharides. *J Chem Soc*. 3338-&.

[69] Maeda, R., Ida, T., Inara, H., Sakamoto, T., 2012. Immunostimulatory Activity of Polysaccharides Isolated from Caulerpa lentillifera on Macrophage Cells. *Biosci Biotech Bioch*. 76, 501-505.

[70] Mao, W. J., Fang, F., Li, H. Y., Qi, X. H., Sun, H. H., Chen, Y., Guo, S. D., 2008. Heparinoid-active two sulfated polysaccharides isolated from marine green algae Monostroma nitidum. *Carbohyd Polym*. 74, 834-839.

[71] Mao, W. J., Li, H. Y., Li, Y., Zhang, H. J., Qi, X. H., Sun, H. H., Chen, Y., Guo, S. D., 2009. Chemical characteristic and anticoagulant activity of the sulfated polysaccharide isolated from Monostroma latissimum (Chlorophyta). *Int J Biol Macromol*. 44, 70-74.

[72] Matsubara, K., Matsuura, Y., Bacic, A., Liao, M. L., Hori, K., Miyazawa, K., 2001. Anticoagulant properties of a sulfated galactan preparation from a marine green alga, Codium cylindricum. *Int J Biol Macromol*. 28, 395-399.

[73] Matsubara, K., Matsuura, Y., Hori, K., Miyazawa, K., 2000. An anticoagulant proteoglycan from the marine green alga, Codium pugniformis. *J Appl Phycol*. 12, 9-14.

[74] Ohta, Y., Lee, J. B., Hayashi, K., Hayashi, T., 2009. Isolation of Sulfated Galactan from Codium fragile and Its Antiviral Effect. *Biol Pharm Bull*. 32, 892-898.

[75] Qi, X. H., Mao, W. J., Gao, Y., Chen, Y., Chen, Y. L., Zhao, C. Q., Li, N., Wang, C. Y., Yan, M. X., Lin, C., Shan, J. M., 2012. Chemical characteristic of an anticoagulant-active sulfated polysaccharide from Enteromorpha clathrata. *Carbohyd Polym*. 90, 1804-1810.

[76] Rao, E. V., Ramana, K. S., 1991. Structural Studies of a Polysaccharide Isolated from the Green Seaweed Chaetomorpha-Anteninna. *Carbohyd Res*. 217, 163-170.

[77] Rao, E. V., Rao, N. V. S. A. V. P., Ramana, K. S., 1991. Structural Features of the Sulfated Polysaccharide from a Green Seaweed, Spongomorpha-Indica. *Phytochemistry*. 30, 1183-1186.

[78] Siddhanta, A. K., Shanmugam, M., Mody, K. H., Goswami, A. M., Ramavat, B. K., 1999. Sulphated polysaccharides of Codium dwarkense Boergs. from the west coast of India: chemical composition and blood anticoagulant activity. *Int J Biol Macromol*. 26, 151-154.

[79] Tabarsa, M., Karnjanapratum, S., Cho, M., Kim, J. K., You, S., 2013. Molecular characteristics and biological activities of anionic macromolecules from Codium fragile. *Int J Biol Macromol*. 59, 1-12.

[80] Uehara, T., Takeshita, M., Maeda, M., 1992. Studies on Anticoagulant-Active Arabinan Sulfates from the Green-Alga, Codium-Latum. *Carbohyd Res*. 235, 309-311.

[81] Wang, X. X., Chen, Y., Wang, J. J., Liu, Z. X., Zhao, S. G., 2014. Antitumor activity of a sulfated polysaccharide from Enteromorpha intestinalis targeted against hepatoma through mitochondrial pathway. *Tumor Biol*. 35, 1641-1647.

[82] Zhang, H. J., Mao, W. J., Fang, F., Li, H. Y., Sun, H. H., Chen, Y., Qi, X. H., 2008. Chemical characteristics and anticoagulant activities of a sulfated polysaccharide and its fragments from Monostroma latissimum. *Carbohyd Polym*. 71, 428-434.

[83] Li, H. Y., Mao, W. J., Zhang, X. L., Qi, X. H., Chen, Y., Chen, Y. L., Xu, J., Zhao, C. Q., Hou, Y. J., Yang, Y. P., Li, N., Wang, C. Y., 2011. Structural characterization of an anticoagulant-active sulfated polysaccharide isolated from green alga Monostroma latissimum. *Carbohyd Polym*. 85, 394-400.

[84] Ramana, K. S., Rao, E. V., 1991. Studies on Indian seaweed polysaccharides. 4. Structural features of the sulfated polysaccharide from a green seaweed, Cladophora socialis. *Phytochemistry*. 30, 259-262.

[85] Bhakuni, D. S., Rawat, D. S., *Bioactive Marine Natural Products*. 2005, India: Anamaya Publishers.

[86] Alves, A., Sousa, R., Reis, R., 2013. A practical perspective on ulvan extracted from green algae. *J Appl Phycol*. 25, 407-424.

[87] Shanmugam, M., Mody, K. H., 2000. Heparinoid-active sulphated polysaccharides from marine algae as potential blood anticoagulant agents. *Curr Sci India*. 79, 1672-1683.

[88] Han, X. F., Lin, B. K., Ru, G. J., Zhang, Z. B., Liu, Y., Hu, Z., 2014. Gene Cloning and Characterization of an alpha-Amylase from Alteromonas macleodii B7 for Enteromorpha Polysaccharide Degradation. *J Microbiol Biotechn*. 24, 254-263.

[89] Collen, P. N., Jeudy, A., Sassi, J. F., Groisillier, A., Czjzek, M., Coutinho, P. M., Helbert, W., 2014. A Novel Unsaturated beta-Glucuronyl Hydrolase Involved in Ulvan Degradation Unveils the Versatility of Stereochemistry Requirements in Family GH105. *J Biol Chem*. 289, 6199-6211.

[90] Quemener, B., Lahaye, M., BobinDubigeon, C., 1997. Sugar determination in ulvans by a chemical-enzymatic method coupled to high performance anion exchange chromatography. *J Appl Phycol*. 9, 179-188.

[91] Barbeyron, T., Lerat, Y., Sassi, J. F., Le Panse, S., Helbert, W., Collen, P. N., 2011. Persicivirga ulvanivorans sp nov., a marine member of the family Flavobacteriaceae that degrades ulvan from green algae. *Int J Syst Evol Micr*. 61, 1899-1905.

[92] Collen, P. N., Sassi, J. F., Rogniaux, H., Marfaing, H., Helbert, W., 2011. Ulvan Lyases Isolated from the Flavobacteria Persicivirga ulvanivorans Are the First Members of a New Polysaccharide Lyase Family. *J Biol Chem*. 286, 42063-42071.

[93] Lahaye, M., Brunel, M., Bonnin, E., 1997. Fine chemical structure analysis of oligosaccharides produced by an ulvan-lyase degradation of the water-soluble cell-wall polysaccharides from Ulva sp. (Ulvales, Chlorophyta). *Carbohyd Res*. 304, 325-333.

[94] Michaud, P., Da Costa, A., Courtois, B., Courtois, J., 2003. Polysaccharide lyases: Recent developments as biotechnological tools. *Crit Rev Biotechnol*. 23, 233-266.

[95] Delattre, C., Michaud, P., Keller, C., Elboutachfaiti, R., Beven, L., Courtois, B., Courtois, J., 2006. Purification and characterization of a novel glucuronan lyase from Trichoderma sp GL2. *Appl Microbiol Biot*. 70, 437-443.

[96] Li, Y. P., Wang, J., Yu, Y., Li, X., Jiang, X. L., Hwang, H., Wang, P., 2013. Production of enzymes by Alteromonas sp A321 to degrade polysaccharides from Enteromorpha prolifera. *Carbohyd Polym*. 98, 988-994.

[97] Costa, L. S., Fidelis, G. P., Cordeiro, S. L., Oliveira, R. M., Sabry, D. A., Camara, R. B. G., Nobre, L. T. D. B., Costa, M. S. S. P., Almeida-Lima, J., Farias, E. H. C., Leite, E. L., Rocha, H. A. O., 2010. Biological activities of sulfated polysaccharides from tropical seaweeds. *Biomed Pharmacother*. 64, 21-28.

[98] Haroun-Bouhedja, F., Ellouali, M., Sinquin, C., Boisson-Vidal, C., 2000. Relationship between sulfate groups and biological activities of fucans. *Thromb Res*. 100, 453-459.

[99] Melo, F. R., Pereira, M. S., Foguel, D., Mourao, P. A. S., 2004. Antithrombin-mediated anticoagulant activity of sulfated polysaccharides: Different mechanisms for heparin and sulfated galactans. *J Biol Chem*. 279, 20824-20835.

[100] Pereira, M. S., Mulloy, B., Mourao, P. A. S., 1999. Structure and anticoagulant activity of sulfated fucans: Comparison between the regular, repetitive, and linear fucans from echinoderms with the more heterogeneous and branched polymers from brown algae. *J Biol Chem*. 274, 7656-7667.

[101] Pomin, V. H., 2010. Structural and functional insights into sulfated galactans: A systematic review. *Glycoconjugate J*. 27, 1-12.

[102] Pomin, V. H., 2014. How to analyze the anticoagulant and antithrombotic mechanisms of action in fucanome and galactanome? *Glycoconjugate J*. 31, 89-99.

[103] Pomin, V. H., 2012. Fucanomics and Galactanomics: Marine Distribution, Medicinal Impact, Conceptions, and Challenges. *Mar Drugs*. 10, 793-811.

[104] Jurd, K. M., Rogers, D. J., Blunden, G., Mclellan, D. S., 1995. Anticoagulant Properties of Sulfated Polysaccharides and a Proteoglycan from Codium Fragile Ssp Atlanticum. *J Appl Phycol*. 7, 339-345.

[105] Li, H. Y., Mao, W. J., Hou, Y. J., Gao, Y., Qi, X. H., Zhao, C. Q., Chen, Y., Chen, Y. L., Li, N., Wang, C. Y., 2012. Preparation, structure and anticoagulant activity of a low molecular weight fraction produced by mild acid hydrolysis of sulfated rhamnan from Monostroma latissimum. *Bioresource Technol*. 114, 414-418.

[106] Lee, J. C., Hou, M. F., Huang, H. W., Chang, F. R., Yeh, C. C., Tang, J. Y., Chang, H. W., 2013. Marine algal natural products with anti-oxidative, anti-inflammatory, and anti-cancer properties. *Cancer Cell Int*. 13.

[107] Chen, H. X., Zhang, M., Xie, B. J., 2004. Quantification of uronic acids in tea polysaccharide conjugates and their antioxidant properties. *J Agr Food Chem*. 52, 3333-3336.

[108] Qi, H. M., Zhang, Q. B., Zhao, T. T., Chen, R., Zhang, H., Niu, X. Z., Li, Z., 2005. Antioxidant activity of different sulfate content derivatives of polysaccharide extracted from Ulva pertusa (Chlorophyta) *in vitro*. *Int J Biol Macromol*. 37, 195-199.

[109] Qi, H. M., Zhao, T. T., Zhang, Q. B., Li, Z., Zhao, Z. Q., Xing, R., 2005. Antioxidant activity of different molecular weight sulfated polysaccharides from Ulva pertusa Kjellm (Chlorophyta). *J Appl Phycol*. 17, 527-534.

[110] Wang, X. M., Zhang, Z. S., Yao, Z. Y., Zhao, M. X., Qi, H. M., 2013. Sulfation, anticoagulant and antioxidant activities of polysaccharide from green algae Enteromorpha linza. *Int J Biol Macromol*. 58, 225-230.

[111] Zhang, Z. S., Wang, X. M., Mo, X. F., Qi, H. M., 2013. Degradation and the antioxidant activity of polysaccharide from Enteromorpha linza. *Carbohyd Polym*. 92, 2084-2087.

[112] Kaeffer, B., Benard, C., Lahaye, M., Blottiere, H. M., Cherbut, C., 1999. Biological properties of ulvan, a new source of green seaweed sulfated polysaccharides, on cultured normal and cancerous colonic epithelial tells. *Planta Med*. 65, 527-531.

[113] Ribeiro, N. A., Abreu, T. M., Chaves, H. V., Bezerra, M. M., Monteiro, H. S. A., Jorge, R. J. B., Benevides, N. M. B., 2014. Sulfated polysaccharides isolated from the green seaweed Caulerpa racemosa plays antinociceptive and anti-inflammatory activities in a way dependent on HO-1 pathway activation. *Inflamm Res*. 63, 569-580.

[114] Rodrigues, J. A. G., Vanderlei, E. D. O., Silva, L. M. C. M., de Araujo, I. W. F., de Queiroz, I. N. L., de Paula, G. A., Abreu, T. M., Ribeiro, N. A., Bezerra, M. M., Chaves, H. V., Lima, V., Jorge, R. J. B., Monteiro, H. S. A., Leite, E. L., Benevides, N. M. B., 2012. Antinociceptive and anti-inflammatory activities of a sulfated polysaccharide isolated from the green seaweed Caulerpa cupressoides. *Pharmacol Rep*. 64, 282-292.

[115] Castro, R., Zarra, I., Lamas, J., 2004. Water-soluble seaweed extracts modulate the respiratory burst activity of turbot phagocytes. *Aquaculture*. 229, 67-78.

[116] Lee, J. B., Ohta, Y., Hayashi, K., Hayashi, T., 2010. Immunostimulating effects of a sulfated galactan from Codium fragile. *Carbohyd Res*. 345, 1452-1454.

[117] Nika, K., Mulloy, B., Carpenter, B., Gibbs, R., 2003. Specific recognition of immune cytokines by sulphated polysaccharides from marine algae. *Eur J Phycol*. 38, 257-264.

[118] Wei, J. T., Wang, S. X., Liu, G., Pei, D., Liu, Y. F., Liu, Y., Di, D. L., 2014. Polysaccharides from Enteromorpha prolifera enhance the immunity of normal mice. *Int J Biol Macromol.* 64, 1-5.

[119] Stadnik, M. J., de Freitas, M. B., 2014. Algal polysaccharides as source of plant resistance inducers. *Trop Plant Pathol.* 39, 111-118.

[120] El Modafar, C., Elgadda, M., El Boutachfaiti, R., Abouraicha, E., Zehhar, N., Petit, E., El Alaoui-Talibi, Z., Courtois, B., Courtois, J., 2012. Induction of natural defence accompanied by salicylic acid-dependant systemic acquired resistance in tomato seedlings in response to bioelicitors isolated from green algae. *Sci Hortic-Amsterdam.* 138, 55-63.

[121] Hernandez-Herrera, R. M., Virgen-Calleros, G., Ruiz-Lopez, M., Zanudo-Hernandez, J., Delano-Frier, J. P., Sanchez-Hernandez, C., 2014. Extracts from green and brown seaweeds protect tomato (Solanum lycopersicum) against the necrotrophic fungus Alternaria solani. *J Appl Phycol.* 26, 1607-1614.

[122] Cluzet, S., Torregrosa, C., Jacquet, C., Lafitte, C., Fournier, J., Mercier, L., Salamagne, S., Briand, X., Esquerre-Tugaye, M. T., Dumas, B., 2004. Gene expression profiling and protection of Medicago truncatula against a fungal infection in response to an elicitor from green algae Ulva spp. *Plant Cell Environ.* 27, 917-928.

[123] Lee, J. B., Koizumi, S., Hayashi, K., Hayashi, T., 2010. Structure of rhamnan sulfate from the green alga Monostroma nitidum and its anti-herpetic effect. *Carbohyd Polym.* 81, 572-577.

[124] Wang, H., Li, Y. L., Shen, W. Z., Rui, W., Ma, X. J., Cen, Y. Z., 2007. Antiviral activity of a sulfoquinovosyldiacylglycerol (SODG) compound isolated from the green alga Caulerpla racemosa. *Bot Mar.* 50, 185-190.

[125] Chiu, Y. H., Chan, Y. L., Li, T. L., Wu, C. J., 2012. Inhibition of Japanese Encephalitis Virus Infection by the Sulfated Polysaccharide Extracts from Ulva lactuca. *Mar Biotechnol.* 14, 468-478.

[126] Alves, A., Sousa, R. A., Reis, R. L., 2013. *In Vitro* Cytotoxicity Assessment of Ulvan, a Polysaccharide Extracted from Green Algae. *Phytother Res.* 27, 1143-1148.

[127] Qi, H.M., Liu, X.L., Wang, K., Liu, D.M., Huang, L.Y., Liu, S.M., Zhang, Q.B., 2013. Subchronic toxicity study of ulvan from Ulva pertusa (Chlorophyta) in Wistar rats. *Food Chem Toxicol.* 62, 573-578.

[128] Holdt, S. L., Kraan, S., 2011. Bioactive compounds in seaweed: functional food applications and legislation. *J Appl Phycol.* 23, 543-597.

In: Sulfated Polysaccharides
Editors: M. Gama, H. Nader, H. de Oliveira Rocha
ISBN: 978-1-63482-975-5

Chapter 7

FUCOIDANS AND FUCANS

Pilar Rupérez*
Metabolism and Nutrition Department. Instituto de Ciencia y Tecnología de Alimentos y Nutrición (ICTAN). Consejo Superior de Investigaciones Científicas (CSIC). Ciudad Universitaria, Madrid, Spain

ABSTRACT

Fucoidans and fucans are the main and typical sulfated-polysaccharides of brown seaweeds (Heterokontophyta, Pheophyceae), missing from other algae or higher plants. These charged polymers play relevant structural roles as cell wall-reinforcing molecules, related to algal survival in the marine environment.Generally, fucoidan refers to a type of homopolysaccharide consisting mainly of α-(1→3)-linked L-fucose with sulfate ester groups, while fucan is used for complex heteropolysaccharides based on sulfated-fucose but with minor amounts of other monosaccharides -including common neutral sugars, uronic acids and amino sugars- or even acetyl group and protein. They are high-molecular weight polymers (100-1,600 kDa), non-gelling agents, which in contrast to other algal sulfated hydrocolloids once extracted from the cell wall are very soluble but do not develop highly viscous solutions. Fucoidans and fucans are readily available from several cheap sources, such as by-products in alginate's extraction process for the food and cosmetic industry. From the nutritional point of view, both are non-digestible polysaccharides which can be regarded as dietary fibre and may potentially exhibit many other useful biological properties. Yet the intimate relationship between polysaccharide structure and function is still barely understood. Their overall bioactivity, mainly anticoagulant-antithrombotic, antioxidant, immunomodulatory, antiviral and antitumoral capacity (among others), seems to be related to sulfate content, average molecular weight (MW), monosaccharide and type of glycosidic linkage, along with the position of sulfate groups on sugar residues. Moreover, structure of fucoidans/fucans as well as their MW are heterogeneous and may differ among species of brown seaweeds. Currently, the search for new marine drugs with novel biological properties has greatly increased interest in these sulfated-polysaccharides for different non-medical (functional foods, cosmetics), medical and pharmaceutical applications.

* E-mail: pruperez@ictan.csic.es

1. Introduction

1.1. Traditional Consumption of Seaweeds

Seaweeds have traditionally been used by many cultures since ancient times, most especially in oriental countries, such as Japan, China and Korea owing to their dietary, medicinal and functional food properties. Until recently, in Western countries their major use has been mainly concentrated on the extraction of algal compounds as phycocolloids, thickening and gelling agents used by the pharmaceutical, cosmetics and food industries [1, 2]. In fact, brown seaweeds are considered as a real treasure-trove of physiologically active polysaccharides [3], because in addition to neutral polysaccharides, such as cellulose and laminaran or polyanionic polysaccharides such as alginic acid, they contain other less known highly sulfated-polysaccharides (SPs): fucoidans and fucans. Whereas alginates are the major products manufactured from brown algae, fucoidans and fucans still remain as underexploited by-products although, together with another algal bioactive components, they may greatly contribute to the therapeutic efficacy of brown seaweeds [3].

1.2. Seaweeds As a Source of Bioactives

The richest source of fucoidans and fucans are marine organisms, including several genera of brown seaweeds such as *Fucus, Laminaria* and *Undaria* from orders Fucales and Laminariales.They are structural polysaccharides of the cell wall which may constitute up to 25–30% algal dry weight, depending on seaweed species and to a lesser extent on harvest season [3]. Similar fucose-based sulfated-polysaccharides have not been found so far, either in land plants or in other terrestrial organisms. However, this type of mucilages is also found in marine invertebrates, for example in echinoderms as sea cucumber and sea urchin [4, 5].

The search of new marine drugs and natural bioactives has renewed interest in fucoidans and fucans. Current interest in fucoidan is global and research occurs all over the world [6]. These polysaccharides possess the great advantage to be readily available from several cheap sources, such as the by-products of alginate extraction for different needs. Only typical fucoidans and fucans from seaweeds, as well as their main biological properties, will be dealt with in this chapter with a especial focus on their use as dietary fibre and prebiotics.

2. Chemical Composition and Structure

Fucoidan from marine brown algae was first isolated and named as "fukoidin" by Kylin in 1913 [7], a century ago. It is now called fucoidan according to IUPAC rules, but it is also known as fucan, sulfated fucan or fucosan [8]. Generally, fucoidan refers to a type of *homopolysaccharide* [9] consisting mainly of α-(1→3)-linked L-fucose with sulfate ester groups, while fucan is used for more complex *heteropolysaccharides* based on sulfated-fucose but with minor amounts (< 10%) of other monosaccharides -including uronic acids, common neutral sugars (galactose, glucose, mannose, xylose, rhamnose) and amino sugars- or even acetyl groups and protein [8, 10]. Berteau and Mulloy [11] preferred to use the name *fucoidan* specifically for sulfated fucans from seaweeds, whereas *sulfated fucan* would refer to polysaccharides from marine invertebrates. For other researchers, fucans can be classified into three major groups: fucoidans, xylofucoglycuronans and glycuronogalactofucans [12]. Fucoidan structure includes sugar composition, sequence and linkage type of sugar residues, sulfate content and position, molecular weight, etc. [8]. The revised average structure for fucoidan was reported by Patankar et al. [13]. Fucoidan structure from *Fucus vesiculosus* and other *Fucus* species seems relatively simple (Figure 1), but most fucoidans from brown seaweeds possess a more complex chemical composition [8].

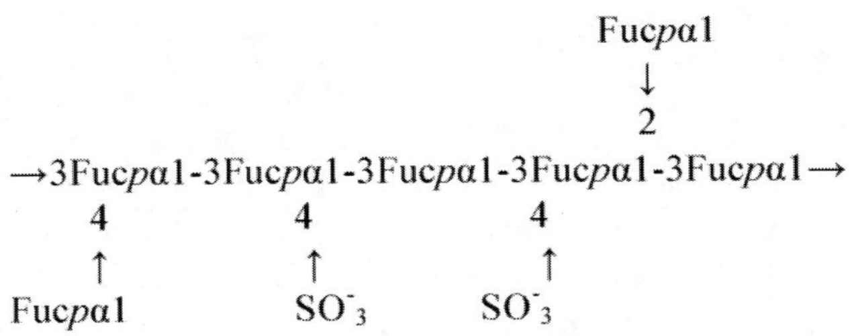

FucoidanType I

Figure 1. Chemical structure of fucoidan Type I from *F. vesiculosus* (according to Patankar et al. [13]).

According to Patankar et al. [13], the core region of fucoidan would be composed primarily of a linear backbone of α-(1→3)-linked L-fucose with sulfate groups substituted at the C4-position on some of the sugar residues. Fucose ($C_6H_{12}O_5$, deoxy-hexose), fucoidan's fundamental subunit is also attached to form branch points, one for every 2-3 fucose residues within the main chain. Sulfate analysis indicated that there is one sulfate group for every 3-4 mol of fucose. The position of sulfate groups is important for the biological activity of SPs. Patankar's main methods used to characterize the structure of fucoidan were infrared spectroscopy and methylation analysis [13]. GC-EIMS analysis of partially methylated alditol acetates from fucoidan indicated predominantly fucose (97%), with trace amounts (< 3%) of galactose, xylose and uronic acid.

This technique also confirmed a highly branched polymer with unsubstituted fucopyranose and fucofuranose at the non-reducing ends [13]. Fucopyranose is the predominant isomer of fucose found in fucoidan. The gentle methylation conditions used in Patankar's study avoided hydrolysis or significant desulfation of fucoidan [13]. The best model for native fucoidan structure was suggested as a branched random coil conformation in aqueous solutions with a large average size of 100 kDa [13].

More recent studies on fucoidan from the common Fucales *Ascophyllum nodosum* have clearly shown that it contains large proportions of α-(1→3)- and α-(1→4)-linked fucose, from which oligosaccharides with a disaccharide repeating structure of alternating α-(1→3) and α-(1→4) glycosidic bonds were prepared by acid hydrolysis (Figure 2). Because of its similarity, fucoidan from *F. vesiculosus* probably displays a very close if not identical structure [14]. According to Cumashi et al. [15], the polysaccharide backbones in fucoidans are known as type I or type II chains (Figures. 1 and 2). The type I chains contain repeating α-(1→3)-linked L-fucopyranose residues, whereas type II chains consist of alternating α-(1→3)- and (1→4)-linked L-fucopyranose residues.

```
                                        Fucpα1
                                          ↓
                                          2
→3Fucpα1-4Fucpα1-3Fucpα1-4Fucpα1-3Fucpα1→              FucoidanType II
   4           4           4
   ↑           ↑           ↑
Fucpα1        SO⁻3        SO⁻3
```

Figure 2. Chemical structure of fucoidan Type II from *Ascophyllum nodosum* (according to Chevolot et al. [14]).

At present the fucoidan preparation from *F. vesiculosus* is commercially available and composed of 44.1% fucose, 26.3% sulfate and 31.1% ash, plus a little aminoglucose content [16]. Pantankar et al. [13] used without distinction both names -fucoidan and fucan- for fucose-containing polysaccharides. Moreover, polysaccharides reported by Li et al. [8], based on sulfated-fucose with different amounts of other monosaccharides, should not strictly be considered as fucoidans but rather as fucans. One difficulty in making comparisons within previous reports lies in the many variations in the composition and molecular weight of the fucoidan fractions used [6]. Thus, data found in the literature may still be controversial and confusing, even regarding the composition and structure of the most carefully studied fucoidan.

A great deal of structural studies show that the chemical composition and structure of fucoidans/fucans from brown algae are very complex and that they can vary according to seaweed species, harvest period and geographical origin. The difference in polysaccharide backbone structures likely reflects the fundamental difference within seaweed families in fucoidan's biosynthetic pathways. More recently, the existence in brown seaweeds of three structural types of fucoidans [17] has been reported:

Type I. Orders Chordariales and Laminariales (i.e., *Laminaria, Ecklonia, Chorda, Kjellmaniella)* produce polysaccharides built of α-(1→3)-linked L-fucopyranose residues, with sulfate groups at C4 and branched at C2 and/or C4 position.

Type II. Order Fucales (i.e., *Fucus, Ascophyllum*) produce polysaccharides built of alternating α-(1→3)- and α-(1→4)-linked L-fucopyranose residues, with sulfate groups at C4 and branched at C2 and/or C4 position.

Type III. Families Alariaceae and Sargassaceae (i.e., *Undaria*, *Sargassum*) produce sulfated-heteropolysaccharides consisting of α-(1→3)- or α-(1→4)-linked L-fucopyranose and D-galactopyranose residues with sulfate groups at the C4 position.

Thus, the structure of fucoidan Type I would be identical to that of *F. vesiculosus* as reported by Patankar et al. [13] (Figure 1), fucoidan Type II would structurally be the same as that reported [14] for *Ascophyllum nodosum* (Figure 2), and Type III would be a galacto-fucan similar to that described for *Undaria* and *Sargassum* [17, 18].

Other researchers distinguish two main fucoidan forms: F-fucoidan, which is composed of sulfate esters of L-fucose and constitutes more than 95% of the fucoidans in brown seaweeds, and U-fucoidan which has approximately 20% glucuronic acid [19].

Average molecular weight (MW) of fucoidans ranges from high to very high (100 to 1,600 kDa) [10, 20, 21]. Once extracted from the cell wall they become very soluble but, unlike other algal SPs, they are neither gelling agents nor develop highly viscous solutions [10] and therefore they are not used as hydrocolloids by the industry [2].

Despite the numerous structural studies on algal fucoidans and the large efforts made, their fine structure remains unclear, mainly due to the absence of a strict regularity, together with the presence of many other minor components (pentose, hexose, uronic acid or protein) and the random sulfation and acetylation. Other possible factors are that the same specific brown seaweed might possess different structural fucoidans or that the extraction mode can exert an effect on their composition and structure [8, 22].

In recent years, high field nuclear magnetic resonance (NMR), mass spectrometry (MS) and tandem MALDI-TOF MS, Fourier transform infrared (FTIR) and FT-Raman spectroscopy techniques have been applied to elucidate the fine structure of fucoidan and other algal polysaccharides [18, 23-27]. Chemical or enzymatic depolymerization of fucoidans also provides an indispensable tool for structural studies and the production of fuco-oligosaccharides [3, 28]. Mild acid hydrolysis of SPs is efficiently used to produce low-MW preparations with improved bioactivity [29]. In addition, gamma irradiation degrades efficiently high-MW into lower-MW fucoidan fractions without considerable removal of the functional sulfate groups [30, 31]. Fucoidan-derived oligosaccharides might display a wide spectrum of biological functions [3, 19]. The increasing availability of specific or monoclonal antibodies capable to detect and quantify structurally different fucoidans will greatly facilitate this type of research [6].

Bioactivity of different fucoidans/fucans has been shown to depend not only on their monosaccharide composition, degree of sulfation, electrical charge density and molecular weight, but also on their three-dimensional structure [6, 25]. In this sense, it is of great interest to have available highly purified and well characterized preparations with which to elucidate their mode of action [21, 32], in order to improve our knowledge of the intimate relationship between structure and bioactivity, and with a view on their potential use as novel marine drugs [33] and functional food ingredients.

3. Natural Sources and Extraction

Fucoidans and fucans are structural SPs from the cell walls of brown seaweeds, mainly found in order Fucales and Laminariales, but which are also present in other algal orders [10, 11]. Fucoidan is considered as a cell wall-reinforcing polymer related to algal protection against the effects of desiccation when seaweeds are exposed at low tide [34]. Some studies show a direct correlation between fucoidan content and the marine depth at which algae grow, so that the closer to the sea surface the greater their fucoidan content [11]. Sulfate is a typical component of marine algal polysaccharides, related to high salt concentration in the environment and with a specific role in ionic regulation [9, 35]. Composition of fucoidan/fucans can also vary depending on taxonomical (algal species), physiological (lifestage), seasonal (harvest period) and environmental factors (geographical origin or cultural area), as well as on other exogenous factors or processing technologies applied to seaweeds for polysaccharide extraction [22, 27, 36-38]. The great structural heterogeneity of fucoidan/fucans could be indicative of polymer molecules being actively metabolized in order to adapt to the ever changing environmental conditions.

Fucoidans and fucans can make up more than 40% d.w. of isolated algal cell walls [9], and typically, they are extracted by treating the raw material with hot aqueous or mild acidic solutions at temperatures ranging from 70–100 ºC for several hours [9, 36]. Purified fucoidans/fucans are highly hygroscopic [9]. *F. vesiculosus* contains the highest concentration of fucoidans (up to 20% d.w.). The fucoidan preparation commercially available is extracted with acid from *F. vesiculosus* [16].

Rupérez et al. [21] extracted sequentially the main polysaccharides of *F. vesiculosus* based on their different solubility in cold and warm water, acid and alkali. According to its chemical composition and infrared spectrum (IR), the acid-soluble fraction (F3) extracted with 0.1 M hydrochloric acid from *Fucus* contains a homogeneous fucoidan with a MW of 1.6×10^6 Da, as calculated by high-performance size exclusion liquid chromatography (HPSEC) with refractive index detection. ^{1}H NMR spectroscopy of this fucoidan confirmed the predominance of fucose, a great proportion of which was released after partial sugar hydrolysis and must be present as fucofuranose (> 50%). F3 has an IR spectrum similar to that of commercial fucoidan and fucoidan preparation previously reported. Fucoidan in F3 shows a broad absorption band at 850 cm^{-1} assigned to sulfate groups at the axial C-4 position of fucose [13, 21]. The large absorption band at 1225-1255 cm^{-1} is also due to sulfate, in accordance with the high sulfate content of fucoidan. This band is common to all SPs and an index of their degree of sulfation, so that the higher the sulfate content the higher its intensity.

The general composition of fucoidan and fucans is dependent on their extraction sequence. Fucans from *F. vesiculosus* and some other brown algae obtained by sequential chemical extraction were reported to be in a continuous spectrum of compounds from highly sulfated homofucans (fucoidans), to high-uronic acid, low-sulfate, fucose-containing polymers [8, 39].

Generally, it is not easy to extract active polysaccharides since the algal cell wall is composed of complex polymers. Therefore, in order to improve the extraction efficiency of the target bioactive compound, novel alternative procedures such as microwave-, ultrasound- and enzyme-assisted techniques are proposed for industrial purposes. The use of ultrasound- and microwave-assisted extraction techniques allows high yields and a low extraction time without modifying the sulfate ester linkage. The enzyme-assisted technique seems advantageous for the large-scale production of fucoidan, but the economic efficiency of the process will depend on the continuous supply of the specific enzymes -lyases- required [36, 40]. Currently, response-surface methodologies (RSM) are being used as an aid to stablish the optimal extraction parameters. The Box–Behnken designs are typically applied to optimize the microwave- or ultrasound-assisted extraction of plant polysaccharides. A more flexible and enhanced variant is offered by d-optimal designs, in which the experimental model and evaluation are software-assisted [36].

4. Biological Properties

Sulfated-polysaccharides, and very especially fucoidan, are within the most extensively researched bioactive compounds from brown seaweeds [12], in a way that scientific publications on fucoidans have increased three-fold between 2000 and 2010 [6]. Fucoidans and fucans possess a wide spectrum of versatile biological properties and potential health effects, namely as: anticoagulant-antithrombotic, antioxidant, antitumoral, antiviral, anti-inflammatory, antiallergic, antiobesity, antihepatopathy, antiuropathy and antirenalpathy, among others. Bioactivity of different fucoidans/fucans is related to sugar composition, degree of sulfation, electrical charge density and average molecular weight, but also to the three-dimensional structure of polymers [6, 25]. Relevant biological properties of algal bioactive compounds, including fucoidan, their applications in functional food and legislation aspects have been reviewed in different seaweed families of interest in Europe [34]. Biological activity of fucoidans in cancer has been reviewed -including apoptosis and growth signalling mechanisms-or as natural bioactive ingredients for functional foods [41, 42]. Other reviews focus on the medicinal, nutraceutical and potential health benefits of fucoidan and other SPs from marine algae [43-46], or provide valuable information on the isolation and purification of fucoidans and other fucose-rich SPs from brown seaweeds, along with their biological activities and potential industrial applications [40]. At the same time, current research interest on the prebiotic potential of fucoidan and other marine algal polysaccharides [12, 47], as well as on their corresponding oligosaccharides is increasing [25]. The main biological properties of fucoidans and fucans from brown seaweeds have been summarized in Table 1.

4.1. Non-toxic

Fucoidan from all brown algal species studied has a non-toxic profile and is well tolerated at the doses assayed [33]. Its use is safe when administered to rats either orally (300 mg/kg bw per day) [48] or intraperitoneally (200 mg/kg). A seaweed-derived nutritional supplement –containing 75 mg fucoidan plus vitamin B6, zinc and manganese-is also found to be safe when given orally to humans in a phase-two clinical trial [6]. Other clinical studies, using orally ingested fucoidan, lack toxicity at doses of 1 g per day up to three months, or 3 g per day for 12 days [6]. Therefore, fucoidan treatment is relatively safe but it should not be administered at doses higher than 900 mg/kg bw per day, because of the anticoagulant effect of fucoidan [48].

4.2. Anticoagulant– Antithrombotic

Their potent anticoagulant action has been the first biological property described for algal fucoidans/fucans and the most widely studied. Although their capacity is lower than that of commercial heparin from animal sources, fucoidan has the advantage to lack side effects due to its vegetable origin [10, 11, 49].

Nishino and Nagumo [50] reported on the sugar constituents and blood-anticoagulant activities of fucose-containing SPs in nine species of brown seaweed. Cumashi et al. [15] studied the anticoagulant, anti-inflammatory, antiangiogenic and antiadhesive activities of nine different fucoidans from brown seaweeds. Jin et al. [51] performed a comparative study on the anticoagulant capacity of eleven fucoidans. Many studies show that the anticoagulant activity of fucoidan is related to its sulfate content, position on sugar residue, molecular weight and sugar composition [8]. Recently, structure-activity relationship of the pro- and anticoagulant effects of fucoidan from *F. vesiculosus* has been reported [52].

Table 1. Main biological properties of fucoidans and fucans from brown seaweeds

Bioactive effect	Seaweed species	References
Anticoagulant-Antithrombotic	*Among others: Fucus vesiculosus, Saccharina latissima, Laminaria digitata, F. serratus, F. distichus, F. evanescens, Ecklonia kurome*	*3, 8, 10, 11, 13, 15, 49-63*
Antioxidant	*Fucus vesiculosus, Saccharina latissima, Laminaria japonica, Sargassum binderi, Padina sp., Ecklonia cava*	*21, 37, 47, 66-70, 104, 111, 112*
Antitumoral	*Saccharina cichorioides, Fucus evanescens, Undaria pinnatifida, S. japonica, Alaria sp.*	*17, 31, 33, 41, 73, 75-79*
Antiviral	*Cystoseira indica, Caulerpa racemosa, Saccharina cichorioides, S. japonica, Dictyota dichotoma*	*80-84*
Immunomodulatory	*Ecklonia cava, Undaria pinnatifida*	*74, 85, 100*
Antiobesity-Antihyperlipidemic	Brown alga *sp, Laminaria japonica*	*86, 87, 90*
Antidiabetic	*Fucus vesiculosus, Ascophyllum nodosum*	*90-92*
Dietary fibre-Prebiotics	*Fucus vesiculosus, Laminaria, Undaria, Himanthalia elongata, Bifurcaria bifurcata, S. latissima (formerly Laminaria saccharina)*	*1, 10, 35, 47, 94, 95, 98, 99*

Oversulfated fucoidans/fucans often present a higher anticoagulant activity than native polysaccharides. Thus, heparin contains three sulfate groups per disaccharide residue, native fucoidan contains one sulfate per fucose unit [13], while the oversulfated fucoidan reported by Qiu et al. [53] may contain two sulfate groups per fucose unit. Consequently, oversulfated fucoidan shows an enhancement of its anticoagulant activity -four times higher upon sulfation- compared to native fucoidan, as measured by an increase in the prothrombin time of normal citrated human plasma.

The position of sulfate groups on sugar residues is also very important for the anticoagulant ability of fucoidan. The activity is related to the concentrations of C-2 sulfate and C-2,3 disulfate, moreover 2,3-disulfated sugar residue is the common structural feature in anticoagulant fucoidans [8, 54].

Nishino and Nagumo [55] found that only fucans with a sulfate to total sugar ratio > 1 displayed a significant anticoagulant capacity. This is in agreement with results from Chevolot et al. [54] indicating that anticoagulant activity requires sulfation at all O-2 and some O-3 positions, whereas sulfation at O-4 seems unnecessary.

Fucoidan requires a long sugar-chain and a certain conformation to bind thrombin, therefore a large molecular weight is needed to achieve anticoagulant activity [8, 51]. Yet it is not clear the minimum-chain length required for anticoagulant capacity, since side reactions cannot be avoided, regardless of the process used to reduce the molecular size of the native polymer [54].

As far as monosaccharides composition (fucose, galactose, uronic acid, etc.) of fucoidan is concerned, some studies show that they may be related to anticoagulant activity [51], but Li et al. [8] suggested that the sulfate group linked to sugar can affect anticoagulant activity, rather than the sugar itself. Uronic acid is not necessary for the anticoagulant ability, but it can act as an enhancer by improving the flexibility of the sugar chain [8].

Regular, linear sulfated α-L-fucans (as well as sulfated α-L-galactans) express their anticoagulant activity not just simply as a function of charge density, but depending critically on the pattern of sulfation and monosaccharide composition [56]. Conformation of sulfated fucan in aqueous solutions has direct implications for their anticoagulant activities [57]. The anticoagulant activity is reported to depend on sulfation degree of α-(1→4) and α-(1→3)-fucosyl units of fucan. Specific substitution patterns are required to allow interactions with particular proteins [58]. Specific sulfation and glycosylation have been recently reported as key factors for the anticoagulation of marine carbohydrates [59].

Fucoidans and other SPs have been tested as anticoagulants of direct action.They are also considered as potential agents to resorb thromboses [3]. Only fucoidans from *Saccharina latissima* (formerly *Laminaria saccharina*), *L. digitata*, *Fucus serratus*, *F. distichus* and *F. evanescens* displayed strong antithrombin activity in a platelet aggregation test [15]. In comparison with heparin, fucoidans are highly antithrombotic without potential haemorrhagic effects in the arterial thrombus model, but this property may differ among algal species due to the variability in molecular weight and structure of fucoidans [60].

Activated partial thrombin time (APTT) and prothrombin time (PT) are the most commonly used clinical laboratory tests to monitor the heparin effect on anticoagulant activity, related either to the intrinsic or the extrinsic coagulation pathways in plasma, respectively. By extension, they are also used to assay the anticoagulant ability of algal fucoidans and fucans. The mechanism of anticoagulant activity for SPs can be shown via thrombin inhibition. Algal sulfated fucans possess potent anticoagulant activity, directly mediated by antithrombin [61] and/or heparin cofactor II, as those from marine invertebrates [56]. These compounds may act only on the intrinsic-but not on the extrinsic pathways of blood coagulation. Structure–function studies reveal that the distribution of negative charges from sulfate groups, rather than the structure of a specific carbohydrate, may constitute a pattern that probably recognizes the biological targets involved in blood coagulation [58]. Mourao [56] has summarized the use and future perspectives of sulfated fucans as anticoagulant and antithrombotic agents, and an analysis of the anticoagulant and antithrombotic mechanisms of action in fucanome and galactanome has been published recently [62].

4.3. Antioxidant

Seaweeds are exposed to adverse environmental conditions, such as light and high oxygen concentrations that lead to the formation of free radicals, and other strong oxidizing agents, however they do not suffer any serious photodynamic damage *in vivo*. Therefore, seaweeds are capable to generate the necessary compounds to protect themselves from external factors such as UV radiation, stress and pollution. This fact suggests that marine algae, like photosynthesizing plants, possess antioxidative mechanisms and compounds which act as antioxidant agents [12].

It has been demonstrated that the antioxidant capability of seaweeds may arise not only from their polyphenols, but also from some other components such as sulfated-polysaccharides [21]. In this regard, SPs from marine algae can act as free radical scavengers and antioxidants for the prevention of oxidative damage. The antioxidant ability of fucoidans/fucans seems to depend on their structural features, such as degree of sulfation (DS), molecular weight, type of predominant sugar and glycosidic branching [63]. Unlike the anticoagulant capacity, the lower the molecular weight of a SPs the more potent its antioxidant ability is. The rationale of this fact is that low-MW SPs may incorporate more efficiently into the cells and may therefore donate protons more effectively than high-MW preparations [63].

Since the antioxidant capacity is due to the combined effect of different antioxidants, several *in vitro* methods are currently used to evaluate the antioxidants of interest in biological systems and food [64], in which samples are commonly extracted with aqueous-organic mixtures (i. e. methanol, ethanol, acetone), and the corresponding antioxidant assays performed in organic solvents. However, under these extraction conditions most polysaccharides would precipitate. Therefore, the antioxidant capacity attributable to fucoidans and fucans must be measured in aqueous solutions. The mechanism leading to the antioxidant properties of charged polysaccharides from seaweeds seems to be multifunctional, and in aqueous solution they can act both as reducing compounds and as radical scavengers [65]. Consequently, the antioxidant power of polysaccharides from *F.vesiculosus* in aqueous solution was estimated from their ability to reduce TPTZ-Fe(III) to TPTZFe(II) complex, as determined by the FRAP assay. The acid soluble fraction (F3), which contained fucoidan, showed the highest reducing power (99.7 and 123.0 μmol Trolox equivalent/g sample d. w. at 4 and 30 min, respectively). Sulfated-polysaccharides from *F.vesiculosus*, and most especially its fucoidan, exhibited antioxidant potential [21].

Freeze-dried raw *F.vesiculosus* powder has higher *in vitro* antioxidant capacity and dietary fibre content than other commercial nutraceuticals derived from this seaweed [66]. This suggests that the technical processes, such as drying at high temperatures or long storage periods, used for the industrial preparation of fucoidan can diminish its antioxidant capability. According to Díaz-Rubio et al. [66], phlorotannins -typical polyphenols of brown seaweeds- seem to be the main responsible agents for the antioxidant capacity as measured in organic extracts of *Fucus* raw powder and in different commercially available fucoidan preparations [66].

In search of macroalgal nutraceuticals a novel approach which takes into account the multifunctional antioxidant activity attributable to both, sulfated-polysaccharides and polyphenols, can actually give a more complete picture of the potential bioactive compounds involved, than the usual approaches based only on algal polyphenols and antioxidant properties [67]. In this regard, the brown seaweed Sugar Kombu (*Saccharina latissima*) is a good source of dietary fibre (DF) and associated compounds [1], and it exhibits antioxidant capacity *in vitro* attributable both to its sulfated-polysaccharides (water-soluble) and polyphenols (alcohol-soluble) [67]. Furthermore, the effect of a DF-rich Sugar Kombu diet on antioxidant ability and biochemical parameters has been evaluated in healthy rats [47]. Several health-promoting effects are observed such as an increase in overall antioxidant status, as well as a decrease in triglycerides and uric acid [47].

Several sulfated fucoidans from the brown seaweed *Laminaria japonica* were evaluated for their antioxidant activity by different *in vitro* methods [68]. Recently, the isolation and antioxidant capacity of fucoidan from selected Malaysian seaweeds has been reported. Among the samples screened, fucoidan from *Sargassum binderi* exhibited the highest antioxidant capacity *in vitro* [69] To test the antioxidant activity *in vivo* of fucoidan from *Eklonia cava* a zebrafish model has been used [70].

SPs and most especially fucoidans/fucans from edible brown seaweeds, byproducts in the industrial extraction of alginates [2], could be used as sources of natural antioxidants with potential applications in the food industry [2, 21]. Nevertheless, at present, the mechanisms by which fucoidans and other sulfated-polysaccharides from marine algae exert their antioxidant power are still unknown. In this sense, it is essential to have available highly purified and well characterized sulfated-polysaccharides with which to elucidate their mode of action [6, 21].

4.4. Antitumoral

Food supplements or drinks containing fucoidan from brown seaweeds have been traditionally administered to cancer patients in Korea, China and Japan. In this regard, the lowest incidence of cancer was found at Okinawa Island in Japan, whose inhabitants consume raw *Laminaria*, which is known to contain a great amount of fucoidan [71, 72].

The antitumoral capacity of fucoidans is exerted by their interference with the initiation (cell transformation), development (colony formation) and progression (metastasis) of cancer growth, through the modulation of various mechanisms including cellular proliferation, differentiation, apoptosis, angiogenesis and metastasis [17]. Fucoidan has been reported to be effective in assays *in vitro* and *in vivo*. Results *in vitro* suggest that the potent apoptotic activity of fucoidan on cancer cells may be cell type specific. Regarding *in vivo* results fucoidan is effective upon oral, intraperitoneal or intravenous administration. The mechanisms underlying the antitumoral activity of fucoidan are diverse and exerted by inducing apoptosis of cancer cells. Fucoidan-mediated apoptosis of cancer cells likely involves up- or down-regulation of multiple signalling pathways. Fucoidan treatment *in vivo* suppresses tumorigenesis and metastasis [73]. As an indirect approach, fucoidans might be capable of defeating different cancers through boosting the immune system [74]. Fucoidan derived from *Undaria pinnatifida* induces apoptosis in human hepatocellular carcinoma [75]. The effect of a sulfated α-(1→3)-L-fucan from the brown alga *Saccharina cichorioides* was studied on resveratrol-induced apoptosis in colon carcinoma cells [76]. Fucoidans from different species show selective antitumoral activity against different types of cancer, which are dependent on unique polysaccharide structure. Low-MW fucoidans (> 2000 Da), mainly composed of highly sulfated-fucose residues, possess higher antitumoral activity than high-MW heterofucans with a low degree of sulfation [17]. There is a close relationship between oversulfation and conformation of low- and high-MW fucoidans and their *in vitro* anticancer activity, so that significant differences were observed between the oversulfated polysaccharide fractions F(5-30K) and F(> 30K), with anticancer activities of 37.3-68.0% and 20.6-35.8%, respectively [77]. Besides degree of sulfation, the anticancer activity of fucoidans seems to depend on their molecular weight, monosaccharide composition and structure of main polymer chain [78]. Fucoidans from reproductive tissue of brown algae have been shown to possess higher antitumoral activity than those from vegetative talli [79].

A recent review discusses fucoidan's structure and its role in apoptosis, invasion, metastasis, angiogenesis and growth signalling mechanisms of cancer [41]. Fucoidan inhibits proliferation of cancer cells by inducing cell cycle arrest and apoptosis, regulating growth signalling molecules and inhibiting metastasis and angiogenesis. Previous results provide strong evidence for further exploring the use of fucoidans from brown seaweeds as novel chemotherapeutic and anticancer drugs against different types of cancer [17]. Combinations of fucoidan or low-MW fractions with other anticancer agents could be very good targets for anticancer therapy [41].

4.5. Antiviral

Antiviral activities and mechanisms of marine polysaccharides and their oligosaccharide derivatives have been reviewed recently [80]. Fucoidans have no direct pathogen killing activity but rather, they inhibit infection via receptor entry blocking and interference with replicative processes [6]. Sulfate seems essential for the antiviral activity of algal polysaccharides, especially the sulfate group located at C-4 of α-(1→3)-linked fucopyranosyl units appears to be particularly important for their antiherpetic activity [81, 82]. Drawbacks to using fucoidan as an orally delivered agent are those of bioavailability [6].

All natural fucoidans can be considered as potential anti-HIV agents, regardless of their carbohydrate backbone and degree of sulfating, since their activity is shown at low concentrations (0.001-0.05 μg/mL). High-MW fucoidans isolated from *Saccharina cichorioides* α-(1→3)-L-fucan and *S. japonica* (galactofucan) are very effective viral inhibitors [83]. Moderate inhibitory effects against herpes simple (HSV-1) and Coxsackie virus (CVB3) are found for galactofucan-rich fractions in *Dictyota dichotoma*, but most of the other polysaccharide fractions tested in the study were inactive against both viruses [84].

4.6. Immunomodulatory

The effect of fucoidan's molecular weight on immune cells has been studied. Spleen cells were cultured in the presence of low- and high-MW fucoidans, respectively. Specifically, viability/death was quantified and the function of treated cells, including cytokine production, was measured. Fucoidans of different MW seem to have differential effects on the viability and function of immune cells. Especially, the high-MW fucoidan fraction exhibits the most potent immunostimulatory activity [85], in the same way that the anticoagulant capacity. Water-soluble sulfated-fucans with immune-enhancing properties have been isolated from *Ecklonia cava* [74].

4.7. Antiobesity– Antihyperlipidemic

Obesity is related to a higher risk of developing diabetes and cardiovascular disease through the accumulation of excess white adipose tissue in the body. Antiobesity effects of fucoidan in an animal model of diet-induced obesity are reported [86]. Fucoidan decreases body weight, fat tissue mass, plasma triglyceride, total- and LDL-cholesterol levels, in part by down-regulating adipogenic transcription factor and other specific target genes. Fucoidan supplementation reduces the level of steatosis induced by high fat diets (HFDs). Besides, it might have the potential to produce antiobesity effects in animals fed HFDs, therefore it may be highly possible to use fucoidan as an antiobesity functional material [86]. The polysaccharide preparation used in the previous study [86] was rather a "fucan" than a "fucoidan", as it is reported to be mainly composed of fucose (64.4±6.0%) and sulfate (31.7±2.2%), but it also contains a considerable amount of galactose (31.9±4.7%) and low mannose (3.6±1.3%). Hypolipidemic effect of fucoidan from *Laminaria japonica* has been shown in hyperlipidemic rats [87]. Low-MW fucoidan is widely used for the treatment of metabolic disorders and could potentially prevent obesity by ameliorating endoplasmic reticulum stress-related leptin resistance [88].

4.8. Antidiabetic

Dietary management of diabetes mellitus involves keeping at normal levels both, blood glucose and blood lipid concentrations, in order to reduce the possibility of associated complications [89]. The effect of fucoidan on insulin secretion and resistance has been studied in overweight or obese adults. Fucoidan administration during a 3-month period decreases diastolic blood pressure and LDL-cholesterol concentrations, increasing insulin secretion and insulin resistance [90]. Potential benefits towards Type 2 diabetes management are attributed to fucoidan which has been shown to inhibit effectively the *in vitro* activity of the two starch-digesting enzymes, α-amylase and α-glucosidase [91]. Low-MW fucoidan protects endothelial function and ameliorates basal hypertension in a diabetic model of rats. Therefore, fucoidan is a potential candidate drug for endothelial protection in diabetic cardiovascular complications [92].

4.9. Dietary Fibre and Prebiotics

From a nutritional point of view, fucoidans and fucans are non-digestible polysaccharides which can be regarded as a source of dietary fibre [10]. Fucoidans cannot be hydrolysed by human digestive enzymes either in the stomach or in the small intestine. Consequently, their intake can result in an increased luminal concentration within the large intestine [17]. Most of the research in this topic has been made either in the whole brown seaweeds or in algal extracts, but few reports deal with purified SPs, such as fucoidans and fucans.

Different species of *Fucus, Laminaria, Undaria* and other common brown seaweeds are reported as good sources of fibre for human consumption with a balanced content of soluble to total dietary fibre [1, 35]. The total indigestible fraction of these seaweeds was further examined under simulated physiological conditions by an *in vitro* enzymatic procedure [93], alternative to AOAC's analysis of dietary fibre. The indigestible fraction value in these seaweeds is slightly higher than the total dietary fibre content by the official AOAC method, and the distribution of the soluble and insoluble fibre fractions is also different by the two procedures. The estimation of the indigestible fraction in edible seaweeds is proposed as a useful approach from the nutritional and commercial point of view [94].

Polysaccharide fractions, including fucoidan, of the dietary fibre from *F. vesiculosus* have been studied in more detail. The acid soluble fraction (F3) has an infrared spectrum similar to that of commercial fucoidan, with a high-MW ($1.6x10^6$ Da), as estimated by molecular exclusion high-performance liquid chromatography [21]. Dietary fibre content and its associated antioxidant capacity have been reported in different commercially available products from *F. vesiculosus* [66].

The prebiotic potential of marine algal polysaccharides with applications in human and animal health is also being increasingly studied [12, 95]. The prebiotic concept is defined as "the selective stimulation of growth and/or activity(ies) of one or a limited number of microbial genus(era)/species in the gut microbiota that confer(s) health benefits to the host" [96, 97]. In early works, sulfated-fucans were reported to be neither degraded nor fermented *in vitro* by the human faecal microbiota [98]. However, the quite unusual fermentation pattern of dietary fibre from brown algae raised the question of the pathways involved in its degradation, disappearance and fermentation. Therefore, it is of great nutritional importance to carry out a detailed research on these specific pathways, as well as on the metabolites released [98]. More recent studies indicate that there are favourable changes in gut microbiota after the ingestion of fucoidan, including an increased population of beneficial *Lactobacilli* [65, 69].

Whole brown seaweeds or purified polysaccharides are given to laboratory and farm animals in order to test their *in vivo* potential as prebiotics [12]. Intake of a brown seaweed rich in dietary fibre and associated compounds -Sugar Kombu (*Saccharina latissima*)- could promote several health benefits *in vivo*. Regarding its prebiotic effect in a rat model, higher caecal weight and total short chain fatty acid contents were evidenced in the seaweed-fed group, together with an improved overall antioxidant status and decreased risk of cardiovascular disease. Fucoidan and other typical polysaccharides from Sugar Kombu - which are an essential part of the dietary fibre complex- reach the large intestine undigested and seem to be responsible for the nutritional benefits observed [47]. McDonnell and coworkers reported the effects *in vivo* of dietary fucoidan in the diet of weaning piglets on performance, selected faecal microbial populations and concentration of short-chain fatty acids. Pigs fed the fucoidan diet increased their *Lactobacilli* population as compared to controls [99].

Furthermore, fucoidans and fucans are a valuable source of potentially prebiotic sulfated fuco-oligosaccharides, which can be prepared from these SPs by chemical or enzymatic hydrolysis.

5. Other Properties

Fucoidans and fucans are now routinely used in biomedical research.

Antiallergic. A common feature of allergic diseases (i.e., asthma and atopic dermatitis) is an elevated serum level of IgE as a response towards common environmental allergens. Peritoneal injection of fucoidan (100 μg per mouse) suppresses the increase of plasma IgE, but no effect is observed when administered orally at higher doses (10 mg per mouse) [6, 100]. Fucoidan may suppress the increase of plasma IgE level in response to ovalbumin immunization, through the inhibition of IgE production and class switching in B cells *in vivo*, suggesting its potential use as a valuable supplement or medicine for the prevention of IgE-mediated allergy [100].

Anti-inflammatory. Inflammation is a normal and protective response to tissue damage or infection [89]. Fucoidan's potential lies in its pleiotropic anti-inflammatory effects. These include inhibition of selectins, complement and enzyme activity [6]. Evaluation of a heterofucan from *Dictyota menstrualis* as anti-inflammatory and anti-nociceptive (against pain) agent with a similar potency to dipyrone has been described [101]. Anti-inflammatory effect of fucoidan from *Ecklonia cava* along with other natural active ingredients from Chinese medicinal plants has been recently reviewed [102].

Antiarthritis. This pathology involves inflammation of the joints and affects mostly the aged people [89]. Fucoidan from *F. vesiculosus* exerts anti-inflammatory action in a zymosan-induced model of acute inflammation in rats [103]. Also, sulfated heterofucan from *Lobophora variegata* exhibits both, potential antioxidant power and anti-inflammatory capacity in acute zymosan-induced arthritis in rats, indicating a possible modulation mechanism in the cellular redox state [104]. The fact that these natural compounds possess minimal toxicity may have important therapeutic implications [103].

Antihepatopathy. Fucoidan and other SPs from macroalgae are potential hepato- or gastroprotective agents. In recent years, studies on the general composition of the bioactive components from algae and their extracts with potential applications for liver and stomach protection, as well as their possible molecular mechanisms of action have notably increased [105]. In fact, liver protection was actually a result of gut protection, by preventing downstream inflammation of the liver [6].

Antileishmaniasis. Leishmania is a protozoan parasite spread by the bites of sandflies. A promising curative effect of fucoidan on visceral leishmaniasis has been reported [106, 107]. Fucoidan prevents endotoxin-induced damage in a mouse model of endotoxemia and increases resistance to lipopolysaccharide (LPS) [107].

Biomaterial for bone tissue engineering. The effect of a low-MW fucoidan on human osteogenic capacity has been recently published [108]. There is also evidence that hydroxyapatite-fucoidan nanocomposites may be used as promising biomaterials for bone tissue engineering. Both materials are non-toxic and biocompatible [109].

Antiulcer. A vegetable extract containing fucoidan has shown activity against the stomach ulcer producer *Helicobacter pylori* [110]. Fucoidan has inhibitory effects on *Helicobacter* infections, largely by inhibiting bacterial adhesion to mucosal surfaces [6].

Neuroprotector. Cognitive impairment and dementia are frequent disorders mostly among elderly people and influence the individual's ability to function independently [89]. Fucoidan improves cognitive impairment induced by infusion of Aβ peptide in rats. Results suggest that fucoidan could ameliorate the learning and memory abilities, and that the mechanisms involved appear to be due to regulating the cholinergic system, reducing oxidative stress and inhibiting cell apoptosis. Fucoidan shows excellent neuroprotective effects against Aβ peptide-induced learning and memory impairment in AD model rats [111].

Antiuropathy and antirenalpathy. Fucoidan exerts a protective role against chronic renal failure and hyperoxaluria in a rat model, by reducing the toxic effects of oxalate on kidney cells through its antioxidant property [112]. In the same way, the mechanisms involved in the protective action of fucoidan derivatives from *L. japonica* in experimental adenine-induced chronic kidney disease seem to be related to their antioxidant activity [113]. *L. japonica* is a popular seafood and traditional medicinal plant in China used for curing edema disease, a symptom of kidney diseases.

Cardioprotector. Fucoidan protects the structural and functional integrity of the myocardial membrane in rats treated with isoproterenol, as inferred from the significant reduction in the high levels of serum marker enzymes in fucoidan administered rats [114]. These recent results could suggest a vital role for fucoidan in cardioprotection.

CONCLUSION

Seaweeds are currently underexploited marine resources. The nutritional value ascribed to brown algae and their sulfated-polysaccharides, along with their non-animal nature, makes them particularly useful in the food and feed industry, as a source of valuable ingredients for human foods or animal feeds, respectively. Moreover, the fact that algal SPs are relatively easy to isolate and possess numerous health benefits gives them the potential to serve as valuable bioactives in functional foods [89]. Among them, fucoidans and fucans are the most promising marine drugs with novel modes of action and a wide spectrum of biological properties. There are also further potential applications for fucoidan products in complementary medicine and topical applications [6]. However, available and carefully controlled algal sources are necessary, together with improved modern extraction techniques and characterization methods, so as to develop reproducible and well defined fucoidan products that can be manufactured to standard quality and suitable for a particular application, for example as ingredients in medical devices or even cross the threshold into the drug development route [6].

The great structural heterogeneity of fucoidans and fucans could possibly be an indication of polymers being actively metabolized as a biological adaptation to the ever changing marine environment.

Regarding macroalgae, recent advances in biotechnological research have stimulated the development of seaweed aquaculture. At present, three genera, two brown (*Laminaria, Undaria*) and one red seaweed (*Porphyra*), constitute 93% of the algal mass cultured for nutritional purposes [89]. In this regard, in-door and/or out-door integrated multitrophic aquaculture (IMTA) would assure a continuous algal supply of functional ingredients with a controlled standard quality.

There is a clear need to reinvestigate the fine structure of pure fucoidan preparations with updated specific methodologies for the structural analysis of polysaccharides, such as high field ^{1}H and ^{13}C NMR, MALDI-TOF MS, FT-infrared or FT-Raman spectroscopy techniques [18, 23-26, 32]. Besides, the increasing availability of monoclonal antibody technology capable to specifically detect and quantify different fucoidans will greatly facilitate this type of research [6].

Fucoidans and fucans can be used as natural sources of novel fuco-oligosaccharides either by chemical, physical and biochemical degradations or by synthesis using chemical processes [28, 115] or specific enzymes [19, 58].

Fucoidan and fucans can be used for the prevention of diseases and healthcare. Yet the way of administration has to be considered carefully, as it may have an influence on their availability and biological activity [100].

In summary, further research on fucoidan's and fucan's biological effects is needed. However to achieve this goal, the availability of deeply purified and well characterized sulfated-polysaccharide preparations, regarding composition, molecular weight and three-dimensional structure is essential in order to establish unequivocally the intimate correlation between structure and bioactivity.

ABBREVIATIONS

GC-EIMS = gas chromatography-electron impact mass spectrometry;
HPSEC= high performance size exclusion chromatography;
MW= Average molecular weight;
SPs = Sulfated polysaccharides

Note

Due to the limited extension of this book chapter some other related and valuable references may have not been included.

REFERENCES

[1] Gómez-Ordóñez, E., Jiménez-Escrig, A., Rupérez, P., 2010. Dietary fibre and physicochemical properties of several edible seaweeds from the northwestern Spanish coast. *Food Research International*. 43, 2289-2294.

[2] Bixler, H., Porse, H., 2011. A decade of change in the seaweed hydrocolloids industry. *Journal of applied phycology*. 23, 321-335.

[3] Kusaykin, M., Bakunina, I., Sova, V., Ermakova, S., Kuznetsova, T., Besednova, N., Zaporozhets, T., Zvyagintseva, T., 2008. Structure, biological activity, and enzymatic transformation of fucoidans from the brown seaweeds. *Biotechnology journal*. 3, 904-915.

[4] Yu, L., Ge, L., Xue, C., Chang, Y., Zhang, C., Xu, X., Wang, Y., 2014. Structural study of fucoidan from sea cucumber *Acaudina molpadioides*: a fucoidan containing novel tetrafucose repeating unit. *Food chemistry*. 142, 197-200.

[5] Mulloy, B., Ribeiro, A.C., Alves, A.P., Vieira, R.P., Mourao, P.A., 1994. Sulfated fucans from echinoderms have a regular tetrasaccharide repeating unit defined by specific patterns of sulfation at the O-2 and O-4 positions. *The Journal of biological chemistry*. 269, 22113-22123.

[6] Fitton, J.H., 2011. Therapies from fucoidan; multifunctional marine polymers. *Marine Drugs*. 9, 1731-60. doi: 10.3390/md9101731. Epub 2011 Sep 30.

[7] Kylin, H., 1913. Zur Biochemie der Meeresalgen. *Hoppe-Seyler´s Zeitschrift für physiologische Chemie*. 83, 171-197.

[8] Li, B., Lu, F., Wei, X., Zhao, R., 2008. Fucoidan: structure and bioactivity. *Molecules*. 13, 1671-1695.

[9] Kloareg, B., Demarty, M., Mabeau, S., 1986. Polyanionic characteristics of purified sulphated homofucans from brown algae. *International journal of biological macromolecules*. 8, 380-386.

[10] Rupérez, P., Gómez-Ordóñez, E., Jiménez-Escrig, A., *Chapter 11 - Biological activity of algal sulphated and non-sulphated polysaccharides*, in *Bioactive Compounds from Marine Foods: Plant and Animal Sources*, Hernández-Ledesma, B., Herrero, M., Editors. 2104, John Wiley & Sons, Ltd. p. 219-248.

[11] Berteau, O., Mulloy, B., 2003. Sulfated fucans, fresh perspectives: structures, functions, and biological properties of sulfated fucans and an overview of enzymes active toward this class of polysaccharide. *Glycobiology*. 13, 29R-40R.

[12] Gupta, S., Abu-Ghannam, N., 2011. Bioactive potential and possible health effects of edible brown seaweeds. *Trends in Food Science and Technology*. 22, 315-326.

[13] Patankar, M.S., Oehninger, S., Barnett, T., Williams, R.L., Clark, G.F., 1993. A revised structure for fucoidan may explain some of its biological activities. *The Journal of biological chemistry*. 268, 21770-21776.

[14] Chevolot, L., Mulloy, B., Ratiskol, J., Foucault, A., Colliec-Jouault, S., 2001. A disaccharide repeat unit is the major structure in fucoidans from two species of brown algae. *Carbohydrate research*. 330, 529-535.

[15] Cumashi, A., Ushakova, N.A., Preobrazhenskaya, M.E., D'Incecco, A., Piccoli, A., Totani, L., Tinari, N., Morozevich, G.E., Berman, A.E., Bilan, M.I., Usov, A.I., Ustyuzhanina, N.E., Grachev, A.A., Sanderson, C.J., Kelly, M., Rabinovich, G.A., Iacobelli, S., Nifantiev, N.E., 2007. A comparative study of the anti-inflammatory, anticoagulant, antiangiogenic, and antiadhesive activities of nine different fucoidans from brown seaweeds. *Glycobiology*. 17, 541-552.

[16] Nishino, T., Nishioka, C., Ura, H., Nagumo, T., 1994. Isolation and partial characterization of a novel amino sugar-containing fucan sulfate from commercial *Fucus vesiculosus* fucoidan. *Carbohydrate research*. 255, 213-224.

[17] Vishchuk, O.S., Ermakova, S.P., Zvyagintseva, T.N., 2013. The fucoidans from brown algae of Far-Eastern seas: anti-tumor activity and structure-function relationship. *Food chemistry*. 141, 1211-1217.

[18] Bilan, M.I., Grachev, A.A., Shashkov, A.S., Thuy, T.T., Van, T.T., Ly, B.M., Nifantiev, N.E., Usov, A.I., 2013. Preliminary investigation of a highly sulfated galactofucan fraction isolated from the brown alga *Sargassum polycystum*. *Carbohydrate research*. 377, 48-57.

[19] Morya, V.K., Kim, J., Kim, E.-K., 2012. Algal fucoidan: structural and size-dependent bioactivities and their perspectives. *Applied microbiology and biotechnology*. 93, 71-82.

[20] Rioux, L.E., Turgeon, S.L., Beaulieu, M., 2007. Characterization of polysaccharides extracted from brown seaweeds. *Carbohydrate polymers*. 69, 530-537.

[21] Rupérez, P., Ahrazem, O., Leal, J.A., 2002. Potential Antioxidant Capacity of Sulfated Polysaccharides from the Edible Marine Brown Seaweed *Fucus vesiculosus*. *Journal of agricultural and food chemistry*. 50, 840-845.

[22] Ale, M.T., Mikkelsen, J.D., Meyer, A.S., 2011. Important Determinants for Fucoidan Bioactivity: A Critical Review of Structure-Function Relations and Extraction Methods for Fucose-Containing Sulfated Polysaccharides from Brown Seaweeds. *Marine drugs*. 9, 2106-2130.

[23] Pereira, L., Sousa, A., Coelho, H., Amado, A.M., Ribeiro-Claro, P.J.A., 2003. Use of FTIR, FT-Raman and 13C-NMR spectroscopy for identification of some seaweed phycocolloids. *Biomolecular Engineering*. 20, 223-228.

[24] Pereira, L., Amado, A.M., Critchley, A.T., van de Velde, F., Ribeiro-Claro, P.J.A., 2009. Identification of selected seaweed polysaccharides (phycocolloids) by vibrational spectroscopy (FTIR-ATR and FT-Raman). *Food Hydrocolloids*. 23, 1903-1909.

[25] Clément, M.-J., Tissot, B., Chevolot, L., Adjadj, E., Du, Y., Curmi, P.A., Daniel, R., 2010. NMR characterization and molecular modeling of fucoidan showing the importance of oligosaccharide branching in its anticomplementary activity. *Glycobiology*. 20, 883-894.

[26] Gómez-Ordóñez, E., Rupérez, P., 2011. FTIR-ATR spectroscopy as a tool for polysaccharide identification in edible brown and red seaweeds. *Food Hydrocolloids*. 25, 1514-1520.

[27] Anastyuk, S.D., Shevchenko, N.M., Dmitrenok, P.S., Zvyagintseva, T.N., 2012. Structural similarities of fucoidans from brown algae *Silvetia babingtonii* and *Fucus evanescens*, determined by tandem MALDI-TOF mass spectrometry. *Carbohydrate research*. 358, 78-81.

[28] Jin, W., Zhang, W., Wang, J., Ren, S., Song, N., Zhang, Q., 2013. Structural analysis of heteropolysaccharide from *Saccharina japonica* and its derived oligosaccharides. *International journal of biological macromolecules*. 62, 697-704.

[29] Pielesz, A., Binias, W., Paluch, J., 2011. Mild acid hydrolysis of fucoidan: characterization by electrophoresis and FT-Raman spectroscopy. *Carbohydrate research*. 346, 1937-1944.

[30] Choi, J.-i., Gu Lee, S., Jong Han, S., Cho, M., Cheon Lee, P., 2014. Effect of gamma irradiation on the structure of fucoidan. *Radiation Physics and Chemistry*. 100, 54-58.

[31] Choi, J.-i., Kim, H.-J., 2013. Preparation of low molecular weight fucoidan by gamma-irradiation and its anticancer activity. *Carbohydrate polymers*. 97, 358-362.

[32] Bilan, M.I., Grachev, A.A., Shashkov, A.S., Kelly, M., Sanderson, C.J., Nifantiev, N.E., Usov, A.I., 2010. Further studies on the composition and structure of a fucoidan preparation from the brown alga *Saccharina latissima*. *Carbohydrate research*. 345, 2038-2047.

[33] Kwak, J.Y., 2014. Fucoidan as a marine anticancer agent in preclinical development. *Marine drugs*. 12, 851-870.

[34] Holdt, S.L., Kraan, S., 2011. Bioactive compounds in seaweed; functional food applications and legislation. *Journal of applied phycology*. 23, 543-597.

[35] Rupérez, P., Saura-Calixto, F., 2001. Dietary fibre and physicochemical properties of edible Spanish seaweeds. *European Food Research and Technology*. 212, 349-354.

[36] Hahn, T., Lang, S., Ulber, R., Muffler, K., 2012. Novel procedures for the extraction of fucoidan from brown algae. *Process Biochemistry*. 47, 1691-1698.

[37] Mak, W., Hamid, N., Liu, T., Lu, J., White, W.L., 2013. Fucoidan from New Zealand *Undaria pinnatifida*: monthly variations and determination of antioxidant activities. *Carbohydrate polymers*. 95, 606-614.

[38] Anastyuk, S.D., Imbs, T.I., Shevchenko, N.M., Dmitrenok, P.S., Zvyagintseva, T.N., 2012. ESIMS analysis of fucoidan preparations from *Costaria costata*, extracted from alga at different life-stages. *Carbohydrate polymers*. 90, 993-1002.

[39] Mabeau, S., Kloareg, B., Joseleau, J.-P., 1990. Fractionation and analysis of fucans from brown algae. *Phytochemistry*. 29, 2441-2445.
[40] Wijesinghe, W.A.J.P., Jeon, Y.-J., 2012. Biological activities and potential industrial applications of fucose rich sulfated polysaccharides and fucoidans isolated from brown seaweeds: A review. *Carbohydrate polymers*. 88, 13-20.
[41] Senthilkumar, K., Manivasagan, P., Venkatesan, J., Kim, S.-K., 2013. Brown seaweed fucoidan: Biological activity and apoptosis, growth signaling mechanism in cancer. *International journal of biological macromolecules*. 60, 366-374.
[42] Vo, T.-S., Kim, S.-K., 2013. Fucoidans as a natural bioactive ingredient for functional foods. *Journal of Functional Foods*. 5, 16-27.
[43] Ngo, D.H., Kim, S.K., 2013. Sulfated polysaccharides as bioactive agents from marine algae. *International journal of biological macromolecules*. 62, 70-75.
[44] Wijesekara, I., Pangestuti, R., Kim, S.-K., 2011. Biological activities and potential health benefits of sulfated polysaccharides derived from marine algae. *Carbohydrate polymers*. 84, 14-21.
[45] Kim, S.K., Li, Y.X., 2011. Medicinal benefits of sulfated polysaccharides from sea vegetables. *Advances in food and nutrition research*. 64, 391-402.
[46] Costa, L.S., Fidelis, G.P., Cordeiro, S.L., Oliveira, R.M., Sabry, D.A., Câmara, R.B.G., Nobre, L.T.D.B., Costa, M.S.S.P., Almeida-Lima, J., Farias, E.H.C., Leite, E.L., Rocha, H.A.O., 2010. Biological activities of sulfated polysaccharides from tropical seaweeds. *Biomedicine and Pharmacotherapy*. 64, 21-28.
[47] Jiménez-Escrig, A., Gómez-Ordóñez, E., Tenorio, M.D., Rupérez, P., 2013. Antioxidant and prebiotic effects of dietary fiber co-travelers from sugar Kombu in healthy rats. *Journal of applied phycology*. 25, 503-512.
[48] Li, N., Zhang, Q., Song, J., 2005. Toxicological evaluation of fucoidan extracted from *Laminaria japonica* in Wistar rats. *Food and Chemical Toxicology*. 43, 421-426.
[49] Jeon, Y.J., Wijesinghe, W.A., Kim, S.K., 2011. Functional properties of brown algal sulfated polysaccharides, fucoidans. *Advances in food and nutrition research*. 64, 163-178.
[50] Nishino, T., Nagumo, T., 1987. Sugar constituents and blood-anticoagulant activities of fucose-contaning sulfated polysaccharides in nine brown seaweed species. *Nippon Nogeikagaku Kaishi*. 61, 361.
[51] Jin, W., Zhang, Q., Wang, J., Zhang, W., 2013. A comparative study of the anticoagulant activities of eleven fucoidans. *Carbohydrate polymers*. 91, 1-6.
[52] Zhang, Z., Till, S., Jiang, C., Knappe, S., Reutterer, S., Scheiflinger, F., Szabo, C.M., Dockal, M., 2014. Structure-activity relationship of the pro- and anticoagulant effects of *Fucus vesiculosus* fucoidan. *Thrombosis and haemostasis*. 111, 429-437.
[53] Qiu, X., Amarasekara, A., Doctor, V., 2006. Effect of oversulfation on the chemical and biological properties of fucoidan. *Carbohydrate polymers*. 63, 224-228.
[54] Chevolot, L., Foucault, A., Chaubet, F., Kervarec, N., Sinquin, C., Fisher, A.M., Boisson-Vidal, C., 1999. Further data on the structure of brown seaweed fucans: relationships with anticoagulant activity. *Carbohydrate research*. 319, 154-165.
[55] Nishino, T., Nagumo, T., 1991. The sulfate-content dependence of the anticoagulant activity of a fucan sulfate from the brown seaweed *Ecklonia kurome*. *Carbohydrate research*. 214, 193-197.

[56] Mourao, P.A., 2004. Use of sulfated fucans as anticoagulant and antithrombotic agents: future perspectives. *Current pharmaceutical design*. 10, 967-981.

[57] Becker, C.F., Guimaraes, J.A., Mourao, P.A., Verli, H., 2007. Conformation of sulfated galactan and sulfated fucan in aqueous solutions: implications to their anticoagulant activities. *Journal of molecular graphics and modelling*. 26, 391-399.

[58] Courtois, J., 2009. Oligosaccharides from land plants and algae: production and applications in therapeutics and biotechnology. *Current Opinion in Microbiology*. 12, 261-273.

[59] Pomin, V.H., Mourao, P.A., 2014. Specific sulfation and glycosylation-a structural combination for the anticoagulation of marine carbohydrates. *Frontiers in cellular and infection microbiology*. 4, 33.

[60] Min, S.K., Kwon, O.C., Lee, S., Park, K.H., Kim, J.K., 2012. An antithrombotic fucoidan, unlike heparin, does not prolong bleeding time in a murine arterial thrombosis model: a comparative study of *Undaria pinnatifida* sporophylls and *Fucus vesiculosus*. *Phytotherapy research : PTR*. 26, 752-757.

[61] Dore, C.M., das, C.F.A.M.G., Will, L.S., Costa, T.G., Sabry, D.A., de Souza Rego, L.A., Accardo, C.M., Rocha, H.A., Filgueira, L.G., Leite, E.L., 2013. A sulfated polysaccharide, fucans, isolated from brown algae *Sargassum vulgare* with anticoagulant, antithrombotic, antioxidant and anti-inflammatory effects. *Carbohydrate polymers*. 91, 467-475.

[62] Pomin, V.H., 2014. How to analyze the anticoagulant and antithrombotic mechanisms of action in fucanome and galactanome? *Glycoconjugate journal*. 31, 89-99.

[63] Li, Y.-X., Kim, S.-K., 2011. Utilization of seaweed derived ingredients as potential antioxidants and functional ingredients in the food industry: An overview. *Food Science and Biotechnology*. 20, 1461-1466.

[64] Frankel, E.N., Meyer, A.S., 2000. The problems of using one-dimensional methods to evaluate multifunctional food and biological antioxidants. *Journal of the Science of Food and Agriculture*. 80, 1925-1941.

[65] Gómez-Ordóñez, E., Jiménez-Escrig, A., Rupérez, P., 2014. Bioactivity of sulfated polysaccharides from the edible red seaweed *Mastocarpus stellatus*. *Bioactive Carbohydrates and Dietary Fibre*. 3, 29-40.

[66] Díaz-Rubio, M.E., Pérez-Jiménez, J., Saura-Calixto, F., 2009. Dietary fiber and antioxidant capacity in *Fucus vesiculosus* products. *International journal of food sciences and nutrition*. 2, 23-34.

[67] Jiménez-Escrig, A., Gómez-Ordóñez, E., Rupérez, P., 2012. Brown and red seaweeds as potential sources of antioxidant nutraceuticals. *Journal of applied phycology*. 24, 1123-1132.

[68] Xue, C.-H., Fang, Y., Lin, H., Chen, L., Li, Z.-J., Deng, D., Lu, C.-X., 2001. Chemical characters and antioxidative properties of sulfated polysaccharides from *Laminaria japonica*. *Journal of applied phycology*. 13, 67-70.

[69] Lim, S.J., Wan Aida, W.M., Maskat, M.Y., Mamot, S., Ropien, J., Mazita Mohd, D., 2014. Isolation and antioxidant capacity of fucoidan from selected Malaysian seaweeds. *Food Hydrocolloids*. 42, 280-288.

[70] Kim, E.A., Lee, S.H., Ko, C.I., Cha, S.H., Kang, M.C., Kang, S.M., Ko, S.C., Lee, W.W., Ko, J.Y., Lee, J.H., Kang, N., Oh, J.Y., Ahn, G., Jee, Y.H., Jeon, Y.J., 2014.

Protective effect of fucoidan against AAPH-induced oxidative stress in zebrafish model. *Carbohydrate polymers*. 102, 185-191.

[71] Kim, S.-K., Bhatnagar, I., *Chapter 7 - Physical, Chemical, and Biological Properties of Wonder Kelp—Laminaria*, in *Advances in food and nutrition research*, Se-Kwon, K., Editor. 2011, Academic Press. p. 85-96.

[72] Willcox, D.C., Scapagnini, G., Willcox, B.J., 2014. Healthy aging diets other than the Mediterranean: A focus on the Okinawan diet. *Mechanisms of Ageing and Development*. 136–137, 148-162.

[73] Lee, H.E., Choi, E.S., Shin, J.A., Lee, S.O., Park, K.S., Cho, N.P., Cho, S.D., 2014. Fucoidan induces caspase-dependent apoptosis in MC3 human mucoepidermoid carcinoma cells. *Experimental and therapeutic medicine*. 7, 228-232.

[74] Cao, R.A., Lee, Y., You, S., 2014. Water soluble sulfated-fucans with immune-enhancing properties from *Ecklonia cava*. *International journal of biological macromolecules*. 67, 303-311.

[75] Yang, L., Wang, P., Wang, H., Li, Q., Teng, H., Liu, Z., Yang, W., Hou, L., Zou, X., 2013. Fucoidan derived from *Undaria pinnatifida* induces apoptosis in human hepatocellular carcinoma SMMC-7721 cells via the ROS-mediated mitochondrial pathway. *Marine drugs*. 11, 1961-1976.

[76] Vishchuk, O.S., Ermakova, S.P., Zvyagintseva, T.N., 2013. The effect of sulfated (1→3)-alpha-l-fucan from the brown alga *Saccharina cichorioides* Miyabe on resveratrol-induced apoptosis in colon carcinoma Cells. *Marine drugs*. 11, 194-212.

[77] Cho, M.L., Lee, B.Y., You, S.G., 2010. Relationship between oversulfation and conformation of low and high molecular weight fucoidans and evaluation of their *in vitro* anticancer activity. *Molecules*. 16, 291-297.

[78] Yang, C., Chung, D., Shin, I.S., Lee, H., Kim, J., Lee, Y., You, S., 2008. Effects of molecular weight and hydrolysis conditions on anticancer activity of fucoidans from sporophyll of *Undaria pinnatifida*. *International journal of biological macromolecules*. 43, 433-437.

[79] Vishchuk, O.S., Tarbeeva, D.V., Ermakova, S.P., Zvyagintseva, T.N., 2012. Structural characteristics and biological activity of Fucoidans from the brown algae *Alaria sp.* and *Saccharina japonica* of different reproductive status. *Chemistry and biodiversity*. 9, 817-828.

[80] Wang, W., Wang, S.X., Guan, H.S., 2012. The antiviral activities and mechanisms of marine polysaccharides: an overview. *Marine drugs*. 10, 2795-2816.

[81] Mandal, P., Mateu, C.G., Chattopadhyay, K., Pujol, C.A., Damonte, E.B., Ray, B., 2007. Structural features and antiviral activity of sulphated fucans from the brown seaweed *Cystoseira indica*. *Antiviral chemistry and chemotherapy*. 18, 153-162.

[82] Ghosh, P., Adhikari, U., Ghosal, P.K., Pujol, C.A., Carlucci, M.a.J., Damonte, E.B., Ray, B., 2004. *In vitro* anti-herpetic activity of sulfated polysaccharide fractions from *Caulerpa racemosa*. *Phytochemistry*. 65, 3151-3157.

[83] Prokofjeva, M.M., Imbs, T.I., Shevchenko, N.M., Spirin, P.V., Horn, S., Fehse, B., Zvyagintseva, T.N., Prassolov, V.S., 2013. Fucoidans as potential inhibitors of HIV-1. *Marine drugs*. 11, 3000-3014.

[84] Rabanal, M., Ponce, N.M., Navarro, D.A., Gómez, R.M., Stortz, C.A., 2014. The system of fucoidans from the brown seaweed *Dictyota dichotoma*: chemical analysis and antiviral activity. *Carbohydrate polymers*. 101, 804-811.

[85] Jang, J.Y., Moon, S.Y., Joo, H.G., 2014. Differential effects of fucoidans with low and high molecular weight on the viability and function of spleen cells. *Food and chemical toxicology: an international journal published for the British Industrial Biological Research Association*. 68, 234-238.

[86] Kim, M.J., Jeon, J., Lee, J.S., 2014. Fucoidan prevents high-fat diet-induced obesity in animals by suppression of fat accumulation. *Phytotherapy research : PTR*. 28, 137-143.

[87] Huang, L., Wen, K., Gao, X., Liu, Y., 2010. Hypolipidemic effect of fucoidan from *Laminaria japonica* in hyperlipidemic rats. *Pharmaceutical biology*. 48, 422-426.

[88] Pan, H., Guo, J., Su, Z., 2014. Advances in understanding the interrelations between leptin resistance and obesity. *Physiology and Behaviour*. 130, 157-169.

[89] Lordan, S., Ross, R.P., Stanton, C., 2011. Marine bioactives as functional food ingredients: potential to reduce the incidence of chronic diseases. *Marine drugs*. 9, 1056-1100.

[90] Hernández-Corona, D.M., Martínez-Abundis, E., Gónzalez-Ortiz, M., 2014. Effect of Fucoidan Administration on Insulin Secretion and Insulin Resistance in Overweight or Obese Adults. *Journal of medicinal food*. 17, 830-832.

[91] Kim, K.T., Rioux, L.E., Turgeon, S.L., 2014. Alpha-amylase and alpha-glucosidase inhibition is differentially modulated by fucoidan obtained from *Fucus vesiculosus* and *Ascophyllum nodosum*. *Phytochemistry*. 98, 27-33.

[92] Cui, W., Zheng, Y., Zhang, Q., Wang, J., Wang, L., Yang, W., Guo, C., Gao, W., Wang, X., Luo, D., 2014. Low-molecular-weight fucoidan protects endothelial function and ameliorates basal hypertension in diabetic Goto-Kakizaki rats. *Laboratory investigation; a journal of technical methods and pathology*. 94, 382-393.

[93] Saura Calixto, F., García Alonso, A, Goñi, I., Bravo, L., 2000. *In Vitro* Determination of the Indigestible Fraction in Foods: An Alternative to Dietary Fiber Analysis. *Journal of agricultural and food chemistry*. 48, 3342-3347.

[94] Rupérez, P., Toledano, G., 2003. Indigestible fraction of edible marine seaweeds. *Journal of the Science of Food and Agriculture*. 83, 1267-1272.

[95] O'Sullivan, L., Murphy, B., McLoughlin, P., Duggan, P., Lawlor, P.G., Hughes, H., Gardiner, G.E., 2010. Prebiotics from Marine Macroalgae for Human and Animal Health Applications. *Marine drugs*. 8, 2038-2064.

[96] Gibson, G.R., Roberfroid, M.B., 1995. Dietary Modulation of the Human Colonic Microbiota: Introducing the Concept of Prebiotics. *The Journal of Nutrition*. 125, 1401-1412.

[97] Roberfroid, M., Gibson, G.R., Hoyles, L., McCartney, A.L., Rastall, R., Rowland, I., Wolvers, D., Watzl, B., Szajewska, H., Stahl, B., Guarner, F., Respondek, F., Whelan, K., Coxam, V., Davicco, M.-J., Léotoing, L., Wittrant, Y., Delzenne, N.M., Cani, P.D., Neyrinck, A.M., Meheust, A., 2010. Prebiotic effects: metabolic and health benefits. *British Journal of Nutrition*. 104, S1-S63.

[98] Michel, C., Lahaye, M., Bonnet, C., Mabeau, S., Barry, J.L., 1996. *In vitro* fermentation by human faecal bacteria of total and purified dietary fibres from brown seaweeds. *The British journal of nutrition*. 75, 263-280.

[99] McDonnell, P., Figat, S., O'Doherty, J.V., 2010. The effect of dietary laminarin and fucoidan in the diet of the weanling piglet on performance, selected faecal microbial populations and volatile fatty acid concentrations. *Animal*. 4, 579-585.

[100] Yanase, Y., Hiragun, T., Uchida, K., Ishii, K., Oomizu, S., Suzuki, H., Mihara, S., Iwamoto, K., Matsuo, H., Onishi, N., Kameyoshi, Y., Hide, M., 2009. Peritoneal injection of fucoidan suppresses the increase of plasma IgE induced by OVA-sensitization. *Biochemical and biophysical research communications*. 387, 435-439.
[101] Albuquerque, I.R., Cordeiro, S.L., Gomes, D.L., Dreyfuss, J.L., Filgueira, L.G., Leite, E.L., Nader, H.B., Rocha, H.A., 2013. Evaluation of anti-nociceptive and anti-inflammatory activities of a heterofucan from *Dictyota menstrualis*. *Marine drugs*. 11, 2722-2740.
[102] Wang, Q., Kuang, H., Su, Y., Sun, Y., Feng, J., Guo, R., Chan, K., 2013. Naturally derived anti-inflammatory compounds from Chinese medicinal plants. *Journal of ethnopharmacology*. 146, 9-39.
[103] Cardoso, M.L., Xavier, C.A., Bezerra, M.B., Paiva, A.O., Carvalho, M.F., Benevides, N.M., Rocha, F.A., Leite, E.L., 2010. Assessment of zymosan-induced leukocyte influx in a rat model using sulfated polysaccharides. *Planta medica*. 76, 113-119.
[104] Paiva, A.A., Castro, A.J., Nascimento, M.S., Will, L.S., Santos, N.D., Araujo, R.M., Xavier, C.A., Rocha, F.A., Leite, E.L., 2011. Antioxidant and anti-inflammatory effect of polysaccharides from *Lobophora variegata* on zymosan-induced arthritis in rats. *International immunopharmacology*. 11, 1241-1250.
[105] Raghavendran, H.R.B., Sathivel, A., Rekha, S., *Chapter 12 - Gastric and hepatic protective effects of algal components*, in *Functional Ingredients from Algae for Foods and Nutraceuticals*, Domínguez, H., Editor. 2013, Woodhead Publishing. p. 416-452.
[106] Sharma, G., Kar, S., Basu Ball, W., Ghosh, K., Das, P.K., 2014. The curative effect of fucoidan on visceral leishmaniasis is mediated by activation of MAP kinases through specific protein kinase C isoforms. *Cellular and molecular immunology*. 11, 263-274.
[107] Kuznetsova, T.A., Besednova, N.N., Somova, L.M., Plekhova, N.G., 2014. Fucoidan extracted from *Fucus evanescens* prevents endotoxin-induced damage in a mouse model of endotoxemia. *Marine drugs*. 12, 886-898.
[108] Pereira, J., Portron, S., Dizier, B., Vinatier, C., Masson, M., Sourice, S., Galy-Fauroux, I., Corre, P., Weiss, P., Fischer, A.M., Guicheux, J., Helley, D., 2014. The *in vitro* and *in vivo* effects of a low-molecular-weight fucoidan on the osteogenic capacity of human adipose-derived stromal cells. *Tissue engineering. Part A*. 20, 275-284.
[109] Jeong, H.-S., Venkatesan, J., Kim, S.-K., 2013. Hydroxyapatite-fucoidan nanocomposites for bone tissue engineering. *International journal of biological macromolecules*. 57, 138-141.
[110] Cai, J., Kim, T.S., Jang, J.Y., Kim, J., Shin, K., Lee, S.P., Choi, E.K., Kim, S.H., Park, M., Kim, J.B., Kim, Y.B., 2014. *In vitro* and *in vivo* anti-*Helicobacter pylori* activities of FEMY-R7 composed of fucoidan and evening primrose extract. *Laboratory animal research*. 30, 28-34.
[111] Gao, Y., Li, C., Yin, J., Shen, J., Wang, H., Wu, Y., Jin, H., 2012. Fucoidan, a sulfated polysaccharide from brown algae, improves cognitive impairment induced by infusion of Aβ peptide in rats. *Environmental Toxicology and Pharmacology*. 33, 304-311.
[112] Veena, C.K., Josephine, A., Preetha, S.P., Varalakshmi, P., Sundarapandiyan, R., 2006. Renal peroxidative changes mediated by oxalate: The protective role of fucoidan. *Life Sciences*. 79, 1789-1795.

[113] Wang, J., Wang, F., Yun, H., Zhang, H., Zhang, Q., 2012. Effect and mechanism of fucoidan derivatives from *Laminaria japonica* in experimental adenine-induced chronic kidney disease. *Journal of ethnopharmacology*. 139, 807-813.

[114] Mayakrishnan, V., Kannappan, P., Abdullah, N., Ahmed, A.B.A., 2013. Cardioprotective activity of polysaccharides derived from marine algae: An overview. *Trends in Food Science and Technology*. 30, 98-104.

[115] Krylov, V.B., Kaskova, Z.M., Vinnitskiy, D.Z., Ustyuzhanina, N.E., Grachev, A.A., Chizhov, A.O., Nifantiev, N.E., 2011. Acid-promoted synthesis of per-O-sulfated fucooligosaccharides related to fucoidan fragments. *Carbohydrate research*. 346, 540-550.

In: Sulfated Polysaccharides
Editors: M. Gama, H. Nader, H. de Oliveira Rocha
ISBN: 978-1-63482-975-5

Chapter 8

CARRAGEENANS

Cinthia Beatrice Silva Telles, Moacir Fernandes Queiroz, Jailma Almeida-Lima and Hugo Alexandre Oliveira Rocha*
Laboratório de Biotecnologia de Polímeros Naturais (BIOPOL), Departamento de Bioquímica, Universidade Federal do Rio Grande do Norte (UFRN), Natal/RN, Brazil

ABSTRACT

Carrageenans are linear, sulfated polysaccharides extracted from red algae and containing repeated disaccharide units composed of β-(1→3)-D-galactopyranose and α-(1→4)-D-galactopyranose or α-(1→4)3,6-anhydro-D-galactopyranose. Carrageenans have a broad rheological spectrum, rangingfrom a viscous thickener to thermally reversible gels, which may be elastic or brittle. Theinternational market for carrageenansis estimated at $450million per year, with applications mainly in the food industry, and cosmetic and pharmaceutical preparations. However, several studies show that carrageenanshave antioxidant, anti-inflammatory, antitumor, immunomodulatory, and anticoagulant activities. In this chapter, we will describe the various properties of this polysaccharide.

Keywords: sulfated polysaccharides; chemical structure; biological activities; red seaweed.

1. INTRODUCTION

The earliest records of the use of macroalgae by humans date from the 4th century in Japan, and the 6th century in China, and in both cases, algae were cited as a food source. Seaweed has high nutritional value, being rich in carotenoids, proteins, dietary fibers, essential fatty acids, vitamins (B-complex and C, D, E, and K), and minerals. Among theapproximately 60 different known elements present in algae, varying concentrations of calcium, phosphorus, sodium, magnesium, iron, copper, manganese, potassium, vanadium,

* Email: hugo-alexandre@uol.com.br.

and iodine (present in significant amounts only in brown algae) have been reported. Algae are used as a food source mainly in Asian countries, where they constitute approximately 25% of the human diet in countries such as Japan, China, and South Korea.

Other important components of algae are the hydrocolloids, which have applications in several human activities. The most abundant hydrocolloids are the fucans/fucoidans and alginates extracted from brown algae and the polysaccharides rich in galactose (agarans and carrageenan), extracted from red algae. It is estimated that 55,000 tons of algal hydrocolloids are produced world wide, from approximately 1 million tons of fresh seaweed.

Carrageenans are sulfated polysaccharides extracted from red algae. Known since the 19th century when they were extracted from the red seaweed *Chondruscrispus,* they were used by the population of the Irish city of "Carraghen" as an emulsifier and gelling agent in homemade food; hence the origin of the name carrageenan [1].

The pure carrageenan is extremely unstable and difficult to obtain; it can be stabilized through binding to one or more cations to form various carrageenan salts, termed the "carrageenans" [2]. Structurally, carrageenans are linear polysaccharides composed of repetitive disaccharide units of β-(1→3)-D-galactopyranose and α-(1→4)-D-galactopyranose or α-(1→4)-3,6-anhydro-D-galactopyranose (Figure 1) [3]. However, as with other sulfated polysaccharides, varieties of this basic structure can be found, and at least 15 different carrageenans were identified [4].

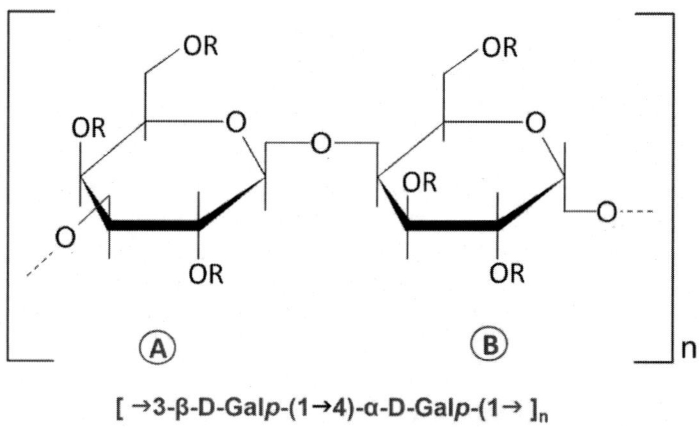

Figure 1. Schematic representation of the repeating β-(1→3)-D-galactopyranose and α-(1→4)-D-galactopyranose unit of carrageenan.

The carrageenans international market is estimated to represent $450million per year [5] and the main families of red algae used inindustrial production of carrageenans are *Gigartinaceae*, *Phyllphoraceae,* and *Solieriaceae* [6]. Though the carrageenans lack nutritional value, they are used inthe food industry as thickening, gelling, and stabilizer agents, as well as in various non-food products, such as pharmaceuticals, cosmetics, textile, and printing [7, 8]. Moreover, in recent decades, studies have shown that in addition to the physicochemical properties, these biopolymers exhibit pharmacological potential as well, especially antioxidant [9], antitumor [10], antiviral [11,12], anticoagulant [13,14], and anti-inflammatory [15] activities. The potential application of carrageenans in products such as drugs may further increase their market value.

2. CARRAGEENANS

2.1. Structure, Classification, and Nomenclature of Carrageenans

Carrageenans are not unique biopolymers, but instead comprise a mixture of water-soluble, sulfated galactans. Specifically, they are composed of alternating units of β-(1→3)-D-galactopyranose (A or G unit) and α-(1→4)-D-galactopyranose (B or D unit) or α-(1→4)-3,6-anhydro-D-galactopyranose (B or DA unit), linked together by a glycosidic bond. The average molecular weight of carrageenans is > 100 kDa, and the sulfate content ranges from 15% to 40% [3].

Carrageenans are classified on the basis of the number and position of sulfate ester groups and content of 3,6-anhydro-α-D-galactopyranose [16] (Table 1). The most common types are traditionally identified by a Greek prefix and classified into four different families: *kappa, lambda, beta*, and *omega.*

I - Family kappa (κ): sulfate groups at C-4;
II - Family Lambda (λ): sulfate groups at C-2;
III - Family Omega (ω): sulfate groups at C-6;
IV - Family Beta (β): have no sulfate groups in the Aunit.

Carrageenans belonging to the *kappa* family contain the A unit 4-*O*-sulfated, while the B unit may vary: 3,6-anhydro-α-D-galactopyranose*(kappa*-carrageenan); 3,6-anhydro-α-D-galactopyranose-2-sulfate (*iota*-carrageenan*)*;α-D-galactopyranose-6-sulfate*(mu*-carrageenan) or α-D-galactopyranose 2,6-disulfate*(nu*-carrageenan). The latter two are biological precursors of *kappa*- and *iota*-carrageenan, respectively. Carrageenans from the *lambda* family are 2-*O*-sulfated in the A unit, and the B unit may be α-D-galactopyranose-2,6-disulfate*(lambda*-carrageenan); α-D-galactopyranose-2-sulfate*(xi*-carrageenan) or 3,6-anhydro-α-D-galactopyranose-2-sulfate (*theta*-carrageenan). Carrageenans which do not have sulfate groups linked to galactopyranose in the A unit belong to the *beta* family, and the B unit may be 3,6-anhydro-α-D-galactopyranose (*beta*-carrageenan); 3,6-anhydro-α-D-galactopyranose-2-sulfate (*alpha*-carrageenan); α-D-galactopyranose-6-sulfate (*gamma*-carrageenan) or α-D-galactopyranose-2,6-disulfate (*delta*-carrageenan). Carrageenans from the *omega* family have both A and B units 6-*O*-sulfated (*psi*-carrageenan) or a 3,6-anhydro-α-D-galactopyranose (*omega*-carrageenan) in the B unit [17].

Since natural carrageenans are a mixture of non-homogeneous polysaccharides, and therefore have a hybrid structure, the term disaccharide-repeating unit refers to the idealized structure. That is why a description based solely on dyads seems to be an oversimplification. Knutsen et al. [17] used an alternative nomenclature using names and letters to describe more complex structures. In this system, the A unit, β-D-galactopyranose, is always identified by the letter, "G. "When the B unit isa α-galactopyranose, it is designated the letter, "D". The 3,6-anhydro-galactopyranose is referred as "DA". The position of the substituent is indicated by the carbon number to which it is connected, followed by "S" in the case of *O*-sulfate group; "M" in the case of *O*-methyl; and "P" in the case of a pyruvic acid.

All carrageenan fractions are water-soluble but insoluble in organic solvents. However, water solubility primarily depends on the number of sulfate groups (very hydrophilic) and their associated cations. The principal cations found in carrageenans are sodium, potassium, calcium, and magnesium, but other ions may also be present at a lower frequency [18].

Commercial carrageenans are typically categorized into 3 main groups:

- Iota (ι): carrageenan 2,4'-sulfate (G4S-DA2S), which form rigid gels in the presence of potassium ions;
- Kappa (κ): carrageenan 4'-sulfate (G4S-DA) form elastic gels in the presence of calcium ions;
- Lambda (λ): carrageenan 2,6,2'-trisulfate (G2S-D2S,6S) do not form gels and are typically used to thicken dairy products.

In addition to these major carrageenan types, two others, called mu (μ) and nu (ν) (G4S-D2S, 6S and G4S-D6S) are often found in commercial samples [19]. They are biological precursors of κ- and ι-carrageenan, respectively, which are formed in vivo by the action of a sulfohydrolase [20]. The κ-, ι-, and λ-carrageenan dimers have 1, 2, and 3 sulfate ester groups respectively (20%, 33%, and 41%), although variations may occur due to differences between algae species [21].

2.2. Carrageenan Isolation

In the original method used until the 1980s carrageenans were extracted using an aqueous solution; the seaweed residues were removed by filtration, and then the carrageenan was recovered from the solution by precipitation with potassium chloride or by with ethanol. This method yields refined or filtered carrageenans, but becomes laborious and expensive if used industrially [22].

The procedure of obtaining semi-refined carrageenans began around the '90s. In this process, the algae are treated with alkaline solutions, which cause cyclization of the precursor units and consequently an increase in carrageenan gel strength. Gel formation prevents dissolution of the carrageenan in solution, while soluble proteins, other salts and carbohydrates dissolve and are removed. The type of base selected will determine the particular type of carrageenan salt produced and has important consequences for the properties of the extract obtained, regarding dispersion, hydration, elasticity, and gel formation. Prolonged treatment with alkali promotes an internal rearrangement that modifies the polysaccharide backbone [23].

The residue is washed several times to remove the alkali and then dried. Despite the lower cost in the production of semi-refined carrageenans, they may contain cellulose resulting in turbid solutions. On the other hand, the refined carrageenans give clear solutions since the pulp is removed during the filtration process.

2.3. Properties of Carrageenan

Carrageenan as a free acid is unstable, so it is not extensively used or studied. However, commercial carrageenans, available as a stable form of sodium, potassium, and calcium salts, or more generally as a mixture of these, are stable, water soluble, and easily obtained, so the properties described for carrageenan normally comes from the characterization of its salts.

Table 1. Carrageenan classification according to the number and position of sulfate ester groups and the content of 3,6-anhydro-α-D-galactopyranose(Adapted from Knutsen et al., [17])

Carrageenan		3-linked (Unit A or G)	4-linked (Unit B or D/DA)	Nomenclature by KNUTSEN et al. [17]
KAPPA	κ (*kappa*)	β-D-Gal*p*-4-sulfate	3,6-anidrogal*p*	G4S – DA
	ι (*iota*)	β-D-Gal*p*-4-sulfate	3,6-anidrogal*p*-2-sulfate	G4S – DA2S
	μ (*mu*)	β-D-Gal*p*-4-sulfate	α-D-Gal*p*-6-sulfate	G4S – D6S
	ν (*nu*)	β-D-Gal*p*-4-sulfate	α-D-Gal*p*-2,6-disulfate	G4S – D2,6S
LAMBDA	λ (*lambda*)	β-D-Gal*p*-2-sulfate	α-D-Gal*p*-2,6-disulfate	G2S – D2,6S
	ξ (*xi*)	β-D-Gal*p*-2-sulfate	α-D-Gal*p*-2-sulfate	G2S – D2S
	θ (*theta*)	β-D-Gal*p*-2-sulfate	3,6-anidrogal*p*-2-sulfate	G2S – DA2S
BETA	β (*beta*)	β-D-Gal*p*	3,6-anidrogalp	G – DA
	α (*alfa*)	β-D-Gal*p*	3,6-anidrogalp-2-sulfate	G – DA2S
	γ (*gamma*)	β-D-Gal*p*	α-D-Gal*p*-6-sulfate	G – D6S
	δ (*delta*)	β-D-Gal*p*	α-D-Gal*p*-2,6-disulfate	G – D2,6S
OMEGA	ω (*omega*)	β-D-Gal*p*-6-sulfato	3,6-anidrogal*p*	G6S – DA
	ψ (*psi*)	β-D-Gal*p*-6-sulfato	α-D-Gal*p*-6-sulfato	G6S – D6S

Physicochemical properties of carrageenans are closely related to the amount of sulfate groups present in their structures, as well as their distribution. Other factors determining the physicochemical properties of carrageenans are the conformation of the monosaccharide units in the chain and cations associated with it [16]. For example, κ- and ι-carrageenans form gels in the presence of potassium or calcium ions, which is not observed for λ-carrageenan [24].

Carrageenans have several applications, depending largely on their rheological properties. Carrageenans are linear, water-soluble polymers. However, only λ-carrageenan and sodium salts of κ- and ι-carrageenans are soluble in cold water. The carrageenans form highly viscous aqueous solutions. This viscosity depends on the concentration, increasing exponentially with increasing concentration. On the other hand, increasing the temperature decreases viscosity. The presence of other solutes, carrageenan type, and molecular weight also influences the viscosity of the carrageenan in solution [25].

Under favorable conditions, some carrageenans such as κ- and ι- can form gels. It has been proposed that the polysaccharide chains of κ- and ι-carrageenan change over cooling from a random coil state into a helical structure (Figure 2). It is noteworthy that not all chains participate in the formation of the helical structure; only the repetitive regions of the molecule participate in this event. Therefore, these tertiary structures are limited to the region of the molecule with structural regularity. The double-helix is disrupted in areas with an irregular sequence, allowing the formation of a double helix with another strand - these are called joining regions or domains. Additional cooling leads to aggregation of junction points to form the 3-dimensional gel structure [26].

Factors such as the number, type, and position of the ester sulfate groups, as well as the presence of loops in the chain have important effects on the jellification. κ-Carrageenans are sensitive to potassium ions and produce hard and brittle gels in aqueous solutions containing this cation. κ-Carrageenan gel presents syneresis (spontaneous extrusion of water through the gel surface when it is at rest) and higher the potassium concentration in the solution, the greater is the syneresis. On the other hand, the ι-carrageenans produce soft and elastic gel in an aqueous solution with calcium salts.

The gel obtained is thermos-reversible and can be subjected to heating and cooling cycles without substantial change in the gel structure (at neutral pH). The gelation and fusion temperatures of the salt/gel depend on the concentration of cations and it has been shown that the increase in the concentration of potassium or calcium salts in aqueous solutions result in increased gelation temperature. However, exposure of gels to high temperatures (above 80°C) affects the stability of the gels [27].

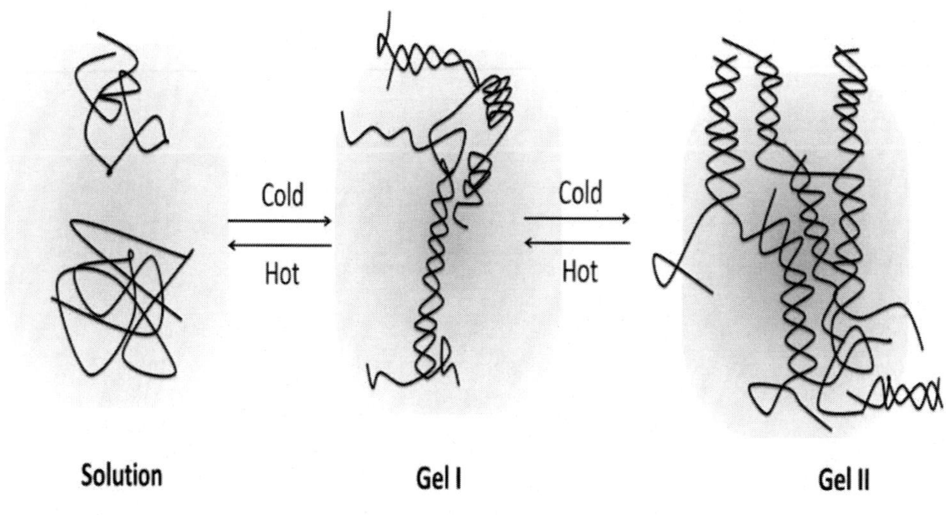

Figure 2. Schematic representation of the gelling carrageenan.

Another important feature of carrageenan is that they can be combined with some proteins, especially milk casein, producing stronger gels. This is the reason why carrageenans are used on a large scale in the dairy industry [28].

2.4. Biological Activities

Carrageenans, due to their abundance, glycidic nature, different chemical structure, and physicochemical properties, have been studied to identify potential end uses in medical and biological sciences [29]. The main outcomes of these studies are described below.

2.4.1. Anticoagulant Activity

The ability to increase time to clot formation was one of the first carrageenans activities evaluated. Several studies have shown that these polymers exhibit anticoagulant activity [30-34], and a recent review states that anticoagulant carrageenans are 15 times less potent than unfractionated heparin. This would make them safer drugs to be administered in patients [29].

In general, the data indicate that among the commercial carrageenans, λ-carrageenans have a greater ability to inhibit blood clotting than κ- and ι-carrageenans. It is stated by several authors that this is due to the greater sulfate content present in λ-carrageenans (about 30%) as compared to other carrageenans (about 20%) [33, 32, 35, 36]. This is corroborated by data presented by Opoku and collaborators [37], which showed that chemically sulfatedκ-carrageenan has higher anticoagulant activity than the native κ-carrageenan. In addition to sulfate, molecular weight and distribution of sugar residues are also suggested to be important characteristics for a carrageenan todisplay anticoagulant activity [29,38,39].

Although some studies have shown that carrageenans act as inhibitors of platelet aggregation [30, 31], the main anticoagulant mechanism of carrageenans is thrombin inhibition. The data indicate that, similar to various heparins, carrageenans associate with antithrombin and thereby promote the inhibition of thrombin [31,40]. However, no studies have identified the structural features necessary for a carrageenan to interact with antithrombin.

Studies have shown that λ-carrageenan also have other anticoagulant mechanisms. It is able to associate with heparin cofactor II and to directly inhibit thrombin, which explains why λ-carrageenans are the most potent anticoagulants [16,41-43].

2.4.2. Antiviral Activity

Together with anticoagulant activity, antiviral capacity is one of the major biological activities of carrageenans (Table 2). These polysaccharides are recognized as selective inhibitors of various enveloped viruses such as retroviruses (HIV-1 and 2), herpes simplex virus (HSV-1 and 2), human cytomegalovirus (HCMV), flaviviruses (dengue virus type 2), hepadnaviruses (hepatitis B virus, HBV), and orthomyxovirus (influenza A virus, IAV), among others [44].

Several negatively-charged compounds have been noted to interfere with the replication of HSV [45-47]. To infect a cell, HSV depends on the interaction of its receptors, present in the viral envelope with molecules present on the cell surface, called Herpes Virus Entry Mediators (HVEM). Some of these viral envelope receptors are glycoproteins called gB, gC, and gD. These glycoproteins are involved in the initial attachment of the virus to the host cell, since they are able to recognize and bind to heparan sulfate proteoglycan molecules and other HVEM receptors, allowing adhesion and subsequent invasion of the cell [48]. Therefore, it is proposed that carrageenans, due to their anionic character, may be able to interact with viral glycoproteins or with molecules of the host cell surface and, thereby, prevent viral attachment to the host cell.

Other studies provide more evidence of the antiherpetic activity of carrageenans. Carlucci and collaborators [49] in a study performed with κ/ι-, λ-carrageenans and the partially cyclized μ/ν-carrageenans from the seaweed, *Gigartinas kottsbergii* noted that all polysaccharides were potent anti-HSV agents, since they prevented the entry of the virus into the host cell. The most potent molecules were the λ- and μ/ν-carrageenans. These data led to the proposition that antiherpetic activity was directly correlated to the amount of α-D-galactose 2,6-disulfate residues in carrageenans. The presence of 3,6-anhydro-α–D-galactose and 3,6-anhydro-α–D-galactose-2-sulfate residues decreased the antiherpetic activity of these polymers. A later study showed that only λ-carrageenan from *G. skottsbergii* had antiviral activity. In this case, it was proposed that the structure of the carrageenan facilitated interaction with the virus, probably with gB andgC, leading to inactivation [50].

Table 2. Antiviral and anticoagulant activity of carrageenans

	Sources	Carrageenans	References
Antiviral activity	*Gigartina skottsbergii*	Lambda, κ-/ι- e ν-/μ carrageenans	[49,50]
		Lambda carrageenans	[53]
		Lambda-carrageenans	[54]
	Schizymenia pacifica	Lambda-carrageenans	[55]
	Stenograme interrupta	Carrageenans	[56]
	Meristiella gelidium	Iota, lamba and kappa-carrageenan	[12,57]
	Commercial	Kappa- carrageenan	[58]
		Iota-Carrageenan	[59]
		Iota-Carrageenan	[60]
		Carrageenan	[51]
		Iota-, lambda- and kappa-carrageenans	[61]
Anticoagulant activity	*Gigartina skottsbergii*	Lambda-carrageenans	[49]
	Phyllophora brodiaei	Lambda- carrageenans	[35]
	Solieria filiformis	Iota-carrageenans	[62]
	Commercial	Lambda andkappa-carrageenans	[14]
		Kappa-carragenana	[38]
		Lambda-carrageenan	[63]

Studies have reported that carrageenans are extremely potent inhibitors of infections caused by a wide variety of sexually transmitted human papilloma virus (HPVs), the etiologic agent responsible for all cases of uterine cancer. The mechanism of action of carrageenans against HPV is very similar to that described for HSV. Carrageenans, because of their resemblance to heparan sulfate, block virus attachment to the host cell surface since they are capable of binding to the viral capsid, and this effect can be 3 times more potent than that of exogenous heparan sulfate [51].

No studies to date correlate carrageenans structure and anti-HPV activity, but it has been reported that more sulfated varieties provide higher efficacy [51].

Studies on sexual lubricant gels containing carrageenans indicate that these may offer protection against HPV infection [52]. A sexual lubricant containing λ/κ-carrageenans has also been studied in a clinical trial to assess the topical microbicide activity against HIV infection in South African women. The results demonstrated that the experimental microbicide gel was safe for use, but did not provide protection against HIV. These results are consistent with *in vitro* studies, which showed that this carrageenan was only active against HIV at concentrations approximately 1000 times higher than that required to inhibit papilloma virus [51].

2.4.3. Antitumor Activity

Carrageenans have cytotoxic activity against a large number of tumor cell lines in culture [64-68]. Some studies using algae extracts have shown that they prevent metastases and this activity was attributed to carrageenans present in the extract. This is because carrageenans associate with the endothelial basal membrane and avoid the passage of cells. However, the mechanism of action is not fully understood [29].

Some articles report carrageenans having antitumor activity, but these carrageenans were chemically altered in order to modulate their properties. He and collaborators showed that κ-carrageenans covalently linked with selenium might inhibit the proliferation of breast tumor cells [69]. The κ-carrageenans super-sulfated oligosaccharides also showed antitumor activity *in vivo* [70]. Hu et al. [71], using κ-carrageenans extracted from the marine bacterium *Cytophaga* MCA-2, produced low molecular weight κ-carrageenan. These derivatives showed much greater tumor inhibition *in vivo* than the parent compounds. In addition, an immunomodulatory effect of these oligosaccharides was observed, which accentuated the antitumor activity.

Several studies have shown antitumor activity of carrageenans *in vivo*, but the authors could not demonstrate the mode of action clearly. It was not clearif this was a direct effect of carrageenan against the tumor, or if the antitumor action was the result of the immune-modulating effect of carrageenans, or if these two effects were additive.

2.4.4. Antioxidant Activity

Red marine algae live in an environment with constant salt and hydrolytic stress, besides being exposed to ultraviolet radiation and high concentrations of oxygen, which are factors that lead to the formation of free radicals and other oxidants. However, these algae do not show any form of severe oxidative damage, even when stored for long periods, indicating that they must have protective mechanisms mediated by enzymes and/or non-enzymatic antioxidants.

Carrageenans may be included in the class of antioxidant compounds found in red algae, since all tested compounds showed some antioxidant activity.

The data presented by Yuan et al., [72] indicated that carrageenans directly influence the formation of superoxide radicals, inhibiting its formation, and this action is positively correlated with the sulfate content in the carrageenans. Later, studies with κ- and ι-carrageenan extracted from the seaweeds, *Eucheuma spinosum* and *Eucheuma cottonii*, respectively, and λ-carrageenan obtained from the algae *Gigartina acicularis* and *Gigartina pistillata* showed that these also possessed strong activity against the sequestration of superoxide and hydroxyl radicals. It was also observed that there was a correlation between antioxidant activity and sulfate content of carrageenans [73].

Several carrageenans extracted from seaweed belonging to the families *Gigartinaceae* and *Tichocarpaceae* were evaluated as antioxidants, and once again, it was observed that carrageenans showed hydroxyl radical (moderate) and superoxide radical (strong) scavenging activity [74]. The data in the literature indicate thatthe sequestration of radicals seems to be the main antioxidant mechanism of carrageenans.

Chemically modification of carrageenans for improved antioxidant activity has also been the target of several studies. Yuan et al. [75] synthesized highly sulfated, phosphorylated, and acetylated κ-carrageenan derivatives and all derivatives showed higher antioxidant activity than native κ-carrageenan, suggesting that chemical modification may improve the antioxidant activity of carrageenan. Additionally, Mou, et al. [76] showed that sulfate oligosaccharides prepared from κ-carrageenans showed significant antioxidant activity.

2.4.5. Inflammatory Activity

Carrageenans are classically used as agents for the induction of experimental inflammation. Carrageenan-induced paw edema is a test widely used to determine anti-inflammatory activity and is a simple and routine animal model for pain assessment at the site of inflammation without any injury or damage to the inflamed paw [77].

λ-carrageenan does not form a gel at room temperature and therefore is the selected carrageenan to induce inflammatory response, which is acute, highly reproducible, and non-immune [78]. Signs of inflammation such as edema, erythema, and hyperalgesia develop immediately after subcutaneous injection as a result of the action of pro-inflammatory factors released by local or infiltrating cells after administration of the carrageenan. Such agents can be generated in situ at the site of injury or by infiltrating cells, such as neutrophils, which easily migrate to sites of inflammation and can generate reactive oxygen species. The development of paw edema after carrageenan administration has been described as age- and weight-dependent and several mediators operate in sequence to produce the inflammatory response. Several mediators are involved in inflammation, including histamine, serotonin, and bradykinin as the first mediators detected in the early phase and prostaglandins (PGs) involved in increased vascular permeability, detectable during the late phase of inflammation induced by carrageenan.

Local and/or systemic inflammation induced by carrageenan is associated with elevated levels of pro-inflammatory cytokines TNF-α, IL-1, and IL-6 [79]. Infiltration of neutrophils to the site of inflammation also contributes to the inflammatory response, with production of, amongst other mediators, oxygen-derived free radicals, such as superoxide anion (O2-) and hydroxyl radicals [80,81]. Furthermore, an increase in the production and release of nitric oxide (NO) at the injured site is also observed [16].

Carrageenans have also been used in studies on the intestinal inflammation process. Oral administration of fragmented carrageenan (carrageenan oligosaccharides), also known as poligeenan, for 20 to 45 days, in animals results in ulcerative lesions associated with clinical and pathological changes that, in some aspects, resemble ulcerative colitis in humans. Clinical observations include weight loss, loose stools, and hidden or visible blood in the stool. Ulcers can be found in all parts of the large intestine, with extensive lesions in the rectum. Microscopically, these similarities include local bleeding of mucosa, cellular infiltration, edema, and ulceration. The ulcerative lesions, however, seem to start in the cecum and extend distally toward the rectum [82]. This activity has been used in intestinal inflammation studies. It is worth mentioning that non-hydrolyzed carrageenans do not promote this effect, due to their high molecular weight, resistance to the pH of the digestive system, and nodegradation by intestinal microorganisms [83].

2.4 6. Anti-Lipidemic Activity

Some studies indicate carrageenans as agents capable of lowering the level of blood cholesterol. For example, Fischer 344 rats fed with diets containing 15% κ/λ-carrageenan derived from the red alga, *Gigartina radula* resulted in a significant reduction in blood cholesterol [84]. Studies with healthy humans who consumed food containing carrageenans as the main source of dietary fiber showed that there was a significant decrease of cholesterol and lipids in the blood. It was proposed that carrageenans show anti-lipidemic action by binding to bile acids and dietary lipids, such as cholesterol, lowering the absorption of those components. As a secondary mechanism, it has been proposed that the carrageenans increase the volume and viscosity of the intestinal contents, thus decreasing the digestion and absorption rate due to a delay in diffusion of digestive enzymes, substrates, and nutrients [85].

2.5. Applications of Carrageenan

Carrageenan, one of the most abundant natural polysaccharides, may be used in the food and the cosmetic industries as a gelling, stabilizing, and thickening agent due to its biocompatibility, biodegradability, high capacity for water-retention, and mechanical strength of the gel formed.

There are several applications of carrageenans in the food industry, such as a food additive in the production of a wide range of processed foods including dairy products, water-based foods, meat products, beverages, condiments, baby formula, and feed [86]. Carrageenan can function as a carrier, emulsifying agent, gelling agent, humectant, stabilizer, and thickener [87], and improve palatability and appearance by interacting with other substances in food [88]. Carrageenans can be used as a fat substitute in processed meats, as it improves moisture retention and restores the taste of processed meats such as burgers. In the case of dairy products, carrageenans increase thickness, texture, and solubility of the food product [86].

A newly explored use of carrageenan in food industry is that of a protective layer in fresh-cut fruit. Carrageenan coatings function as a barrier to gases, reducing the rate of 'breathing', which delays the discoloration of fruits [89]. Recent studies have shown that carrageenan can retain fruit texture beyond the expiration date, and moreover, provides antibacterial protection when used as acoating of edible fruit such as bananas [90] and mango [91].

Carrageenans are also used to produce antibiotics such as tetracycline and chlortetracycline, wherein κ-carrageenans are used to immobilize *Streptomyces aureofaciens* in an attempt to increase production of antibiotics.

Another interesting use of carrageenan is in the cleaning of industry effluent. A system developed by the co-immobilization of carrageenans with thebacterial species *Nitrosomonas europaea* and *Pseudomonas* spp., results in efficient nitrogen removal by leveraging the oxygen gradient within the trapping beads [92,93]. Furthermore, due to their acidic nature, carrageenans can bind to heavy metals and thus remove them from contaminated water. This property can also be harnessed for the development of drugs for elimination of metals from the body or targeted delivery of metal ions for treatment [94].

2.6. Carrageenan Toxicity

Tobacman [95] conducted an evaluation of the effects of carrageenan, with special attention to the occurrence of ulcers and cancers in association with exposure to carrageenan. Tobacman showed that exposure to both non-degraded as well as degraded carrageenan was associated with the occurrence of intestinal ulcers and cancer. This association can be attributed to the contamination of non-degraded carrageenan by low molecular-weight components produced by the spontaneous metabolism of carrageenans by acid hydrolysis under normal digestion conditions, or via interactions with intestinal bacteria.

However, other studies show that carrageenans are resistant to the pH of the gastrointestinal tract and are not degraded by the bacteria present in the gastrointestinal tract of humans, and are therefore, not absorbed, as reviewed by McKim [96]. In addition, more recent toxicological studies performed with different animal models,using chronic doses (800 times greater than that consumed daily by a human) and sub-chronic doses, showed that the use of carrageenans does not affect the immune system, and does not promote histopathological, clinical, biochemical, or hematological changes in these animals. There is also no evidence that carrageenans are genotoxic or mutagenic [83].

Conclusion

Carrageenans are a family of sulfated polysaccharides with hydrophilic properties, obtained from various species of red seaweed (Rhodophyta). Chemically, they are highly sulfated galactan polymers, with a strongly anionic character. Different carrageenans have been investigated for various applications, some of which are relevant to the food and cosmetics industries. Carrageenan obtained from different seaweed species were reported to have a potential and specific biological effects such as anticoagulant, antiviral, antitumor, antioxidant, and immunomodulatory activity. A greater understanding of the structure and analysis of chemical modifications allows the use of carrageenans and the application of its mechanisms against major diseases. Therefore, these polysaccharides may be useful in the development of low-cost, effective drugs in the future.

References

[1] Pereira, L.; Mesquita, J. F. 2004. Population studies and carrageenan properties of Chondracanthusteedei var. lusitanicus (Gigartinaceae, Rhodophyta). *J .Appl. Phycol.* 16, 369–383.

[2] Perez, R ; Kaas, R.; Campello, F.; Arbaut, S.; Barbaroux, O. 1992. La culture des algues marines dans le monde. Plouzane: Ifremer.

[3] Usov A. I. 1984. NMR spectroscopy of red seaweed polysaccharides: agars, carrageenans, and xylans. *Bot .Mar.* 27, 189–202,

[4] Lahaye M. 2001. Developments on gelling algal galactans, their structure and physico-chemistry. *J. Appl. Phycol.* 13, 173–84.

[5] Sahu, N.; Meena, R.; Ganesan, M. 2011. Effect of graftingon the properties of *kappa*-carrageenan of the red seaweed *Kappaphycusalvarezii* (Doty) Doty ex Silva. *Carbohydr. Polym.* 84, 584–92.

[6] Saito, R. M. ; Oliveira, E. C. 1990. Chemical screening of Brazilian marine algae producing carrageenans. *Hydrobiologia* 204-205, 585-588.

[7] Campo, V. L.; Kawano, D. F.; Silva, D. B. Carvalho, I. 2009. Carrageenans: Biological properties chemical modifications and structural analysis – a review. *Carbohydr. Polym.* 77, 167-180.

[8] Robledo, D.; Freile-Pelegrin, Y. 2010. Prospects for thecultivation of economically important carrageenophytes in Southeast Mexico. *J. Appl. Phycol.* 3, 127-134.

[9] Zhang, Z.; Zhang, Q.; Wang, J.; Song, H. ; Zhang, H.; Niu, X. 2010. Regioselective syntheses of sulfated porphyrans from *Porphyrahaitanensis* and their antioxidant and anticoagulant activities *in vitro*. *Carbohydr. Polym.* 79, 1124-1129.

[10] Zhou, G. F.; Sheng, W. X.; Yao, W. H.; Wang C. 2006. Effect of low molecular lambda-carrageenan from *Chondrusocellatus*on antitumor H-22 activity of 5-Fu. *Pharmacol. Res.* 53, 129-134.

[11] Duarte, M. E. R.; Cauduro, J. P.; Noseda, D. G.; Noseda, M. D.; Gonçalves, A. G.; Pujol, C. A.; Damonte, E. B. ; Cerezo A. S. 2004. The structure of the agaran sulfate from *Acanthophoraspicifera* (Rhodomelaceae, Ceramiales) and its antiviral activity. Relation between structure and antiviral activity in agarans. *Carbohydr. Res.* 339, 335-347.

[12] Talarico, L. B.; Damonte, E. B. 2007. Interference in dengue virus adsorption and uncoating by carrageenans. *Virology* 363, 473– 485.

[13] Farias, W. R. L.; Valente, A-P. ; Pereira, S. P.; Mourão P. A. S. 2000. Structure and anticoagulant activity of sulfated galactans. *J. Biol. Chem.* 275, 29299-29307.

[14] Silva, F. R. F; Dore, C. M. P. G.; Marques, C. T. ; Nascimento, M. S. ; Benevides, N. M. V.; Rocha, H. A. O. ; Chavante, S. F. ; Leite, E. L. 2010. Anticoagulant activity, paw edema and pleurisy induced carrageenan: Action of major types of commercial carrageenans. *Carbohydr. Polym.* 79, 26-33.

[15] Zacharopoulos, V. R.; Phillips, D. M. 1997. Vaginal formulations of carrageenan protect mice from herpes simplex virus infection. *Clin Diagn Lab Immunol.* 4, 465–468.

[16] Necas, J; Bartosikova, L. 2013. Carrageenan: a review. *Vet. Med. – Czech.* 58, 187–205.

[17] Knutsen, S. H.; Mylabodsky, D. E.; Larsen, B.; Usov, A. I. 1994. A modified system of nomenclature for red algal galactans. *Bot. Mar.* 37, 163-169.

[18] Pardonche, P. E. 1985. Aplicação de extratos de algas marinhas na indústria de alimentos. Boletim técnico CECA Produtos Químicos S/A (p. 15). São Paulo.

[19] Van De Velde, F. , Peppelman, H. A. , Rollema, H. S. , & Tromp, R. H. 2001. On the structure of kappa/iota-hybrid carrageenans. *Carbohydr. Res.* 331, 271–283.

[20] De Ruiter, G. A., Richard, O., Rudolph, B., Genicot, S., Kloareg, B., Penninkhof, B., Richard, O. 2001. Sulfohydrolases, corresponding amino acid and nucleotide sequences, sulfohydrolase preparations, processes, and products thereof. World Patent application: WO0068395.

[21] De Ruiter, G. A. ; Rudolph, B. 1997. Carrageenan biotechnology. *Trends Food Sci. Technol.* 8, 389–395.

[22] Stanley, N. 1987. Production and utilization of products from commercial seaweeds. In D. J. McHugh (Ed.), *FAO fisheries technical paper* (Vol. 288) (pp. 116–146). Rome: FAO.

[23] Imeson, A. 2011. *Food Stabilisers, Thickeners and Gelling Agents*, John Wiley & Sons, ISBN 978-1-4443-6033-2.

[24] Michel, A-S; Mestdagh, M. M. ; Axelos, M. A. V. 1997. Physico-chemical properties of carrageenan gels in presence of various cations. *International Journal of Biological Macromolecules* 21, 195–200.

[25] Lai, V. M. F.; Wong, P. A-L.; Lii, C-Y. 2000. Effects of cation properties on sol-gel transition and gel properties of κ-carrageenan. *J. Food Sci.* 65, 1332–1337.

[26] Rees, D. A. 1977. *Polysaccharide shapes.* Ch. 4 in Polysaccharide Shapes, p. 62-73. Chapman and Hall, London.

[27] Stanley, N. 2011. FAO Corporate document repository. Chapter 3: Production, properties and uses of carrageenan. FMC Corporation, Marine Colloids Division 5 Maple Street, Rockland Maine 04841, USA.

[28] Bobbio, P. A.; Bobbio, F. O. 2001. Química do processamento e alimentos. 3.ª Edição. Varela Livraria. 143p.

[29] Prajapati, V. D.; Maheriya, P. M.; Jani, G. K.; Solanki, H. K. 2014. Carrageenan: A natural seaweed polysaccharide and its applications. *CarbohydrPolym.* 105, 97-112.

[30] Hawkins, W. W.; Leonard, V. G. 1963. Antithrombic Activity of carrageenin in Human Blood. *Can. J. Biochem. Physiol.* 41, 1325.

[31] Kindness G.; Long W. F.; Williamson, F. B.; Boyd J. 1979. Effects of carrageenans on the aggregation of human blood platelets. *Thromb Res.* 15, 3–15.

[32] Shanmugam, M.; Mody, K. H. 2000. Heparinoid-active sulphated polysaccharides from marine algae as potential blood anticoagulant agents. *Curr. Sci.* 79, 1672-1683.

[33] Silva, F. R. F.; Dore, C. M. P. G.; Marques, C. T. ; Nascimento, M. S.; Benevides, N. M. B.; Rocha, H. A. O. ; Chavante, S. F.; Leite, E. L. 2010. Anticoagulant activity, paw edema and pleurisy induced carrageenan: action of major types of commercial carrageenans. *Carbohydr. Polym.* 79, 26-33.

[34] Sebaaly C., Karaki N., Chahine N., Evidente A., Yassine A., Habib J., Kanaan H. 2012. Polysaccharides of the red algae "Pterocladia" growing on the Lebanese coast: Isolation, structural features with antioxidant and anticoagulant activities. *J. App. Pharm. Sci.* 2, 1-10.

[35] Sen, A. K., Das, A. K., Banerji, N., Siddhanta, A. K., Mody, K. H., Ramavat, B. K.; Chauhan, V. D.; Vedasiromoni, J. R.; Ganguly, D. K. 1994. A new sulfated polysaccharide with potent blood anti-coagulant activity from the red seaweed *GrateloupiaIndica. Int. J. Biol. Macromol.* 16(5), 279-280.

[36] Efimov, V. S.; Usov, A. L.; Ol'skaya, T. S. ; Baliunis, A. Y.; Rozlin, M. Y. 1983. Comparative study of anticoagulant activity of sulphate polysaccharides obtained from red sea algae. *Farmakol. Toksikol.* 46, 61-67.

[37] Opoku, G.; Qiu, X.; Doctor, V. 2006. Effect of oversulfation on the chemical and biological properties of kappa carrageenan. *Carbohydr. Polym.* 65, 134–138.

[38] Nishino T., Aizu, Y.; Nagumo, T. 1991. Antithrombin activity of a fucan sulfate from the brown seaweed Eckloniakurome. *Thromb Res.* 62, 765-774.

[39] Nishino T., Aizu, Y.; Nagumo, T. 1991. The relationship between the molecular weight and the anticoagulant activity of two types of fucan sulfates from the brown seaweed Eckloniakurome. *Agric. Biol. Chem.* 55, 791-796.

[40] Kindness, G.; Long, W. F.; Williamson, F. B. 1979. Enhancement of antithrombin III activity by carrageenans. *Thromb Res.* 15, 49–60.

[41] Kindness, G.; Long, W. F.; Williamson, F. B. 1980. Inhibition by antithrombin III of carrageenan and xilan SP 54-induced aggregation of human blood platelets. *Biochem. Soc. Trans.* 8, 84-85.

[42] Kindness, G., Williamson, F. B., Long, W. F. 1979. Effect of polyanetholesulphonic acid and xylansulphate on antithrombin III activity. *Biochem Biophys Res Commun.* 13, 1062-1068.

[43] Wunderwald, P.; Schrenk, W. J ; Port, H. 1979. Antithrombin BM from human plasma: an antithrombin binding moderately to heparin. *Thromb Res.* 15, 49–60.

[44] Baba, M.; Snoeck, R; Pauwels, R.; de Clercq, E. 1988. Sulfated polysaccharides are potent and selective inhibitors of various enveloped viruses, including herpes simplex virus, cytomegalovirus, vesicular stomatitis virus, and human immunodeficiency virus. *Antimicrob Agents Chemother.* 32, 1742–1745.

[45] Gerber, P.; Dutcher, J. D.; Adams, E. V.; Sherman, J. H. 1958. Protective effect of seaweed extracts for chicken embryos infected with influenza B or mumps virus. *Proc. Soc. Exp. Biol. Med.* 99, 590–593.

[46] Ehresmann, D. W. Deig, E. F.; Hatch, M. T.; Disalvo, L. H.; Vedros, N. A. 1977. Antiviral substances from California marine algae. *J. Phycol.* 13, 37–40.

[47] Richards, J. T., Kern, E. R.; Glasgow, L. A.; Jr., J. C. O.; Deign, E. F.; Hatch, M. T. 1978. Antiviral activity of extracts from marine algae. *Antimicrob Agents Chemother.* 14, 24–30.

[48] Herold, B. C.; Visalli, R. J.; Susmarski, N.; Brandt, C. R.; Spear, P. G. 1994. Glycoprotein C-independent binding of herpes simplex virus to cells requires cell surface heparansulphate and glycoprotein B. *J. Gen. Virol.* 75, 1211-1222.

[49] Carlucci, M. J.; Pujol, C. A.; Ciancia, M.; Noseda, M. D; Matulewicz, M. C.; Damonte, E. B.; Cerezo, A. S. 1997. Antiherpetic and anticoagulant properties of carrageenans from the red seaweed Gigartinaskottsbergii and their cyclized derivatives: correlation between structure and biological activity. *Int. J. Biol. Macromol.* 20, 97-105.

[50] Carlucci, M. J.; Scolaro, L. A.; Damonte, E. B. 1999. Inhibitory action of natural carrageenans on herpes simplex virus infection of mouse astrocytes. *Chemotherapy* 45, 429–436.

[51] Buck, C. B.; Thompson, C. D; Roberts, J. N.; Müller, M.; Lowy, D. R.; Schiller, J. T. 2006. Carrageenan is a potent inhibitor of papillomavirus infection. *PLoSPathog.* 2, 69.

[52] Roberts, J. N.; Buck, C. B.; Thompson, C. D.; Kines, R.; Bernardo, M.; Choyke, P. L.; Lowy, D. R.; Schiller, J. T. 2007. Genital transmission of HPV in a mouse model is potentiated by nonoxynol-9 and inhibited by carrageenan. *Nature Med.* 13, 857–861.

[53] Carlucci, M. J.; Scolaro, L. A.; Noseda, M. D.; Cerezo, A. S. ; Damonte, E. B. 2004. Protective effect of a natural carrageenan on genital herpes simplex virus infection in mice. *Antiviral Res.* 64, 137–141.

[54] Diogo, J. V.; Novo, S. G.; González, M. J.; Ciancia, M.; Bratanich, A. C. 2015. Antiviral activity of lambda-carrageenan prepared from red seaweed (Gigartinaskottsbergii) against BoHV-1 and SuHV-1. *Res. Vet. Sci.* 98, 142-4.

[55] Nakashima, H.; Kido, Y.; Kobayashi, N.; Motoki, Y.; Neushul, M.; Yamamoto, N. 1987. Purification and chracterization of an avian myeloblastosis and human immunodeficiencyvirus reverse transcriptase inhibitor, sulfated polysaccharides extracted from sea algae. *Antimicrob. Agents Chemother.* 31, 1524-1528.

[56] Cáceres, P. J.; Carlucci, M. J.; Damonte, E. B.; Matsuhiro, B.; Zúñiga, E. A. 2000. Carrageenans from Chilean samples of Stenogramm- einterrupta (Phyllophoraceae): structural analysis and biological activity. *Phytochemistry* 53, 81–86.

[57] Talarico, L. B., Noseda, M. D., Ducatti, D. R., Duarte, M. E., Damonte, E. B. 2011. Differential inhibition of dengue virus infection in mammalian and mosquito cells by iota-carrageenan. *J. Gen. Virol.* 92, 1332–1342.

[58] Ya-Huang, C.; Yi-Lin, C.; Li-Wen, T.; Tsung-Lin, L.; Chang-Jer, W. 2012. Prevention of human enterovirus 71 infection by kappa carrageenan. *Antiviral Res.* 95, 128-134.

[59] Grassauer, A.; Weinmuellner, R.; Meier, C.; Pretsch, A.; Prieschl-Grassauer, E.; Unger, H. 2008. Iota-Carrageenan is a potent inhibitor of rhinovirus infection. *Virol. J.* 5,107.

[60] Leibbrandt, A.; Meier, C.; König-Schuster, M.; Weinmüllner, R.; Kalthoff, D.; Pflugfelder, B.; Graf, P.; Frank-Gehrke, B.; Beer, M.; Fazekas, T.; Unger, H.; Prieschl-Grassauer, E.; Grassauer, A. 2010. Iota-carrageenan is a potent inhibitor of influenza A virus infection. *PLoS One* 5, e14320.

[61] Girond, S.; Crance, J. M.; Van Cuyck-Gandre, H.; Renaudet, J.; Deloince, R. 1991. Antiviral activity of carrageenan on hepatitis A virus replication in cell culture. *Res. Virol.* 142, 261–270.

[62] Araújo, I. W. F.; Rodrigues, J. A. G.; Vanderlei, E. S. O.; Paula, G. A.; Lima, T. B.; Benevides, N. M. B. 2012. Iota-carrageenans from *Solieriafiliformis* (Rhodophyta) and their effects in the inflammation and coagulation. *Acta Sci.* 34, 127-135.

[63] Sokolova, E. V.; Byankina, A. O.; Kalitnik, A. A.; Kim, Y. H.; Bogdanovich, L. N.; Solov'eva, T. F.; Yermak, I. M. 2014. Influence of red algal sulfated polysaccharides on blood coagulation and platelets activation in vitro. *J. Biomed. Mater. Res. A.* 102, 1431-1438.

[64] Yuan, H.; Song, J.; Li, X.; Li, N.; Dai, J. 2004. Immunomodulation and antitumor activity of κ-carrageenan Oligosaccharides. *Pharmacol. Res.* 50, 47–53.

[65] Yuan, H.; Song, J.; Li, X.; Li, N.; Dai, J. 2006. Immunomodulation and antitumor activity of κ-carrageenan oligosaccharides. *Cancer Lett.* 243, 228–234.

[66] Zhou, G.; Sun, YP.; Xin X.; Zhang, Y.; Li, L.; Xu, Z. 2004: In vivo antitumor and immunomodulation activities of different molecular weight lambda-carrageenans from Chondrusocellatus. *Pharmacol. Res.* 50, 47–53.

[67] Zhou, G.; Sheng W.; Yao, W.; Wang, C. 2006. Effect of low molecular-carrageenan from Chondrusocellatus on antitumor H-22 activity of 5-Fu. *Pharmacol. Res.* 53, 129–134.

[68] Yuan, H.; Song, J. 2005. Preparation, structural characterization and in vitro antitumor activity of kappa-carrageenan oligosaccharide fraction from Kappaphycus striatum. *J. Appl. Phycol.* 17, 7–13.

[69] He, G.; Cheng, C.; Lu, R. 1997. Studies on biological effects of kappaselenocarrageenan on human breast cancer cell line BCaP-37. *Zhonghua Yu Fang Yi Xue Za Zhi.* 31, 103–106.

[70] Mou, H.; Jiang, X.; Guan, H. 2003. A kappa-carrageenan derived oligosaccharide prepared by enzymatic degradation containing anti-tumor activity. *J. Appl. Phycol.* 15, 297–303.

[71] Hu, X.; Jiang, X.; Aubree, E.; Boulenguer, P.; Critchley, A. T. 2006. Preparation and in vivo antitumor activity of k-carrageenan oligosaccharides. *Pharm. Biol.* 44, 646-650.

[72] Yuan, H.; Song, J.; Zhang, W.; Li, X.; Li, N.; Gao, X. 2006. Antioxidant activity and cytoprotective effect of kappa-carrageenan oligosaccharides and their different derivatives. *Bioorg. Med. Chem.* 16, 1329-1334.

[73] Rocha de Souza, M. C.; Teixeira Marques, C.; Guerra Doré, C. M.; Ferreira da Silva, F. R.; Oliveira Rocha, H. A. ; & Lisboa Leite, E. 2007. Antioxidant activities of sulfated polysaccharides from brown and red seaweeds. *J. Appl. Phycol.* 19, 153e160.

[74] Sokolova, E. V.; Barabanova, A. O.; Homenko, V. A.; Solov'eva, T. F.; Bogdanovich, R. N.; Yermak, I. M. 2011. In Vitro and Ex Vivo Studies of Antioxidant Activity of Carrageenans, Sulfated Polysaccharides from Red Algae. *Bull. Exp. Biol. Med.* 150, 398-401.

[75] Yuan, H.; Zhang, W.; Li, X.; Lü, X.; Li, N.; Gao, X.; Song, J. 2005. Preparation and in vitro antioxidant activity of kappa-carrageenan oligosaccharides and their oversulfated, acetylated, and phosphorylated derivatives. *Carbohydr. Res.* 340, 685-692.

[76] Mou, H.; Jiang, X.; Liu, Z. 2007. Method for manufacturing carrageenan-derived oligosaccharide sulfate with antioxidation activity. China Patent Office, Patent No. CN1986824.

[77] Sini, J. M.; Yaro, A. H.; Ayanwuyi, L. O.; Aiyelero, O. M.; Mallum, S. M.; Gamaniel, K. S. 2010. Antinociceptive and anti-inflammatory activities of the aqueous extract of the root bark of Combretumsericeum in rodents. *Afr. J. Biotechnol.* 9, 8872–8876.

[78] Adeyemi, O. O.; Okpo, S. O.; Ogunti, O. O. 2002. Analgesic and anti-inflammatory effects of the aqueous extract of leaves of Perseaamericana mill (lauraceae). *Fitoterapia* 73, 375-380.

[79] Cuzzocrea, S.; Sautebin, L.; De Sarro, G.; Costantino ,G.; Rombola, L.; Mazzon, E.; Lalenti, A.; De Sarro, A.; Ciliberto, G.; Di Rosa, M.; Caputi, A. P.; Thiemermann, C. H. 1999. Role of IL-6 in the pleurisy and lung injury caused by carrageenan. *J. Immunol.* 163, 5094–5104.

[80] Salvemini, D.; Wang, S-Q.; Wyatt, P. S.; Bourdon, D. M.; Marino, M. H.; Manning, P. T.; Currie, M. G. 1996. Nitric oxide: a key mediator in the early and late phase of carrageenan-induced rat paw inflammation. *Br. J. Pharmacol.* 118, 829–838.

[81] Posadas, I.; Bucci, M.; Roviezzo, F.; Rossi, A.; Parente, L.; Sautebin, L.; Cirino, G. 2004. Carrageenan-induced mouse paw oedema is biphasic, age-weight dependent and displays differential nitric oxide cyclooxygenase-2 expression. *Br. J. Pharmacol.* 142, 331–338.

[82] Watt, J.; Marcus, R. 1971. Carrageenan-induced ulceration of the large intestine in the guinea pig. *Gut* 12, 164–171.

[83] Weiner, M. L. 2014. Food additive carrageenan: Part II: A critical review of carrageenan in vivo safety studies. *Crit. Rev. Toxicol.* 44,244-269.

[84] Reddy, B. S.; Watanabe, K.; Sheinfil, A. 1980. Effect of dietary wheat bran, alfalfa, pectin and carrageenan on plasma cholesterol and fecal bile acid and neutral sterol excretion in rats. *J. Nutr.* 110, 1247–1254.

[85] Panlasigui, L. N.; Baello, O. Q.; Dimatangal, J. M.; Dumelod, D. B. 2003. Blood cholesterol and lipid-lowering effects of carrageenan on human volunteers. *Asia Pac. J. Clin. Nutr.* 12, 209–214.

[86] McHugh, D. J. 2003. A guide to the seaweed industry. FAO Fisheries Technical Paper-T441. Food and agriculture organization of the United Nations, Rome. Available online at http://www. fao. org/docrep/006/ y4765e/y4765e0a. htm

[87] Codex Alimentarius Commission. 2010. GSFA Online: Food Additive Details, Carrageenan (407). Available online at http://www. codexalimentarius. net/gsfaonline/additives/details. html?id=49

[88] Piculell, L. 2006. Chapter 8: GellingCarrageenans. In A. M. Stephen, G. O. Phillips, and P. A. Williams (Eds.) Food polysaccharides and their applications, 2nd edition pp. 239-288. CRC Press. Boca Raton, FL.

[89] Baeza, R. 2007. Comparison of technologies to control the physiological, biochemical and nutritional changes of fresh cut fruit. A report for completion of the Masters of Science degree. Kansas State University: Manhattan, Kansas. Available online at http://krex. k-state. edu/dspace/bitstream/2097/494/1/RitaBaeza2007. pdf.

[90] Bico, S. L. S.; Rapaso, M. F. J.; Morais, R. M. S. C.; Morais, A. M. M. B. 2009. Combined effects of chemical dip and/or carrageenan coating and/or controlled atmosphere on quality of fresh-cut banana. *Food Control.* 20, 508-514.

[91] Plotto, A.; Narciso, J.; Rattanapanone, N.; Baldwin, E. A. 2010. Surface treatments and coatings to maintain fresh-cut mango quality in storage. *J. Sci. Food Agric.* 90, 2333-2341.

[92] Dos Santos, V. A. P. M.; Bruijnse, M.; Tramper, J.; Wijffels, R. H. 1996. The magic-bead concept: An integrated approach to nitrogen removal with co-immobilized micro-organisms. *Appl. Microbiol. Biotechnol.* 45, 447–453.

[93] Dos Santos, V. A. P. M.; Tramper, J.; Wijffels, R. H. 1996. Dynamic modelling of an integrated nitrogen removal system using Co-immobilized microorganisms. *Biotechnol. Progr.* 11, 486–493.

[94] Carrageenans as a New Source of Drugs with Metal Binding Properties. Khotimchenko, Y. S.; Khozhaenko, E. V.; Khotimchenko, M. Y.; Kolenchenko, E. A.; Kovalev, V. V. 2010. *Mar. Drugs.* 8, 1106-1121.

[95] Tobacman, J. K. 2001. Review of harmful gastrointestinal effects of carrageenan in animal experiments. *Environ. Health Perspect.* 109, 983–994.

[96] McKim, J. M. 2014. Food additive carrageenan: Part I: A critical review of carrageenan in vitro studies, potential pitfalls, and implications for human health and safety. *Crit. Rev. Toxicol.* 44, 211-243.

In: Sulfated Polysaccharides
Editors: M. Gama, H. Nader, H. de Oliveira Rocha
ISBN: 978-1-63482-975-5

Chapter 9

AGARANS: SULFATED PRECURSORS AND DERIVATIVES FROM AGAROSE, AND RELATED SULFATED GALACTANS

***Marina Ciancia*[1,2,*] *and María Cristina Matulewicz*[2]**

[1]Cátedra de Química de Biomoléculas, Departamento de Biología Aplicada y Alimentos, Facultad de Agronomía, Universidad de Buenos Aires, Buenos Aires, Argentina
[2]CIHIDECAR (CONICET-UBA), Departamento de Química Orgánica, Facultad de Ciencias Exactas y Naturales, Universidad de Buenos Aires, Ciudad Universitaria, Buenos Aires, Argentina

ABSTRACT

Agarans are sulfated galactans biosynthesized by red seaweeds; they are constituted by alternating 3-linked β-D-galactose units and 4-linked α-L-galactose units with different degree of cyclization to give 3,6-anhydro-α-L-galactose. The neutral, completely cyclized agaran is known as agarose and is the major component of the products industrially obtained mainly from seaweeds of the orders Gelidiales, Anhfeltiales, and Gracilariales. Its gelling properties have a wide range of industrial uses. The biological precursor structure of agarose is usually known as porphyran, being found in important amounts in seaweeds of the genus *Porphyra* (Bangiales). The ideal structure of this galactan is composed of alternating 3-linked β-D-galactose and 4-linked α-L-galactose 6-sulfate units. The latter residues are cyclized by specific enzymes in the cell wall regulating the physical characteristics of the amorphous matrix. The same reaction, but in the presence of alkali, is used to improve the rheological properties of extracts from different sources with commercial purposes. Besides, water extracts from agarophytes usually comprise not only major amounts of agarose, which is only obtained with hot water or hot water solutions, but also minor quantities of galactans with related structures but with a certain degree of substitution with sulfate ester groups, methyl ethers, pyruvate ketals, D-xylose and/or 4-*O*-methyl-L-galactose side chains, and different percentages of 3,6-anhydrogalactose, depending on the source and sample work up. These kind of polymers constitute the major sulfated galactans from Ceramiales, while

* Email: ciancia@agro.uba.ar.

those obtained from calcareous seaweeds of the order Corallinales are characterized by important amounts of single stubs of D-xylose, a complex methylation pattern, and by the absence of 3,6-anhydrogalactose units. The presence of substituents, as well as non-cyclized units in the backbone, precludes helix formation and gelling in aqueous systems.

1. INTRODUCTION

Agarans are sulfated galactans biosynthesized by red seaweeds. They are constituted by alternating 3-linked β-D-galactose and 4-linked α-L-galactose units with different degree of replacement of the latter units by 3,6-anhydro-α-L-galactose. The neutral, completely cyclized agaran is called agarose (Figure 1) and it is the major component of the products industrially obtained mainly from seaweeds of the orders Gelidiales, Anhfeltiales, and Gracilariales, commercially known as agar [1]. Uses of agarose are related to formation of thermoreversible ion-independent gels, at low concentration in water, with large hysteresis [2].

The world's first source of agar, from the middle of the seventeenth century, was *Gelidium* from Japan, but by the beginning of the twentieth century demand for the phycocolloid exceeded of the supply of this alga. Since then *Gracilaria* has played an important role in the production of agar. Currently agars are obtained from genera *Gelidium, Gelidiella, Pterocladia, Gracilaria, Gracilariopsis,* and *Ahnfeltia* of red algae and marketed as 'natural agar' in squares or strips or as 'industrial agar' in powder form. The development of production processes through alkaline elimination of sulfates increases greatly the agarose content and therefore has allowed a good quality food agar to be obtained from *Gracilaria,* which comprises more than 90% of industrial agars. Besides, bacteriological and pharmacological grade agar is still obtained mostly from *Gelidium* and also in small quantities from *Pterocladia* [1].

The gel-forming ability and solubility of agar polysaccharides rely on the relative hydrofobicity of the basic repeating disaccharide, agarobiose, and its substitution by hydrophobic (methoxyl) and polar (sulfate, pyruvate) groups. Agar gels are formed if helical conformation of agar polysaccharides is possible and if these helices can aggregate. Whether these helices are two intertwined left-handed [3-5] or extended single helices [6-8] is still a matter of controversy, it is possible that both forms exist and which is obtained could depend on the experimental conditions. Nevertheless, it is known that upon cooling below the gelation temperature, the coils reorder to form helices that subsequently aggregate into a three-dimensional network of thick bundles and large pores of water to form a gel. Due to the presence of thick bundles, agarose gels are typically rigid and turbid, and show considerable degree of hysteresis between the melting and setting temperatures. The formation of thick bundles leads to the thinning of polymer chains among the bundles, as a result of which the local viscosity is reduced, which in turn markedly affects the molecular diffusion in gels.For this reason, the network structure of bundles is particularly important in separation technologies, such as in the use of agarose in gel electrophoresis [5].

According to the US Pharmacopoeia and the Food Chemicals Codex, agar is defined as a hydrophilic colloid, insoluble in cold water but soluble in boiling water. Its 1.5% (w/w) aqueous solutions are clear and form, between 32° and 39° C, firm gels which do not melt below 85° C. Quality of agars is primarily dependent upon the gelling capacity and the texture of the gel they produce, and it is usually evaluated in terms of practical use and very often the requirements of the application determine the chemical 'tailoring' of agar preparations.

In the cell wall, the type and extent of chemical substitution and molecular weight and molecular weight distribution widely affect the physical properties of agar polymers influencing the functionality of the whole algal thallus. Such effects on the properties are usually preserved when these galactans are isolated from the algal matrix. Consequently, the yield and the chemical and physical properties of agar preparations reflect seaweed source, environmental and physiological conditions, as well as the extraction and recovery procedures [9].

Commercial agarophytes also synthesize related galactan structures, which can be in part in some cases converted into gelling products by alkaline treatment (Figure 1). Besides, seaweeds of other orders, like Ceramiales and Corallinales, produce galactans which are structurally related with agarose but do not gel. These polymers are usually discarded by the agar industry, but they have varied potential applications, mostly due to their biological activity. All these products are generically known as agaroids or agarans.

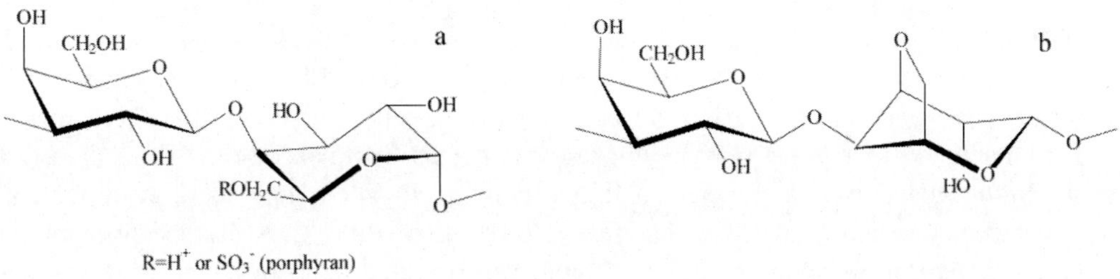

Figure 1. Structure of the basic backbone of non cyclized agarans (a) and of agarose (b).

2. Agarans

As it was previously mentioned, the repeating units of agarans are constituted by alternating 3-linked β-D-galactopyranose and 4-linked α-L-galactopyranose residues. This backbone is usually masked by different substituents:

a) The 3-linked and 4-linked units can carry methyl groups on different positions.
b) The 3-linked units can carry 4-*O*-methyl-α-L-galactopyranosyl groups on the 6-position.
c) The 3-linked and 4-linked units can carry β-D-xylopyranosyl side chains in different positions.
d) The 3-linked units can carry pyruvate cyclic ketals at the 4- and 6-positions (1`-carboxyethylidene groups).
e) The 3-linked and 4-linked units can carry sulfate groups on different positions.
f) The 4-linked units can be partially or wholly converted to 3,6-anhydro forms by enzymatic elimination of sulfate from the 6-position. According to the initial

substitution pattern, this conversion may afford agarose (Figure 1) or agarose derivatives.

The accumulation of ionic groups influences very much the solubility of polysaccharides and may greatly diminish their gel–forming properties.

Due to the complexity and diversity of substitution patterns, it has not been possible to classify the agarans onto "ideal structures" as it has been done for carrageenans. Using fractional extraction or ion-exchange chromatography agaran structures, which could eventually afford chemotaxonomic information, were isolated. This review will describe the latest literature on the agaran structures from several seaweeds belonging to different families. An extensive review on polysaccharides of the red algae has been recently published [10] which is strongly recommended for a more general picture about seaweed galactans.

2.1. Agarans from the Order Ceramiales

2.1.1. Rhodomelaceae

This family comprises most of the structural studies carried out on agarans.

Within the genus *Laurencia* [11], the structure of the agarans from *L. papillosa*, *L. cruciata*, *L. pedicularioides*, and *L. majuscula* of the west coast of India were reported; this characterization was mainly carried out by ^{13}C NMR spectroscopy. The first three species were found to be of similar type having as main diads β-D-galactose 2-sulfate-(1→4)-3, 6-anhydro-α-L-galactose, the latter unit partially methylated in the 2-position, and β-D-galactose-(1→4)-α-L-galactose 6-sulfate. However, in the ^{13}C NMR spectrum of the agaran of *L. majuscula* no signal at *ca.* 59 ppm was detected indicating that no methylated units were present; the diads found were: β-D-galactose-(1→4)-3,6-anhydro-α-L-galactose, β-D-galactose 2-sulfate→3,6-anhydro-α-L-galactose, and β-D-galactose 6-sulfate→3,6-anhydro-α-L-galactose.

A more detailed chemical and spectroscopic analysis was carried out on the agarans from *L. obtusa* and *L. filiformis* [12]. Those from *L. obtusa* had 3-linked β-D-galactose units in part sulfated on 2-position or methylated on 6-position, while the 4-linked units were mostly 3,6-anhydro-α-L-galactose and α-L-galactose 6-sulfate, some of the latter units were substituted with β-D-xylose on 3-position, precluding alkaline cyclization. The 3-linked β-D-galactose units of the agarans from *L. filiformis* were mostly sulfated on 2-position, but ~ half of these residues also carried the 4,6-*O*-(1`-carboxyethylidene) group. The 4-linked 3,6-anhydro-α-L-galactose units were methylated or substituted in part with single stubs of β-D-xylose on 2-position. This was the first time that substitution with xylose of 3,6-anhydro-α-L-galactose was reported. Besides, α-L-galactose 2-sulfate carrying single stubs of β-D-xylose on 3-position were also detected.

The agaran of *L. thyrisifera* contained a galactose:3,6-anhydrogalactose molar ratio of 1.00:0.24 and a very high level of xylose (14 mol%); in the infrared spectrum only a signal at 825 cm^{-1} was observed indicating sulfate substitution on the 2-position. Methylation analysis and ^{13}C NMR spectroscopy was consistent with a complex substitution pattern with partial sulfation on the 2-or 6-position of the 3-linked units, non-methylated 3,6-anhydrogalactose and single stubs of xylose on the 3-position of 4-linked α-galactose residues [13].

The application of partial reductive hydrolysis was especially effective in the structural investigation of an agaran from *L. nipponica* [14], allowing to describe the structures found after in other species of this genus. This procedure is described later in the chapter.

A very complex pyruvylated and sulfated agaran was also isolated from *Palisada flagelifera* [15]. This seaweed is a member of the *Laurencia* J. V. Lamouroux complex. It has been previously reported [16] that *Chondrophycus flagellifera* and *Chondrophycus papillosus* (nomenclature transferred to *Palisada flagellifera* and *Palisada papillosa*) biosynthesized sulfated xylomannans. After precipitation of the sulfated xylomannans from *P. flagellifera* with potassium chloride, the polysaccharides recovered from supernatants of extracts obtained at room temperature (FCS) and at 80 °C (FHS) were fractionated by anion-exchange chromatography.The homogeneous subfraction FCS-3, eluted with 0.6 M NaCl, was selected for further studies. Analyses indicated that the β-D-galactose units were mostly pyruvylated, 2-sulfated, and 6-*O*-methylated, with the latter partially 2- and 2,4-sulfated. Minor amounts of β-D-galactose units 2-, 6- and 2,6-sulfated, 6-glycosylated and non-substituted were also detected. On the other hand, the α-units were mainly 3,6-anhydro.L-galactose or 3,6-anhydro.2-*O*-methyl-L-galactose; but also L-galactose 6-sulfate units, substituted on the 3-position by xylose, galactose, and/or 2,3-di-*O*-methylagalactose residues or sulfate groups which precluded cyclization to 3,6-anhydrogalactose, were found. The structural complexity of this agaran was still increased by the presence of 2- and 3,6-sulfated α-L-galactoses, with the latter additionally 2-*O*-methylated.

Villanueva et al. [17] studied the agar from *L. flexilis* from northern Philippines. Recent molecular phylogenetic analysis revealed that this alga is an atypical species in the genus *Laurencia* and is an intermediate species between *Laurencia* and *Chondrophycus* [18]. The gel strength of the native and alkali-treated extracts was similar indicating the presence of minor amounts α-L-galactose 6-sulfate units. Chemical analysis showed very low sulfate and high 3,6-anhydrogalactose contents; FTIR gave a small band at 845 cm^{-1} consistent with sulfate ester at 4-position of 3-linked β-D-galactoseresidues. ^{13}C NMR spectroscopy revealed that the main diad was 6-*O*-methyl-β-D-galactose→3,6-anhydro-L-galactose, but also minor sulfation at 4-position of 3-linked galactose units was found. The high gelling temperature of the native and alkali-treated extracts (46-48 °C) was attributed to the high level of methyl ether substitution of the samples.

Structural determination of the polysaccharides of the cold- and hot-water extracts from *Acanthophora spicifera*, as well as those of their desulfated and alkali-treated derivatives was reported [19]. The agarans were isolated and further fractionated in a similar way to that previously described for the polysaccharides from *Palisada flagellifera* [15].The seaweed was extracted with water at 25 °C and the extracts were combined; the residue was re-extracted with water at 70 °C and the first two extracts were worked up separately. Addition of potassium chloride to their aqueous solutions until a 2 M concentration produced in each case a precipitate.The supernatants C (25 °C), H (first extraction with hot water) and W (second extraction with hot water) were dialyzed and freeze-dried. The resulting extracts were refractionated by anion-exchange chromatography. The fractions studied were those obtained with higher yield in each case (C3, eluted with 0.75 NaCl; H-3, eluted with 1.0 M NaCl; W4, eluted with 1.0 M NaCl). Structural analysis of these fractions showed that major amounts of the β-D-galactose units were sulfated on the 2-position and that some of these residues were pyruvylated; minor 2,6-disulfation, 6-*O*-methylation or non-substitution of β-D-galactose was also detected. The α-units were mainly constituted by 3,6-anhydro-L-galactose and its precursor, L-galactose 6-sulfate, together with lesser amounts of 3,6-anhydro-L-galactose 2-sulfate, L-galactose 2,6-disulfate, L-galactose 2,3,6-trisulfate, L-galactose 2,6-disulfate 3-xylose, 2-*O*-methyl-L-galactose, and non-substituted L-galactose.

The polysaccharides from *Bostrychia montagnei* were investigated [20,21]. In a first step, the seaweed was treated with methanol and the pigment free powder was then successively extracted with water (CW) at room temperature and at 85 °C. CW was fractionated by anion-exchange chromatography eluting with water and increasing sodium chloride concentrations. The major fractions eluted with water (B1), 1.25 M NaCl (B4), 1.5 M NaCl (B5), and 4.0 M NaCl (B6) were subjected to structural analysis. This seaweed biosynthesizes agarans with a linear backbone of 3-linked β-D-galactose and 4-linked 3,6-anhydro-α-L-galactose and α-L-galactose. The D-units are mainly sulfated on the 4-position, with lesser sulfation on the 6-position and minor at the 2-position; substitution on the 6-position with methyl and xylose single stubs was also found. The L-units are mainly methylated in the 2-position of 3,6-anhydrogalactose and sulfated on the 3-position of galactose; in addition, minor amounts of 2,3- and 3,6-disulfated, and 2-*O*-methyl or 2-*O*-glycosyl 3-sulfated galactose were detected.

The structures of the agarans of some species of the genus *Polysiphonia* have been determined: *P. lanosa* [22], *P. morrowii* [23,24], *P. strictissima* [25], *P.abscissoides* [25], *P. atterima* [26], and *P. nigrescens* [27].

With the exception of the studies reported for the polysaccharides of *P. lanosa*, *P. morrowii*, and *P. nigrescens*, the rest were mainly carried out by ^{13}C NMR spectroscopy of the native and alkali-treated polysaccharides.

The gel-forming galactan isolated from *P. morrowii* was subjected to β-agarase from *Pseudomonas atlantica* and the mixture of oligosaccharides obtained was analyzed by ^{13}C NMR spectroscopy [23,24] (see also later). The polysaccharide was shown to contain: a) blocks composed of agarobiose residues, partly 6-*O*-methylated and 6-sulfated, which are sensitive to enzymolysis; b) enzyme resistant blocks of agarobiose 6-sulfate residues with some disaccharide units containing L-galactose 6-sulfate instead of 3,6-anhydro-L-galactopyranose residues. These latter units were hardly detected in the ^{13}C NMR spectrum of the native polysaccharide.

Besides, a detailed study was carried out on the polysaccharides from *P. nigrescens* [27]. The seaweed was sequentially extracted with water at room temperature, 70 and 90 °C. The extracts were analyzed and the major one, isolated at 70 °C was fractionated by anion-exchange chromatography eluting with water and solutions of increasing sodium chloride concentration. Structural analysis, carried out by methylation and NMR spectroscopy (1D and 2D experiments), showed that four of the major fractions were partially cyclized agarans highly substituted on the 6-position of the β-unit with sulfate, although methyl ether and single stubs of β-D-xylose were found in minor proportions. A fifth fraction comprising 6-sulfated agarose was also isolated.

In the ^{13}C NMR spectra of the agarans from *P. strictissima* and *P. atterima* [25,26] low contents of non-substituted α-L-galactose residues were observed.

Thus, the agarans from seaweeds of the genus *Polysiphonia* showed similar substitution patterns and much simpler to those found for seaweeds of the *Laurencia* complex.In addition, it is very clear that the substitution pattern is very different between agarans from both genera, showing no chemotaxonomic pattern at a family level.

The structures of other agarans from members of the Rhodomelaceae were also studied, including those from *Chondria macrocarpa* [28,29], *Cladhymenia oblongifolia* [30], and two *Lophurella* species [31].

Seasonal variations were found in the structure of the polysaccharides from *Chondria macrocarpa*. Particularly, a complete structural analysis was carried out on a March extract [28] and it was found the β-units were sulfated in the 2- or the 6-position, or both. The α-units were 3,6-anhydro-L-galactose, L-galactose 6-sulfate or L-galactose having both single stubs of β-D-xylose attached to the 3-position and a sulfate group attached to the 6-position. Miller [31] reported for a November extract, mainly analyzed by ^{13}C NMR spectroscopy, a similar substitution pattern but with higher amounts of precursor units and significant amounts of pyruvate ketal not found in the original chemical analysis of the polysaccharide.

The agaran from *Cladhymenia oblongifolia* [30], analyzed by ^{13}C NMR spectroscopy, showed a significant level of sulfation. The β-units were mainly sulfated on the 2-position but moderate amounts of sulfation on the 6-position and minor amounts of pyruvate ketal were also found. The α-unis were 3,6-anhydrogalactose and galactose; the former were non-sulfated but sulfation on the 2-position was predominant in the latter, together with moderate levels on the 6-position and low levels on the 3-position.

Similarly, Miller [31] determined the structure of the agarans from *Lophurella kookeriana* and *Lophurella caespitosa*, both collected in spring. Both agarans were partially sulfated at the 2- and 6-positions of both the 3- and 4-linked units. The polysaccharide from *L. hookeriana* comprised higher amounts of L-galactose 6-sulfate residues than that of *L. caespitosa*. However, within the error range of NMR assessments, the polysaccharide from *L. caespitosa* had approximately twice the degree of sulfation in each of the remaining positions where sulfation occurred than did the polysaccharide from *L. hookeriana*.

Other agarans from Rhodomelaceae whose structures were studied comprise those from *Osmundaria colensoi*, *Bryocladia ericoides*, *Lenormandia chauvinii, Lenormandia augustifolia* [13], and *Streblocladia glomerulata* [25].

2.1.2. Wrangeliaceae

The monosaccharide composition of the biomass from *Neoptilota asplenioides* (formerly included within the Ceramiaceae) was determined by reductive hydrolysis and it was found, that, besides galactose and 3,6-anhydrogalactose, 3,6-anhydro-2-*O*-methylgalactose and 6-*O*-methylgalactose were present [32].Thus, it was concluded that this seaweed biosynthesized a highly methylated agarose.

The agaran from *Griffithsia antarctica* (formerly included within the Ceramiaceae) [33], studied by ^{13}C NMR spectroscopy and methylation analysis, was almost completely sulfated at the 4-position of the 3-linked units, while approximately half of the 4-linked units were 3,6-anhydrogalactose or galactose 6-sulfate, and the other half sulfated at the 3-position, and additionally, with 2-xylose, 2-sulfate or 6-sulfate as substituents. Besides, the author mentioned that sugar analysis of the natural extract also showed the presence of small amounts of 3,6-anhydro-2-*O*-methylgalactose and 6-*O*-methylgalactose.

The water-soluble polysaccharides from *Georgiella confluens* were analyzed in detail [34]. They were fractionated with cetrimide and the complexed material was further subjected to fractional solubilization in solutions of increasing sodium chloride concentration. The initially extracted polysaccharide and the major fraction isolated, soluble in 0.5 M NaCl, were studied. These were sulfated xylogalactans naturally rich in methylated sugar residues, comprising 3,6-anhydro-2-*O*-methyl-L-galactose, 2-*O*-methyl-L-galactose and 6-*O*-methyl-D-galactose. Structural analysis was carried out by methylation, ethylation, desulfation-ethylation, desulfation-methylation, Smith degradation, ^{13}C NMR spectroscopy and determination of the absolute configuration of monosaccharides by reductive amination. The results indicated the presence of an agaran backbone with a low 3,6-anhydrogalactose content, bearing xylose side-chains at the 4-position of the β-D-galactose and sulfated mainly at the 3-position of the α-L-galactose units, with a lesser amount of sulfation at the 4-position of β-D-galactose.

2.1.3. Ceramiaceae

Miller studied the agarans from *Cereamium rubrum* [29] and *Ceramium uncinatum* [33]. IR spectroscopy and ^{13}C NMR analysis indicated that the backbone of *C. rubrum* consisted of 3-linked galactose partially 6-sulfated or 2-sulfated and 4-linked units in a 2:1 ratio of 3,6-anhydrogalactose and galactose; 4-linked galactose was only partially 6-sulfated. Low levels of xylose substitution possibly at more than one location were detected. The structure of the agaran from *C. uncinatum* was similar but the 3-linked units were non-substituted or 6-sulfated. Monosaccharide composition indicated low levels of 3,6-anhydro-2-*O*-methylgalactose and 6-*O*-methylgalactose.

2.1.4. Delesseriaceae

The polysaccharide extract from *Myriogramme denticulata* was examined by ^{13}C NMR spectroscopy and it consisted largely of one diad repeat unit, comprising 3-linked β-D-galactose 4-sulfate and 4-linked α-L-galactose 2,3-disulfate [35].

In the same way, the polysaccharides from *Hymenena palmata*, *Hymenena palmata* f. *marginata*, *Hymenena variolosa*, and *Acrosorium decumbens* were analyzed [36]. The backbones of all polysaccharides comprised mainly diads of 3-linked β-D-galactose and 4-linked α-L-galactose, and these were heavily sulfated. The polysaccharide from *H. palmata* f. *marginata* was similar to that from *M. denticulata*. The polysaccharides from the remaining samples appeared to be very similar to each other and had a higher degree sulfation. The author reported as predominant diad β-D-galactose 2,4,6-trisulfate-(1→4)-α-L-galactose 2,3-disulfate. Sugar analysis of the four samples gave 3,6-ahydrogalactose contents of 4-12 mol%.

It is noteworthy that Usov & Klochkova [32]detected 3,6-anhydrogalactose and 3,6-anhydro-2-*O*- methylgalactose by reductive hydrolysis of the biomass from *H. ruthenica* (P. *et* R.) A.D. Zinova.

2.2. Agarans from Other Orders

2.2.1. Bangiales, Bangiaceae

The chemical composition of polysaccharides extracted from *Porphyra capensis* growing at different sites along the coast of Namibia and South Africa and collected at different times (mainly in November and January) was determined [37]. Considering their similarity, the authors selected the polysaccharide collected at Cape Town for FT-IR and ^{1}H and ^{13}C NMR spectral analysis. They also subjected this sample to alkaline treatment. Structural analysis was consistent with a linear backbone of alternating 3-linked β-D-galactose and 4-linked α-L-galactose 6-sulfate or 3,6-anhydro-α-L-galactose units (1.2:1 ratio determined by ^{1}H NMR spectroscopy). In the ^{1}H NMR spectrum a strong signal at 3.40 ppm was attributed to the methyl group of 6-*O*-methyl-D-galactose and the estimated degree of substitution was 0.64 for the D-galactose residue.

The structure of an agaran from *Porphyra haitanensis* was similar to that reported for *P. capensis* but partial methylation was observed at the 6-position of 3-linked β-D-galactose and at the 2-position of 3,6-anhydro-α-L-galactose [38].

2.2.2. Gracilariales, Gracilariaceae

The chemical analysis combined with FT-IR spectroscopy of a cold water extracted polysaccharide from *Gracilaria corticata* collected from the coast of Western India was carried out [39]. This polysaccharide consisted of a high molecular weight agaran, mainly sulfated at the 4-position of 3-linked β-D-galactose and at the 6-position of 4-linked α-L-galactose. For agarans isolated from the same seaweed but from the south-east of Madagascar the main physical properties and especially their rheological behavior in aqueous solution in the presence of sodium chloride and potassium chloride was determined [40].The seaweeds were stirred successively in ethanol/water 80, 60, 40, and 20% (v/v) at 70 °C and later in water at 100 °C. Particularly, the fraction extracted with ethanol/water 60% had low methoxyl and pyruvate contents and a great ability to form relatively strong physical gels in the presence of potassium chloride.

2.2.3. Corallinales, Corallinaceae

The composition and structure of the sulfated xylogalactans from the calcareous red seaweed *Corallina officinalis* was reported [41,42]. Structural analysis, carried out by ethylation, desulfation-methylation and ethylation analysis, indicated an agaran backbone with the 3-linked β-D-galactose units mainly substituted on the 6-position with single stubs of β-D-xylose or sulfate and also with minor amounts of 4-*O*-methyl-L-galactose. The 4-linked α-L-galactose residues carried sulfate or methoxyl groups on the 2- or 3-position, but never both together in the same unit. These agarans were called corallinans.

The structure of an agaran from *Joculator maxima* Manza was studied by ^{13}C NMR spectroscopy, controlled Smith degradation, specific 6-*O*-desulfation and methylation [43]. It was shown that the β-D-galactose units were substituted on the 6-position with single stubs of β-D-xylose while sulfates were located on the 6-position of the β-unit and on the 2- or 3-position of the α-L-galactose units. Minor amounts of the latter units were 2-methylated.

Usov et al. [44] reported the monosaccharide composition of the biomass from several calcareous red alga; glucose, galactose, xylose, and uronic acids were shown to be the main monosaccharide components, but no 3,6-anhydrogalactose was detected. It was suggested that these monosaccharides were the constituents of floridean starch, sulfated xylogalactan, and a polyuronide. The corresponding polysaccharides were isolated from *Corallina pilulifera*. The polyuronide was proved to be alginic acid and the xylogalactan was practically devoid of 3,6-anhydrogalactose both before and after alkaline treatment. The structure of the xylogalactan was studied by methylation analysis before and after desulfation or Smith degradation, as well as by NMR spectroscopy (1D and 2D experiments) [45]. The polysaccharide was shown to consist of an agaran backbone; the β-units were substituted on the 6-position with single stubs of β-D-xylose, whereas sulfate and methyl groups mainly occupied the 2-position of the α-unit and in lesser extent the 6-position of the β-unit.

Studying the influence of the extraction method on yields and compositions of xylogalactans from the Corallinales, significant amounts of 3,6-anhydrogalactose were detected in *Jania rubens* and small amounts of 2-*O*-methyl-D-galactose in *Bossiella orbigniana* [46].

Even though the xylogalactans from *J. rubens* showed the characteristic substitution pattern of the corallinans previously studied [47], some structural features, not previously detected, were reported: the presence of 3,6-anhydro-L-galactose (partly sulfated or methylated on the 2.position) and single stubs of 2,3-di- and 3-*O*-methyl-D-galactose and 3-*O*-methyl-L-galactose.

The main polysaccharides from *Lithothamnion heterocladum* shared the general characteristics of corallinans [48], even though this seaweed was crustose, whereas those previously studied were articulated seaweeds [49]. In addition, significant amounts of single stubs of 2,3-di-*O*-methyl-D-galactose and of 4-linked α-L-galactose glycosylated on the 3- or 6-position were found.

These results confirm that within a common pattern in the xylogalactans of Corallinales, each species has minor characteristics of their own [48].

3. Oligosaccharides Obtained from Agarans

Difficulties found to study agarans due to their high molecular weight and complex structure drove researchers to develop several partial degradation methods to obtain oligosaccharides, as representative as possible of the native structure. This is a difficult task due to the acid labile nature of 3,6-anhydrogalactose residues, as well as sulfate esters and pyruvate ketals, which are main factors determining the differences between agarans from various sources.

Due to the extension of the literature on the subject, two very successful degradation procedures will be reviewed, and only selected work on this subject will be described in detail.

3.1. Oligosaccharides Obtained by Hydrolysis of Agarans with Specific Enzymes

The use of specific enzymes to produce oligosaccharides from agarans is a powerful tool in structural analysis. Agarases are the enzymes which catalyze the hydrolysis of agarose sequences in a polysaccharide. They are classified into α-agarases and β-agarases according to the cleavage pattern. α-Agarases cleave α-(1→3)-linkages to produce agarooligosaccharides of series related to agarobiose, while β-agarases cleave β-(1→4)-linkages to produce neoagarooligosaccharides of series related to neoagarobiose (Figure 2). Several agarases have been isolated from different genera of bacteria found in seawater and marine sediments, as well as engineered microorganisms and they have a wide variety of applications [50]. They have been used to hydrolyze agar to produce oligosaccharides, which exhibit important physiological and biological activities beneficial to the health of human being. Besides, they have been used to isolate protoplasts from seaweeds and to recover DNA from agarose gel; also their use for bioconversion of seaweed galactans into biofuel or other industrial chemicals has been suggested [51]. However, since the late sixties, they have been used to investigate the composition and structure of cell wall polysaccharide of agarophytes. A purified extracellular β-agarase from a *Cytophaga* species was used to hydrolyze agarose, porphyran and alkali-treated porphyran [52, 53]. The oligosaccharide mixture from agarose was separated by gel filtration chromatography into the series of neoagarooligosaccharides, with predominance of the tetrasaccharide. Enzyme action on alkali-treated porphyran gave, in addition, oligosaccharides containing 6-*O*-methyl-D-galactose units. From the composition of these oligosaccharides it was deduced that action of the enzyme on a D-galactosidic linkage is four to five times faster than on the 6-*O*-methyl-D-galactosidic linkage. Enzyme action on native porphyran gave a similar series of oligosaccharides but in smaller yield, much of the polysaccharide being either not degraded or only degraded to a series of large, highly sulfated oligosaccharides. This procedure showed that 6-*O*-methyl groups were distributed randomly on half the D-galactose units in the porphyran chain, but the 6-sulfate groups on L-galactose units occurred in blocks. Similar enzymatic treatments were also used to study in detail the structure of agar from *Gelidium cartilagineum* [54] and the differences in structure between agars from several *Gracilaria* spp., which had different amounts of 6-*O*-methylgalactose, sulfate, pyruvate and degree of cyclization of the α-L-galactose 6-sulfate units, and their alkaline modified counterparts [55].

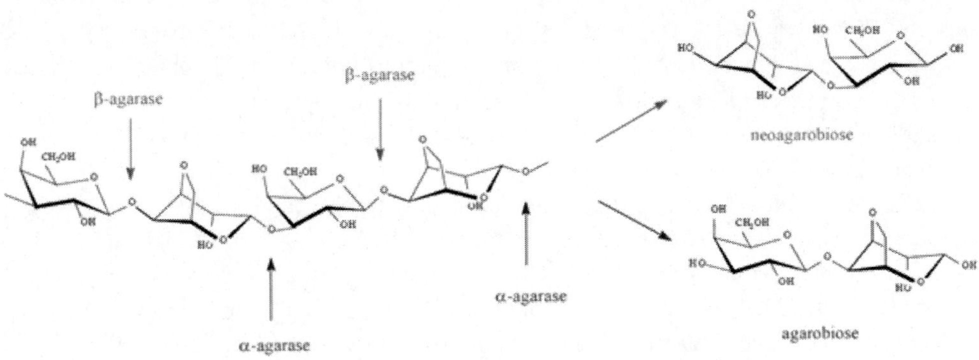

Figure 2. Treatment of agarose with agarases.

Morrice et al. [56] applied the β-agarase I from *Pseudomonas atlantica* to study the substitution pattern of the galactan from *Porphyra umbilicalis*. Two major neutral oligosaccharides were shown by ^{13}C-NMR spectroscopy to be 6^3-*O*-methyl-neoagarotetraose and 6^3,6^5-di-*O*-methyl-neoagarohexaose. The anionic oligosaccharides revealed two novel monosulfated tetrasaccharides, α-L-galactose 6-sulfate-(1→3)-β-D-galactose-(1→4)-3,6-anhydro-α-L-galactose-(1→3)-β-D-galactose and its 6^3-*O*-methylated derivative.

The ^{13}C-NMR data from the sulfated tetrasaccharides provided a novel reference which was used to characterize higher, partially sulfated fragments. The fraction of higher molecular weight had an average degree of polymerization of 40. This study showed that porphyran comprises 49 % sulfated disaccharide units and these were calculated to occur in stretches averaging 2.0-2.5 contiguous units.

Oligosaccharides of the agarobiose series obtained by treatment of agars with α-agarase were first described by Yaphe and coworkers [57,58].

Treatment of a commercial agarose with α-agarase with different degree of contamination with β-agarase from *Alteromonas agarlithycus* produced even and odd agarooligosaccharides that were studied by ^{13}C NMR spectroscopy [59].

More recently a separation and quantification method for neoagarose and agarose oligosaccharides (the latter in this case were obtained by acid hydrolysis) was developed. The interest in the above oligomer forms was due to the fact that some of them exhibit a variety of physiological activities. The potential of further applications of each different degree of polymerization of neoagarose oligosaccharide and agarose oligosaccharide could be discovered if qualitative and quantitative methods were established to prepare and determine the identity of individual oligomers [60].

Porphyran from the red seaweed *Porphyra umbilicalis* was degraded with α-porphyranase A derived from the marine flavobacterium *Zobellia galactanivorans*: The new series of sulfated oligosaccharides produced from porphyran were characterized by NMR, HPAEC and size-exclusion chromatography. α-Porphyranase A afforded oligosaccharides of the α-L-galactose6-sulfate-(1→3)-β-D-galactose (L6S-G) series and the L6S-G disaccharide was the major final product. The newly identified α-porphyranase A was further used in combination with β-agarase B to revise porphyran polysaccharide structure [61].

The complexity of porphyran is due to the co-occurrence of two repeating moieties, which are 3,6-anhydro-α-L-galactose (1→3) β-D-galactose (LA-G, neoagarobiose) and L6S-G, as well as by additional decoration with methyl groups on the hybrid copolymer chain. A low rate of porphyran degradation using β-agarase B was observed, and showed that an abundant enzyme-resistant fraction of high molecular weight remains, containing almost exclusively L6S-G moieties. This means that most of the LA-G moieties occur in the oligosaccharides and in the form of neoagarotetraose and hybrid LA-G/L6S-G oligosaccharides, as previously reported. In summary, blocks of agarose, if present, should be very short in porphyran, since the amount of pure agarose oligosaccharides was low compared to that of hybrid LAG/L6S-G oligosaccharides. Overall, using a combination of two enzymes, a β-agarase and a α-porphyranase, a more complete view of porphyran structure was obtained [61].

3.2. Oligosaccharides Obtained by Partial Reductive Hydrolysis of Agarans

Methods used for the study of monosaccharide composition by reductive hydrolysis and subsequent derivatization for the analysis by GC and GCMS were developed independently by two groups of researchers [62, 14] based on studies carried out by Garegg et al. [63]. Reductive hydrolysis was also used to study monosaccharide composition by high-performance anion exchange chromatography, without derivatization of the mixtures of alditols [64]. The importance of this methodology is based on the fact that by the standard procedure [65] using sodium borohydride to obtain alditols from monosaccharides, 3,6-anhydrogalactose residues are lost due to their acid labile nature. Reductive hydrolysis using

4-methylmorpholine-borane complex (MMB) as reductive agent allows reduction to take place in the acid medium of hydrolysis. In this way, 3,6-anhydrogalactose is converted to the acid stable 3,6-anhydrogalactitol, as soon as it is hydrolyzed. These methods comprise two hydrolysis steps, a first one in milder conditions in which only the 3,6-anhydrogalactosidic linkages are broken and a second one, in standard conditions.

Mild acid hydrolysis in the presence of MMB was used to obtain oligosaccharide alditols for detailed structural studies. A good example was the structural investigation of a sulfated galactan isolated from *Laurencia nipponica* [66]. After gel chromatography of the hydrolysate, a high molecular fraction and a mixture of neutral and charged oligosaccharides were obtained. Subsequent ion-exchange and reversed-phase chromatography gave rise to a series of individual oligosaccharides, from which agarobiitol and its 2′-sulfate were the main components. Two reduced sulfated tetrasaccharides in which 3,6-anhydrogalactose was not present, and one pentasaccharide were isolated. The tetrasaccharides had identical sugar chains and differed only in the position of one sulfate group (at 2-position of internal or terminal β-D-galactose residue, respectively). The pentasaccharide had the same backbone and sulfation pattern with an additional β-D-xylose residue located at position 3 of a 4-linked α-L-galactose 6-sulfate residue. The high-molecular weight fraction differed considerably from the parent polysaccharide in composition, since it had no 3,6-anhydrogalactose and was enriched in xylose, methylated galactoses and sulfate. On the basis of the amount of products isolated after partial reductive hydrolysis, the proportion of different repeating units in the parent polysaccharide was calculated. It was clear from these data that the polysaccharide from *Laurencia nipponica* has a hybrid backbone containing both agaran and agarose type segments, where sulfate groups occupy mainly 2-position of 3-linked β-D-galactose residues [66].

Other very interesting work was carried out by Gonçalvez et al. [67-69], who obtained oligosaccharides from the complex pyruvylated agaran 2-sulfate from *Acanthophora spicifera* [19], previously submitted to alkaline treatment to give a substituted agarose. The following disaccharide alditols were obtained: β-D-galactose-(1→4)-3,6-anhydro-L-galactitol; 4,6-*O*-(1`-carboxyethylidene)-β-D-galactose-(1→4)-3,6-anhydro-L-galactitol; β-D-galactose2-sulfate-(1→4)-3,6-anhydro-L-galactitol and 4,6-*O*-(1`-carboxyethylidene)-β-D-galactose 2-sulfate-(1→4)-3,6-anhydro-L-galactitol and analyzed by ESIMS and NMR spectroscopy [67]. These disaccharide alditols were used with others obtained from other carrageenophytes and agarophytes, or synthesized to develop a separation method using capillary electrophoresis [68], and then to explore the possible use of ESIMS differential fragmentation to distinguish between positional isomers [69].

CONCLUSION

Agarans are a family of sulfated galactans biosynthetized by red seaweeds, structurally related to agarose, but showing a tremendous variation of structures. These structural differences could be given by evolutional and environmental factors, and would be related with the functions of the amorphous matrix that they form in the cell wall [70]. This diversity of structures is very difficult to study in a systematic way and only very disperse information is available. However it gives rise to great possibilities concerning the applications of agarans in very different areas ranging from their classical use as hydrocolloids to many potential pharmacological uses. Moreover, we think that the information gathered in this field should be inspiration for the synthesis of new drugs.

As it was previously described, within some genera similar substitution patterns occur, however it is difficult to assess, for some others, given the very limited literature data, if such substitutions have taxonomic significance. It should be kept in mind that the amount of precursor structures is often not a taxonomic marker, as there is ample evidence in the literature of seasonal variation in precursor/anhydrogalactose levels for many algae [71].

REFERENCES

[1] Armisen, R. 1995. World-wide use and importance of *Gracilaria. Journal Applied Phycology.*7, 231-243.

[2] Lahaye, M.; Rochas, C. 1991. Chemical structure and physico-chemical properties of agar. *Hydrobiologia,* 221, 137-148.

[3] Arnott, S.; Fulmer, A.; Scott, W.E. Dea; I.C.M. Moorhouse, R.; Rees, D.A. 1974. The agarose double helix and its function in agarose gel structure. *Journal Molecular Biology.* 90, 269–284.

[4] Norton, I. T.; Goodall, D. M.; Austen, K. R. J.; Morris, E. R. 1986. Dynamics of molecular organization in agarose sulphate. *Biopolymers.* 25, 1009–1029.

[5] Dai, B.; Matsukawa, S. 2013. Elucidation of gelation mechanism and molecular interactions of agarose in solution by ^{1}H NMR. *Carbohydrate Research.* 365, 38–45.

[6] Foord, S. A.; Atkins, E.D.T. 1989. New X-ray diffraction results from agarose: Extended single helix structures and implications for gelation mechanism. *Biopolymers.* 28, 1345–1365.

[7] Jimenez-Barbero, J.; Bouffar-Roupe, C.; Rochas, C.; Pérez, S. 1989. Modelling studies of solvent effects on the conformational stability of agarobiose and neoagarobiose and their relationship to agarose. *International Journal of Biological Macromolecules.* 11, 265-272.

[8] Guenet, J.M.; Brulet, A.; Rochas, C. 1993. Agarose chain conformation in the sol state. *International Journal of Biological Macromolecules.* 15, 131–132.

[9] Murano, E. 1995. Chemical structure and quality of agars from *Gracilaria. Journal Applied Phycology.* 7, 245-254.

[10] Usov, A.I. 2011. Polysaccharides of the red algae. *Advances Carbohydrate Chemistry Biochemistry.* 65, 115-217.

[11] Siddhanta, A. K.; Goswami, A. M.; Shanmugam, M.; Mody, K.H.; Ramavat, B.K.; Mairh, O.P. 2002. Sulphated galactams of marine red alga *Laurencia* spp. (Rhodomelacea, Rhodophyta) from the west coast of India. *Indian Journal Marine Sciences.* 31, 305-309.

[12] Canelón, D.J.; Ciancia, M.; Suárez, A.I.; Compagnone, R.S.; Matulewicz, M.C. 2014. Structure of highly substituted agarans from the red seaweeds *Laurencia obtusa* and *Laurencia filiformis. Carbohydrate Polymers.*101, 705-713.

[13] Miller, I. J.; Falshaw, R.; Furneaux, R.H. 1993. The chemical structures of polysaccharides from New Zealand members of the Rhodomelaceae. *Botanica Marina.* 36, 203-208.

[14] Usov, A. I. 1993. Sulfated polysaccharides of red seaweeds. *Hydrobiologia.* 260/261, 641-645.

[15] Ferreira, L. G.; Noseda, M. D. Gonçalves, A.G.; Ducatti, D.R. B.; Fujii, M. T.; Duarte, M.E.R. 2012. Chemical Structure of the complex pyruvylated and sulfated agaran from the red seaweed *Palisada flagellifera* (Ceramiales, Rhodophyta). *Carbohydrate Research.* 347, 83-94.

[16] Cardoso, M.A.; Noseda, M.D.; Fujii, M. T.; Zibetti, R.G. M.; Duarte, M.E.R. 2007. Sulfated xylomannans isolated from red seaweeds *Chondrophycus papillosus* and *C. flagelliferus* (Ceramiales) from Brazil. *Carbohydrate Research.* 342, 2766-2775.

[17] Villanueva, R. D.; Romero, J.B.; Ragasa, A.L. R.; Montaño, M.N.E. 2010. Agar from the red seaweed, *Laurencia flexilis* (Ceramiales, Rhodophyta) from northern Philippines. *Phycological Research.* 58, 151-156.

[18] Abe, T.; Kurihara, A.; Kawaguchi, S.; Terada, R.; Masuda, M. 2006. Preliminary report on the molecular phylogeny of the *Laurencia* complex (Rhodomelaceae). *Coastal Marine Science* 30, 209-213.

[19] Duarte, M.E. R.; Cauduro, J. P.; Noseda, D. G.; Noseda, M.D.; Gonçalves, A. G.; Pujol, C.A.; Damonte, E. B.; Cerezo, A.S. 2004. The structure of the agaran sulfate from *Acanthophora spicifera* (Rhodomelaceae, Ceramiales) and its antiviral activity. Relation between structure and antiviral activity in agarans. *Carbohydrate Research.* 339, 335–347.

[20] Noseda, M. D.; Tulio, S.; Duarte, M.E.R. 1999. Polysaccharides from the red seaweed *Bostrychia montagnei*: chemical characterization. *Journal Applied Phycology.* 11, 35-40.

[21] Duarte, M.E. R.; Noseda, M. D.; Cardoso, M. A.; Tulio, S.; Cerezo, A.S. 2002. The structure of a galactan sulfate from the red seaweed *Bostrychia montagnei.Carbohydrate Research.* 337, 1137-1144.

[22] Batey, J. F.; Turvey, J.R. 1975. The galactan sulphate of the red alga *Polysiphonia lanosa. Carbohydrate Research.* 43, 133-143.

[23] Usov, A. I.; Ivanova, E. G.; Shashkov, A.S. 1983. Polysaccharides of algae. XXXIII: Isolation and ^{13}C NMR spectral study of some new gel-forming polysaccharides from Japan Sea red seaweeds. *Botanica Marina.* XXVI, 285-294.

[24] Usov, A. I.; Ivanova, E.G. 1987. Polysaccharides of algae. XXXVII. Characterization of hybrid structure of substituted agarose from *Polysiphonia morrowii* (Rhodophyta, Rhodomelaceae) using β-agarase and ^{13}C NMR spectroscopy. *Botanica Marina.* 30, 365-370.

[25] Miller, I. J. ; Furneaux, R.H. 1997. The structural determination of the agaroid polysaccharides from four New Zealand algae in the order Ceramiales by means of ^{13}C NMR spectroscopy. *Botanica Marina.* 40, 333-339.

[26] Miller, I.J. 2003a. Evaluation of the structures of polysaccharides from two New Zealand members of the Rhodomelaceae by ^{13}C NMR spectroscopy. *Botanica Marina.* 46, 386-391.

[27] Prado, H. J.; Ciancia, M.; Matulewicz, M.C. 2008. Agarans from the red seaweed *Polysiphonia nigrescens* (Rhodomelaceae, Ceramiales). *Carbohydrate Research.* 343, 711-718.

[28] Furneaux, R. H. ; Stevenson, T.T. 1990. The xylogalactan sulfate from *Chondria macrocarpa* (Ceramiales, Rhodophyta). *Hydrobiologia.* 204/205, 615-620.

[29] Miller, I. J.; Blunt, J.W. 2002. Evaluation of the structure of the polysaccharides from *Chondria macrocarpa* and *Ceramium rubrum* as determined by ^{13}C NMR spectroscopy. *Botanica Marina.* 45, 1-8.

[30] Miller, I. J.; Blunt, J.W. 2000. New ^{13}C NMR methods for determining the structure of algal polysaccharides. Part 3. The structure of the polysaccharide from *Cladhymenia oblongifolia*. *Botanica Marina.* 43, 263-271.

[31] Miller, I.J. 2002. The structure of the polysaccharides from two *Lophurella* species as determined by ^{13}C NMR spectroscopy. *Botanica Marina.* 45, 373-379.

[32] Usov, A. I.; Klochkova, N.G. 1992. Polysaccharides of algae. 45. Polysaccharide composition of red seaweeds from Kamchatka coastal waters (Northwestern Pacific) studied by reductive hydrolysis of biomass. *Botanica Marina.* 35, 371-378.

[33] Miller, I.J. 2003b. Evaluation of the structures of polysaccharides from two New Zealand members of the Ceramiaceae. *Botanica Marina.* 46, 378-385.

[34] Kolender, A. A.; Matulewicz, M.C. 2002. Sulfated polysaccharides from the red seaweed *Georgiella confluens*. *Carbohydrate Reearch.* 337, 57-68.

[35] Miller, I.J. 2001. Evaluation of the structure of the polysaccharide from *Myriogramme denticulata* as determined by ^{13}C NMR spectroscopy. *Botanica Marina.* 44, 253-259.

[36] Miller, I.J. 2005. Evaluation of the structures of polysaccharides from three taxa in the genus *Hymenena* and from *Acrosorium decumbens* (Rhodophyta, Delesseriaceae). *Botanica Marina.* 48, 148-156.

[37] Zhang, Q.; Qi, H.; Zhao, T.; Deslandes, E.; Ismaeli, N. M.; Molloy, F.; Critchley, A.T. 2005. Chemical characteristics of a polysaccharide from *Porphyra capensis* (Rhodophyta). *Carbohydrate Research.* 340, 2447-2450.

[38] Zhang, Q.; Li, N.; Liu, X.; Zhao, Z.; Li, Z.; Xu, Z. 2004. The structure of a sulfated galactan from *Porphyra haitanensis* and its in vivo antioxidant activity. *Carbohydrate Research.* 339, 105-111.

[39] Mazumder, S.; Ghosal, P .K.; Pujol, C. A.; Carlucci, M. J.; Damonte, E.B.; Ray, B. 2002. Isolation, chemical investigation and antiviral activity of polysaccharides from *Garcilaria corticata* (Gracilariaceae, Rhodophyta). *International Journal of Biological Macromolecules.* 31, 87-95.

[40] Andriamanantoanina, H.; Chambat, G.; Rinaudo, M. 2007. Fractionation of extracted Madagascar *Gracilaria corticata* polysaccharides: structure and properties. *Carbohydrate Polymers.* 68, 77-88.

[41] Cases, M. R.; Stortz, C. A.; Cerezo, A.S. 1992. Methylated, sulphated xylogalactans from the red seaweed *Corallina officinalis*. *Phytochemistry* 31, 3897-3900.

[42] Cases, M.R., Stortz, C.A., Cerezo, A.S. 1994. Structure of the ´corallinans´ - sulfated xylogalactans from *Corallina officinalis*. *International Journal of Biological Macromolecules.* 16, 93-97.

[43] Takano, R.; Hayashi, J.; Hayashi, K.; Hara, S.; Hirase, S. 1996. Structure of a water-soluble polysaccharide sulfate from the red seaweed *Joculator maximus* Manza. *Botanica Marina.* 39, 95-102.

[44] Usov, A. I.; Bilan, M. I.; Klochkova, N.G. 1995. Polysaccharides of algae. 48. Polysaccharide composition of several calcareous red algae. Isolation of alginate from *Corallina pilulifera* P. et R. (Rhodophyta, Corallinaceae). *Botanica Marina.*38, 43-51.

[45] Usov, A. I.; Bilan, M.I.; Shashkov, A.S. 1997. Structure of a sulfated xylogalactan from the calcareous red alga *Corallina pilulifera* P. et R. (Rhodophyta, Corallinaceae). *Carbohydrate Research.*303, 93-102.

[46] Navarro, D. A.; Stortz, C.A. 2002. Isolation of xylogalactans from the Corallinales: influence of the extraction method on yields and compositions. *Carbohydrate Polymers.* 49, 57-62.

[47] Navarro, D. A.; Stortz, C.A. 2008. The system of xylogalactans from the red seaweed *Jania rubens* (Corallinales, Rhodophyta). *Carbohydrate Research.*343, 2613-2622.

[48] Navarro, D. A.; Ricci, A. M.; Rodríguez, M. C.; Stortz, C.A. 2011. Xylogalactans from *Lithothamnion heterocladum*, a crustose member of the Corallinales (Rhodophyta). *Carbohydrate Polymers.* 84, 944-951.

[49] Bilan, M.I., Usov, A.I. 2001. Polysaccharides of calcareous algae and their effect on the calcification process. *Russian Journal Bioorganic Chemistry.* 27, 2-16.

[50] Fu, X. T.; Kim, S.M. 2010. Agarase: review of major sources, categories, purification method, enzyme characteristics and applications. *Marine Drugs.*8, 200-218; doi:10.3390/md8010200

[51] Gupta, V.; Trivedi, N.; Kumar, M.; Reddy, C.R.K.; Jha, B. 2013. Purification and characterization of exo-β-agarase from an endophytic marine bacterium and its catalytic potential in bioconversion of red algal cell wall polysaccharides into galactans. *Biomass Bioenerg.* 49, 290-298.

[52] Duckworth, M.; Turvey, J.R. 1969a. The action of a bacterial agarase on agarose, porphyran and alkali-treated porphyran. *Biochemical Journal.* 113, 687-692.

[53] Duckworth, M.; Turvey, J.R.1969b. The specificity of an agarase from a *Cytophaga* species. *Biochemical Journal.* 113, 693-696.

[54] Duckworth, M.; Yaphe, W. 1971. The structure of agar. Part II. The use of a bacterial agarose to elucidate structural features of the charged polysaccharides in agar. *Carbohydrate Research.* 16, 435–445.

[55] Duckworth, M.; Hong, C. K.; Yaphe, W. 1971. The agar polysaccharides of *Gracilaria* species. *Carbohydrate Research.* 18, 1–9.

[56] Morrice, L.; McLean, M.W.; Long, W. F.; Williamson, F.B. 1983. Porphyran primary structure. An investigation using β-agarase I from *Pseudomonas atlantica* and ^{13}C-NMR spectroscopy. *European Journal Biochemistry.* 133, 671-684.

[57] Hamer, G. K.; Bhattacharjee, S. S.; Yaphe, W. 1977. Analysis of the enzymic hydrolysis products of agarose by ^{13}C-n.m.r. spectroscopy. *Carbohydrate Research.* 54, C7-C10.

[58] Young, K.S., Bhattacharjee, S.S.; Yaphe, W. 1978. Enzymic cleavage. *Carbohydrate Research.* 66, 207-212.

[59] Rochas, C.; Potin, P.; Kloareg, B. 1994. NMR spectroscopic investigation of agarose oligomers produced by an α-agarase. *Carbohydrate Research.* 253, 69-77.

[60] Kazlowski, B.; Pan, C. L.; Ko. Y.T. 2008. Separation and quantification of neoagaro- and agaro-oligosaccharide products generated from agarose digestion by β-agarase and HCl in liquid chromatography system. *Carbohydrate Research.* 343, 2443-2450.

[61] Correc, G.; Hehemann, J.H.; Czjzek, M.; Helbert, W. 2011. Structural analysis of the degradation products of porphyran digested by *Zobellia galactanivorans* β-porphyranase A. *Carbohydrate Polymers.* 83, 277–283.

[62] Stevenson, T. T.; Furneaux, R.H. 1991. Chemical methods for the analysis of sul sulphated galactans from red algae. *Carbohydrate Research.* 210, 277-298.

[63] Garegg, P. J.; Lindberg, B.; Konradsson, P.; Kvarnström, I. 1988. Hydrolysis of glycosides under reducing conditions. *Carbohydrate Research.*176, 45-148.

[64] Jol, C. N. ; Neiss, T. G.; Penninkhof, B.; Rudolph, B.; De Ruiter, G.A. 1999. A novel high-performance anion-exchange chromatographic method for the analysis of carrageenans and agars containing 3,6-anhydrogalactose. *Analytical Biochemistry.* 268, 213-222.

[65] Morrison, I. M. 1988. Hydrolysis of plant cell walls with trifluoroacetic acid. *Phytochemistry, 27,* 1097–1100.

[66] Usov, A. I.; Elashvili, M.Ya. 1991. Polysaccharides of algae. 44. Investigation of sulfated galactans from *Laurencia nipponica* Yamada (Rhodophyta, Rhodomelaceae) using partial reductive hydrolysis. *Botanica Marina.* 34, 553-560.

[67] Gonçalves, A.G., Ducatti, D.R.B.; Duarte, M.E. R.; Noseda, M.D. 2002. Sulfated and pyruvylated disaccharide alditols obtained from a red seaweed galactan: ESIMS and NMR approaches. *Carbohydrate Research.* 337, 2443–2453.

[68] Gonçalves, A. G.; Ducatti, D.R. B.; Paranha, R. G.; Duarte, M.E. R.; Noseda, M.D. 2005. Positional isomers of sulfated oligosaccharides obtained from agarans and carrageenans: preparation and capillary electrophoresis separation. *Carbohydrate Research.* 340, 2123–2134.

[69] Gonçalves, A. G.; Ducatti, D.R. B.; Grindley, T. B.; Duarte, M.E. R.; Noseda, M.D. 2010. ESI-MS Differential fragmentation of positional isomers of sulfated oligosaccharides derived from carrageenans and agarans. *Journal of the American Society for Mass Spectrometry.* 21, 1404–1416.

[70] Kloareg, B.; Quatrano, RS. 1988. Structure of the cell wall of marine algae and ecophysiological functions of the matrix polysaccharides. *Oceanography and Marine Biology Annual Review.* 26, 259-315.

[71] Miller, I.J. 1997. The chemotaxonomic significance of the water soluble red algal polysaccharides. *Recent Research Developments in Phytochemistry.* 1, 531-565.

In: Sulfated Polysaccharides
Editors: M. Gama, H. Nader, H. de Oliveira Rocha

ISBN: 978-1-63482-975-5

Chapter 10

SULFATED POLYSACCHARIDES FROM UNUSUAL NATURAL SOURCES

Marília Medeiros Fernandes Negreiros, Jailma Almeida-Lima and Hugo Alexandre Oliveira Rocha *

Laboratório de Biotecnologia de Polímeros Naturais (BIOPOL),
Departamento de Bioquímica,
Universidade Federal do Rio Grande do Norte (UFRN), Natal/RN, Brazil

ABSTRACT

Sulfated polysaccharides (SPs) are mainly found in the animal kingdom and marine seaweeds. Apart from these groups of living organisms, studies have shown that organisms from other taxa also synthesize SPs. Some of these data show that plants synthesize mainly sulfated galactans, while the SPs found in microorganisms are generally heteropolysaccharides that contain xylose, galactose, arabinose, mannose, fucose, glucuronic acid, galacturonic acid, rhamnose, or glucose. In plants, the SPs are localized in the cell walls, mostly in the rhizomes and roots, indicative of their relationship with nutrient absorption and a possible structural function. However, the physiological function of SPs in other organisms remains unclear. Similar to the SPs found in animals and seaweeds, the SPs from other organisms have been studied in light of the information obtained on their pharmacological activities, such as anticoagulant, antioxidant, antitumor, anti-inflammatory, and antiviral properties. In this chapter, we have reviewed the research regarding the bioactivity of sulfated polysaccharide-rich extracts and structurally uncharacterized sulfated polysaccharides from plants, bacteria, cyanobacteria, and microalgae. In addition, the present chapter also reviews the recent progress made in the research related to structural features of SPs.

Keywords: sulfated galactans; plants; microorganisms; biological activities

---

* Corresponding Author Email: hugo-alexandre@uol.com.br

1. SULFATED POLYSACCHARIDES SYNTHESIZED IN PLANTS

The first scientific evidence of sulfated polysaccharides (SPs) synthesized in animals and macroalgae dates back to the middle of the last century, and a quick search of the literature shows a considerable number of publications related to the synthesis of SPs in these two groups of organisms. A major finding is that all the species of animals and seaweeds studied to date have exhibited synthesis of SPs. On the other hand, the number of publications relating to the synthesis of SPs in plants is quite low, with the first data reported about 20 years ago.

More specifically, the first report of the synthesis of SPs in plants dates back to 1995, where it was demonstrated that it was possible to purify SPs from *Artemisia princeps* leaves. These plant SPs showed the property of preventing the clotting of human blood plasma by increasing the inhibitory effects of heparin cofactor II (HC-II) toward thrombin. Further, it was observed that incubation of these SPs with various glycosaminoglycan lyase enzymes abolished their anticoagulant effect, which led the authors of the study to propose that *Artemisia* SPs are glycosaminoglycan-like molecules [1]. However, these data need to be carefully interpreted, as until now there have been no structural studies suggesting that *Artemisia* SPs are sulfated polysaccharides or glycosaminoglycans.

The presence of SPs in plants was reported again only 10 years later. In this case, the plant studied was *Utricularia aurea*, also known as the golden bladderwort, a fresh water carnivorous plant found commonly in Asia and Australia. The SP from this plant consist of glucuronic acid, fucose, and sulfates (65%, 15.3%, and 2.1%, respectively). Due to the presence of fucose, the *Utricularia* SP is called as fucoidan. It was also demonstrated that this polymer has cytotoxic activity against tumor cells [2]. However, the authors did not present any evidence in this study that the sulfated groups are linked to sugar residues in the *Utricularia* SP.

Gerencer and colleagues [3] reported the purification of an acid polysaccharide from *Chelidonium majus* L. using the 9-aminoacridine precipitation method, followed by ion exchange and gel filtration chromatography techniques. This polysaccharide was observed to prevent infection with human immunodeficiency virus type 1 (HIV-1) in AA2 and H9 cells (human CD4+ T-cell lines), as well as the cell-to-cell virus spread in H9 cells that were continuously infected with HIV-1. Moreover, on the basis of elemental analysis and mass spectrometry results, the authors have suggested that this acid polysaccharide is a 3.8 kDa low-sulfated glycosaminoglycan molecule. However, appropriate techniques for the structural analysis of these sulfated polysaccharides need to be developed to confirm these data since, no research group has managed to prove that plants synthesize glycosaminoglycans.

Further, using several steps of purification, a sulfated acetamide-polysaccharide was obtained from the roots of *Radix hedysari*, a traditional Chinese medicinal plant. Antioxidant assays demonstrated that *R. hedysari* SP has strong 2,2-diphenyl-1-picrylhydrazyl (DPPH) and hydroxyl radical scavenging activities. However, in this study, the authors neither mentioned the histological localization of *R. hedysari* SP nor did they discuss about its physiological functions. In addition, after several structural analyses, including 2D NMR analysis, the authors were unable to identify the position of the sulfate group in *R. hedysari* SP [4].

Ilex latifolia is another plant that produces sulfated polysaccharides. This plant is a component of Chinese traditional tea called Kudingcha, which has been used in the formulae for treating several conditions such as obesity, hypertension, cardiovascular diseases, and hyperlipidemia [5]. This plant produces four sulfated galactoarabinan-rich fractions that possess antioxidant and hepatoprotective properties. The presence of sulfate groups was confirmed using Fourier transform infrared spectroscopy (FT-IR) [5].

In a more detailed study, Aquino and colleagues [6] demonstrated that SPs were synthesized by three different species of seagrasses, i.e., *Ruppia maritima*, *Halodule wrightii*, and *Halophila decipiens* (all three SPs are sulfated galactans) [6], as well as by two species of mangroves, i.e., *Avicennia schaueriana* and *Rhizophora mangle* (both SPs are sulfated arabinogalactans] and the mangrove pteridophyte *Acrostichum aureum* (SP is a sulfated glucan) [7]. In the studies focused on the SPs from *R. maritima*, it was observed that this plant synthesized a sulfated homogalactan that was composed of the following repeating tetrasaccharide units: [3-β-D-Gal-2(OSO_3)-1→4-α-D-Gal-1→4-α-D-Gal-1→3-β-D-Gal-4(OSO_3)-1→] (Figure 1). Histochemical analysis of these SPs showed that the sulfated galactan was present in the plant cell walls, mostly in rhizomes and roots, which is indicative of their involvement in the absorption of nutrients and a possible structural function. Moreover, when *R. maritima* was exposed to several artificial environments containing different concentrations of salt, the data showed that the concentration of sulfated polysaccharides and the degree to which they were sulfated were positively correlated with the salinity. However, no data regarding the pharmacological activities of sulfated galactans from *R. maritima* are available.

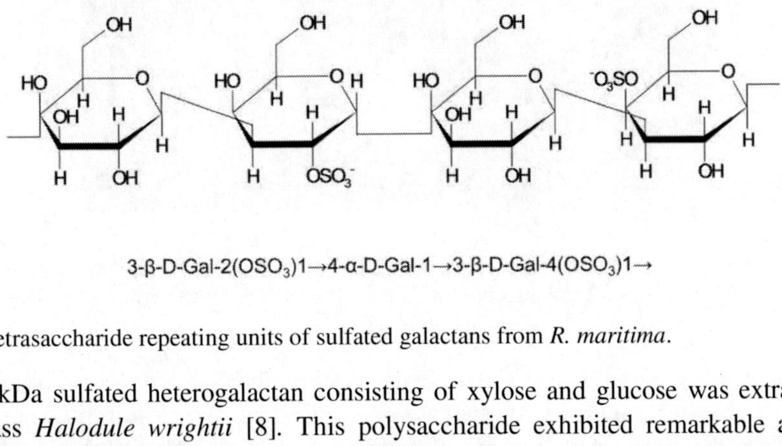

Figure 1. Tetrasaccharide repeating units of sulfated galactans from *R. maritima*.

A 11-kDa sulfated heterogalactan consisting of xylose and glucose was extracted from the seagrass *Halodule wrightii* [8]. This polysaccharide exhibited remarkable antioxidant properties in different *in vitro* assays. It scavenged over 41.4% of the DPPH radicals at a concentration of 500 μg/mL and the scavenging pattern observed was dose-dependent. In this study, the authors also showed that, in order to obtain higher reducing activity from the *Halodule* SP, which was equivalent to 50% of ascorbic acid, the required concentration for this compound was only 2.5-fold higher than that for vitamin C. Moreover, the *Halodule* SP showed outstanding anticoagulant activity that was found to be 2.5-fold higher than that of low molecular weight heparin in the activated partial thromboplastin time (APTT) test [8].

The sulfated polysaccharides were also extracted from water hyacinth *(Eicchornia crassipes*). Galactose, glucose, and arabinose were found to be the main monosaccharides in

E. crassipes SPs. The histological and gel electrophoresis analyses showed that *E. crassipes* SPs are primarily found in the roots (epidermis and vascular bundle), as well as in the rhizomes, leaves, and petioles. When the anticoagulant activity for these SPs was investigated, only the SPs from the roots and rhizomes caused prolongation of coagulation time to double of the baseline value and the SPs from the leaves and petioles showed no anticoagulant activity. This finding indicates that at least two different types of SPs are synthesized in *E. crassipes*. Since the *E. crassipes* root portion displayed the highest number of SPs, it was subjected to scanning electron microscopy (SEM) and energy-dispersive X-ray analysis (EDXA) in order to identify the SP localization in this portion of the plant. As shown in Figure 2, three locations in the roots, i.e., epidermis, cortex, and vascular bundle, were analyzed using the aforementioned methods. Sulfur was mainly found in the epidermis and vascular bundles, confirming the presence of sulfated polysaccharides in *E. crassipes* roots [9].

Figure 2. Scanning electron microscopy images for *E. crassipes* roots: White arrow, arrowhead, and yellow arrow indicate the points of collection of data for the EDXA. White arrow – epidermis; arrowhead – cortex; and yellow arrow – vascular bundles.

Moreover, the authors suggested that the role of SPs in *E. crassipes* roots may be related to nutrition and water capture, as in the case of *R. maritima SPs*. The results obtained from EDXA showed the presence of sulfur in the epidermis and cell wall, which are the important regions for water and nutrient capture in plants, thus reinforcing the argument about the function of SPs. However, the role of SPs in the leaves and petioles of *E. crassipes* is not clearly understood. It has been suggested that SPs present in the seaweeds may protect against the water loss due to sun exposure during low tide. Thus, based on this, the SPs found in the leaves and petioles of *E. crassipes* may have a similar function. Furthermore, *E. crassipes* is a freshwater plant and hence is different from seagrasses. *E. crassipes* is found in a non-saline environment, which indicates that freshwater plants may produce SPs that do not have any role related to salinity stress and that other factors may promote their production [9].

Table 1. Monosaccharide composition and biological properties of plant SPs

Plant Species	Sugar residues	Properties	Structural studies	References
Artemisia princeps	ND	anticoagulant	ND	[1]
Utricularia aurea	GlcA and Fuc	Tumor cell cytotoxicity	ND	[2]
Chelidonium majus	ND	Anti-HIV	Elemental analysis and mass spectrometry	[3]
Radix hedysari	Rha, Glu, Ara, Gal, and 2-acetamido-2-deoxy-D-galactose	DPPH and hydroxyl radical scavenging	Elemental analysis, partial acid hydrolysis, FT-IR, GC, GC–MS, 1D and 2D NMR	[4]
Ilex latifolia	Ara, Gal and traces of Xyl and Glu	Antioxidant and hepato-protective	Partial acid hydrolysis, FT-IR	[5]
Ruppia maritima	Gal	ND	GC–MS, GLC, electrophoresis, 1D and 2D NMR	[6]
Halophila decipiens	Gal	ND	GLC, electrophoresis	[6]
Avicennia schaueriana	Ara and Gal	ND	GLC, electrophoresis	[6]
Rhizophora mangle	Ara and Gal	ND	GLC, electrophoresis	[6]
Acrostichum aureum	Glu	ND	GLC, electrophoresis	[6]
Halodule wrightii	Gal, Glu, and Xyl	Antioxidant and anticoagulant	Partial acid hydrolysis, FT-IR, HPLC, electrophoresis	[8]
Eicchornia crassipes	Gal, Glu, and Ara	anticoagulant	Partial acid hydrolysis, FT-IR, SEM, HPLC, EDXA, electrophoresis	[9]

Gal – Galactose; GlcA – Glucuronic acid; Glu – Glucose; Fuc – Fucose; Rha – Rhamnose; Xyl – Xylose; FT-IR - Fourier transform infrared spectroscopy; GC – Gas chromatography; GC-MS - Gas Chromatography–mass spectroscopy; GLC – Gas-liquid chromatography; HPLC – High-performance liquid chromatography; NMR – Nuclear magnetic resonance; SEM – Scanning electron microscopy; EDXA – Energy-dispersive X-ray analysis; ND – Not determined.

During the preparation of this chapter, several references on sulfated polysaccharides from terrestrial plants were reviewed. However, the SPs in these studies were obtained after artificial sulfation of polysaccharides. Therefore, such references were not considered for the discussion in this chapter. Table 1 presents a summary of the data analyzed during the preparation of this chapter.

In the aforementioned table, we refer to the articles that have reported the presence of SPs in land plants. However, these references did not show any data that clearly indicate that these plants actually synthesize SPs (Table 1). However, recently two research groups have used suitable techniques to demonstrate the presence of SPs in two species of land plants [4, 5]. Hence, the question arises: do other land plants also synthesize SPs? However, this question cannot be answered with certainty. On the other hand, sea grasses, mangroves, and freshwater plants synthesize SPs, which in some cases present biological activities that indicate the potential of these plants as sources of bioactive SPs.

2. Sulfated Polysaccharides Synthesized by Microorganisms

Both autotrophic and heterotrophic unicellular organisms synthesize homo- and heteropolysaccharides. These polysaccharides could be stored intracellularly (i.e., endopolysaccharides) or secreted outside the cells (i.e., exopolysaccharides). In the case of sulfated polysaccharides, the literature shows that a vast majority of microorganisms that synthesize these molecules are autotrophic (phototrophic) in nature and that these microbial SPs are mainly exopolysaccharides.

2.1. SPs from Cyanobacteria (Blue-Green Algae)

The main monosaccharides found in cyanobacterial SPs are glucose (more than 90% of the polymer), followed by galactose, rhamnose, and mannose. Further, smaller proportions of ribose, xylose, arabinose, fucose, glucuronic acid, and galacturonic acid as well as some methylated or amino sugars can also be found in these SPs.

The function of some of these sulfated endopolysaccharides found in cyanobacteria has not been understood yet. Furthermore, various functions have been assigned to the sulfated exopolysaccharides, and it is suggested that they could be responsible for homogeneous distribution of extracellular structures to improve light and nutrient inputs. Moreover, the SP molecules act as antioxidant that prevents oxidative damage to nitrogenase, a nitrogen-fixing enzyme. The SPs may also participate in the process of immobilization of the microbial cells, which facilitates the colonization of surfaces (adhesion) and symbiotic relationships with other organisms. Further, the SPs also seem to protect cyanobacteria from dehydration [10].

Matsunag and colleagues [11] showed that the cyanobacterium, *Aphanocapsa halophytia,* produced 116 mg of sulfated polysaccharides/mg of the algae, when they were immobilized on high diffusion optical fibers. However, when these cyanobacteria were immobilized in calcium alginate, the output reduced to 11.7 mg of sulfated polysaccharides/mg of algae. These data show that light intensity could influence SP production.

Moreover, the composition and production of SPs by cyanobacteria may also be affected by various other factors such as pH, temperature, incubation time [10], mineral salts, vitamins, plant growth hormones, and light conditions [12].

Other more subtle factors may affect the type and amount of sulfated polysaccharides synthesized by cyanobacteria. Two of the strains of *Synechocystis*, *Synechocystis* PCC 6714 and PCC 6803, synthesize SPs. However, *Synechocystis* PCC 6803 is known to exhibit a mobile phase during its lifecycle, whereas PCC 6714 strain does not.

This factor seems to change the monosaccharide composition of SPs synthesized by the PCC 6803 strain, as it lacks the monosaccharide composition found in PCC 6714 SPs. It is noteworthy that the amount of sulfate in both the SPs is very similar; however, the sulfate content was found to be much lower in the SPs after 2 months of cultivation than in the SPs obtained after 15 days of cultivation [13].

2.1.1. SPs from Arthrospira Platensis (Spirulina Platensis)

Arthrospira platensis (formerly known as *Spirulina platensis*) is a cyanobacterium whose SPs have been studied in detail. This cyanobacterium is found in salt-rich lakes in Africa and Central and South America. *Arthrospira platensis* synthesizes the sulfated endopolysaccharide calcium spirulan (Ca-Sp). Several studies have shown the monosaccharide composition of Ca-Sp. The data from these studies indicate the presence of rhamnose, ribose, mannose, fructose, galactose, xylose, glucose, glucuronic acid, and galacturonic acid in Ca-Sp. The proportion of different monosaccharide components varies in different reports, and certain aforementioned monosaccharides were also absent. However, there is a consensus between these studies that Ca-SP consists mainly of rhamnose (50–60%). In addition, the main disaccharide repeating units in Ca-SP are composed of O-rhamnosyl-3-O-methylrhamnose (Figure 3) and O-hexuronosyl-rhamnose. The sulfate groups in this endopolysaccharide are linked to the C2 and C4 positions in rhamnose and O-methyl rhamnose, respectively.

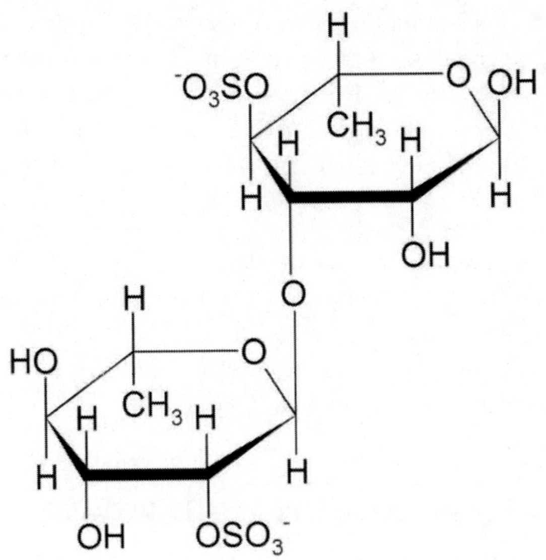

O-rhamnosyl-3-O-methylrhamnose

Figure 3. Sulfated disaccharide repeating units found in calcium spirulan.

When calcium is replaced with sodium in the SPs from spirulan, it is called sodium spirulan (Na-Sp), which is an anti-HIV agent and is more potent in comparison to Ca-Sp. This is probably because sodium promotes better interaction of sulfate groups with the virus particles [14]. Ca-Sp shows low anticoagulant activity; however, several articles have reported that this SP presents antiviral activity against several human viruses (Table 2). These reports suggest that this polysaccharide acts by selectively inhibiting the entry of viruses into several cells, including HeLa cells (human cervix adenocarcinoma cells), Vero cells (kidney epithelial cells extracted from an African green monkey), MDCK cells (kidney epithelial cells extracted from a dog), and MT-4 cells (human T-cell leukemia). Ca-Sp also prevents binding of *Helicobacter pylori* on the gastric mucin optimally at pH 2.0, without affecting the viability of either bacteria or gastric epithelial cells. In addition, Ca-Sp reduces the bacterial load in BALB/c mice around 90% [15].

Another study showed that Na-Sp also inhibits endothelial cell proliferation. This activity of Na-Sp was observed to be strongest in comparison to that by several glycosaminoglycans, such as heparan sulfate, heparin, dermatan sulfate, and chondroitin sulfate. Furthermore, structural studies indicate that the inhibitory activity of Na-Sp requires polymerization of sulfated O-rhamnosyl-3-O-methylrhamnose repeating units [16].

2.1.2. SPs from Other Cyanobacteria

Sacran is a fucose-rich sulfated polysaccharide synthesized by *Aphanothece sacrum*. This high-molecular weight SP (1.6×10^7 Da) exerts anti-inflammatory effects by improving skin barrier function and reducing Th2 cytokine production [17]. In addition, it inhibits viral adsorption by the receptor of the host cells involved in the replication process of herpes simplex virus type 2 (HSV-2) and influenza virus type A (IFV-A and H1N1) [18]. Another species from this genus, *A. halophytica*, produces a highly sulfated exopolysaccharide in its late logarithmic growth phase. This SP has been reported to show antiviral activity against influenza virus type A (H1N1) *in vivo*. The mechanism of its action might be mediated, at least in parts, by modulating the host immune system. However, further studies are necessary to confirm this hypothesis [19].

Several species from the genera *Anabaena*, *Nostoc*, *Gloeothece*, *Cyanothece*, *Synechocystis*, and *Phormidium* synthesize small amounts of SPs. They are heteropolysaccharides composed of different proportions of arabinose, fucose, galactose, glucose, mannose, rhamnose, ribose, xylose, acetyl-galactose, and acetyl-glucose. Several factors could be responsible for influencing the production of SPs in these species, and the synthesis of sulfated polysaccharides may often not occur [20].

2.2. SPs from Microorganisms from Eukarya (Eukaryota)

In the domain Eukaryota, the microorganisms that synthesize sulfated polysaccharides are found in the taxon Diaphoretickes. We found only one report on the sulfated oligosaccharides for fungi [21]. In addition, we found three reports describing the sulfated polysaccharides synthesized by fungi [22, 23, 24]. However, appropriate techniques for the structural analysis of sulfated polysaccharides need to be performed for these polymers to confirm these data.

2.2.1. Diaphoretickes

SPs were found in three supergroups of the taxon Diaphoretickes i.e., Cryptophyceae, Archaeplastida, and SAR (Stramenopiles, Alveolates, and Rhizaria).

2.2.1.1. Cryptophyceae

Only one species of this supergroup was identified for producing SPs. *Cryptomonas obovata* synthesizes a sulfated exopolysaccharide composed of fucose (42%), *N*-acetyl-galactosamine (26%), rhamnose (15%), and a small proportion of other monosaccharides. The amount of SP is affected by the presence of nitrates. Hence, a low amount of nitrates stimulates SP synthesis because the cell proliferation is inhibited at the expense of the synthesis of several kinds of molecules, including SPs [25]. However, there are no data on the physiological functions or pharmacological properties of these SPs (Table 2).

2.2.1.2. Archaeplastida

In the supergroup Archaeplastida, the species that synthesize SPs are found in the taxon Rhodophyta. Most of the studies have been carried out on the species from the genus *Porphyridium*.

Complex structures comprising cell wall-bound and soluble sulfated polysaccharide fractions encapsulate the cells of the red microalga *Porphyridium* sp. The soluble SPs dissolve in the culture medium, making it more viscous. Therefore, several studies have evaluated these compounds as gelling, thickening, stabilizing, or emulsifying agents. In addition, SPs from this particular and other red microalgae showed anti-inflammatory [26], antitumor [27], antioxidant [28], and antiviral [29] activities (Table 2).

The properties of the SPs from red microalgae depend mainly on their molecular weight, distribution of the sulfate groups, and monosaccharide composition. However, these structural features could change due to several factors. Keidan and colleagues [30] observed that a particular strain of *Porphyridium* sp. synthesized an SP that was slightly different from those synthesized by other strains. In addition, studies involving *Porphyridium* sp. (UTEX 637) showed that, in the first few hours of its cell cycle, this alga synthesized an SP with a molecular weight of 5×10^5 Da, which was further polymerized after 4–6 h of the cycle to produce a larger (2×10^6 Da) polysaccharide. However, after approximately 8 h of the cell cycle, a full-size polysaccharide was produced ($6–7 \times 10^6$ Da). The chemical composition of the SP undergoes a change during the cell cycle, wherein xylose was found to be the predominant sugar in the early phase, whereas the concentrations of glucose, galactose, and glucuronic acid peaked after 8–12 h. Other minor sugars appear at a later stage in the life cycle (12–14 h). The sulfate groups were found to show a peak concentration at 2–4 h [31]. The amount of nitrogen present in the medium also influenced the production of SPs, and higher amounts of SPs were produced in the presence of low levels of nitrogen [32].

The data obtained so far suggest that further efforts are required to understand the factors that favor the production of certain polysaccharides over others in a red alga culture. This information would allow the optimization of the synthesis of a molecule with certain defined features and therefore expand the potential use of red algal polysaccharides for various economic applications.

2.2.1.3. SAR (Stramenopiles, Alveolates, and Rhizaria)

For the supergroup SAR, we found reports regarding sulfated polysaccharides produced by organisms from the taxa Stramenopiles (Bacillariophyta) and Alveolata (Dinophyta).

2.2.1.3.1. Stramenopiles

Two species of Stramenopiles, *Phaeodactylum tricornutum* and *Navicula directa,* synthesize sulfated polysaccharides. However, little is known about the physiological functions of sulfated polysaccharides produced by *P. tricornutum.*

The available data only suggest that they are composed of mannose, glucuronic acid, and sulfate groups (7.5%) [33]. Further studies need to be conducted to confirm the identity of *P. tricornutum* polysaccharides.

N. directa synthesizes a 220-kDa sulfated polysaccharide composed of fucose (26.6%), xylose (25.0%), galactose (20.7%), mannose (13.1%), rhamnose (8.7%), and sulfate (8.0%). The SP produced by this diatom has been named naviculan; it shows antiviral activities against herpes simplex type 1 and 2 and influenza A viruses because of the inhibition of virus binding and penetration into the host cells. In addition, naviculan interferes with the formation of cell-cell fusion between HIV gp160- and CD4-expressing HeLa cells (Table 2) [34].

2.2.1.3.2. Alveolates

Cochlodinium polykrikoides is a planktonic dinoflagellate species with a distinctive spiral-shaped cingulum. This microorganism synthesizes two kinds of sulfated exopolysaccharides composed of mannose, galactose, and glucose. Both *Cochlodinium* polysaccharides show anticoagulant and antiviral activities (Table 1). In addition, their antiviral activity was observed at concentrations that were not markedly inhibitory for the blood coagulation process (Table 2) [35].

Gyrodinium impudicum is a non-toxic, chain-forming, red tide dinoflagellate. The strain KG03, originally obtained from a red tide bloom in Korea, synthesizes a 10^6 Da sulfated exopolysaccharide called p-KG03. Structural studies suggest that this polymer is a homogalactan containing uronic acid (2.7%) and sulfate (10%) [36].

p-KG03 increases the macrophage tumoricidal activity because it induces nitric oxide synthesis in a NFκB- and JNK-dependent manner, as well as stimulates the release of interleukins (IL-1 and IL-6) and TNF-α in macrophages, thus preventing tumor cell growth both *in vitro* and *in vivo* [34] [35]. p-KG03 also shows antiviral activity against HIV (very low) and encephalomyocarditis virus (EMCV). However, the mechanisms responsible for these activities are not clearly understood [36]. In addition, p-KG03 inhibits influenza type A virus replication in MDCK renal cells when added within 6 h after viral infection. This antiviral activity of p-KG03 is possible because it is able to interact with viral particles blocking the adsorption and internalization steps [40].

2.3. SPs from Prokaryotes

In bacteria, polysaccharides are mainly found in the cell wall as an essential constituent of lipopolysaccharides (LPS), as capsular material, or as exopolysaccharides that form dispersed slime. Sulfated polysaccharides are rare in prokaryotes and have been reported in very few genera so far; however, they are important for the interaction between bacteria and their habitat. These polysaccharides participate in events such as cell attachment and adhesion to surfaces and protection against toxic molecules, desiccation, and damage.

Rhizobia are symbiotic bacteria that elicit the formation of specialized organs on leguminous plants, for example, root nodules, in which they fix nitrogen. They produce lipochitooligosaccharides called Nod factors in response to plant-derived inducer molecules. Certain kinds of rhizobia synthesize sulfated Nod factors, for example, *Rhizobium meliloti*, which produces a sulfated tetrasaccharide composed of β-1,4-N-acetyl-D-glucosamine. In this tetrasaccharide, one of the residues is acylated with a C16 bis-unsaturated fatty acid. The sulfate group is linked to C-6 position on the reducing end of the N-acetylglucosamine backbone [41]. Other rhizobia, such as *Mesorhizobium loti* and *Sinorhizobium meliloti*, produce not only sulfated Nod factors but also a sulfate-modified form of LPS. These LPS are very complex polymers composed of several sugar residues, including mannose, galactose, fucose, glucose, xylose, and arabinose. However, their structure has not been elucidated yet [42]. *Azospirillum brasilense* produces sulfated LPS composed of O-repeating units made up of rhamnose and galactose residues, with a 3,4-linked fucose at the branching point and a xylose residue present in a ratio of 1:1 with fucose. The sulfate groups are linked at C-6 and C-2 positions of galactose and rhamnose residues, respectively [43]. The role of this sulfated cell-surface polysaccharide has been poorly understood. It has been suggested that this polysaccharide is required for optimum nodule formation [42].

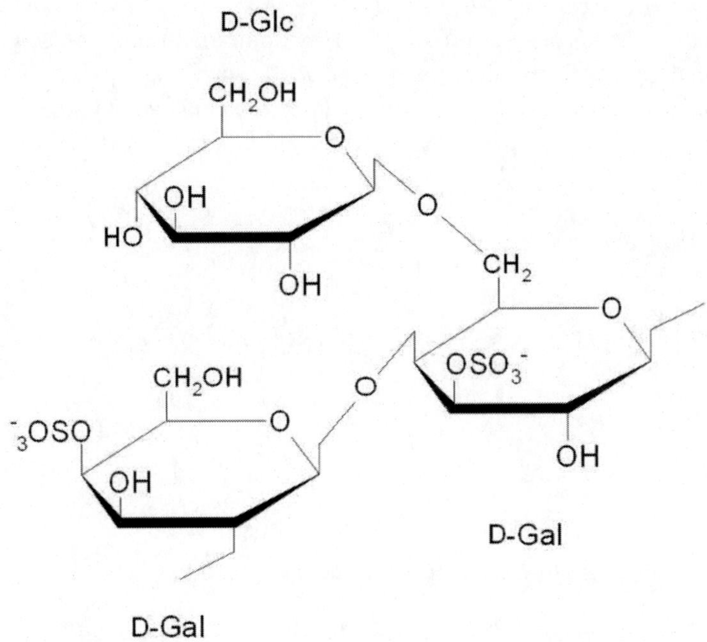

Figure 4. Sulfated trisaccharide units of SP extracted from *Pseudomonas* sp.

Mycobacterium tuberculosis is the causative agent of human tuberculosis, a disease that causes approximately 3 million deaths per year. This bacterium produces several types of sulfated glycolipids, which are located on the cell surface and are considered as the primary determinants for the intracellular survival and virulence of this organism. The major sulfated glycolipid in *Mycobacterium* is a family of trehalose-2-sulfate esters that are acylated with different acyl fatty acids (phthioceranate, hydroxyphthioceranate, palmitate, and stearate) at several positions in the oligosaccharides [44].

The marine *Pseudomonas* sp. produces a sulfated exopolysaccharide composed of a repeating unit consisting of a trisaccharide [-2)-β-D-Gal(4SO_4)(1-4)[-β-D-Glc(1-6)]-β-D-Gal(3SO_4)(1-] (Figure 4). This polymer shows cytotoxic activity against 24 different human cancer cell lines and was found to be more effective against central nervous system cancer (SF-295, SF-539, and SNB-75) and lung cancer (NCI-H522, DMS273, and DMS114) (Table 2) [45].

Staphylococcus epidermidis usually has a benign relationship with its host. However, currently, it is considered as an important opportunistic pathogen in relation with prosthetic joints, vascular grafts, surgical sites, central nervous system shunts, and cardiac device infections. Several *S. epidermidis* strains show resistance toward many antibiotics (such as penicillins, aminoglycosides, and quinolones). In addition, they form a biofilm, which is an important virulence factor in *Staphylococcus*. Several exopolysaccharides are involved in the biofilm function of *Staphylococcus*, including a 20-kDa sulfated polysaccharide. This polysaccharide is composed of glucosamine (which is in part N-sulfated to ~25%), glucose, and traces of xylose and fucose. However, the proportion of these sugar residues depends on the type of *Staphylococcus* strain [46].

Certain studies on the bacteria from deep-sea hydrothermal vents revealed the presence of sulfated exopolysaccharides in some species, such as *Alteromonas infernus*, *Pseudoalteromonas* strain HYD721 [47], and *Alteromonas macleodii* subsp. *fijiensis* biovar *deepsane* [48] (Figure 5). However, the physiological functions of these SPs have not been reported.

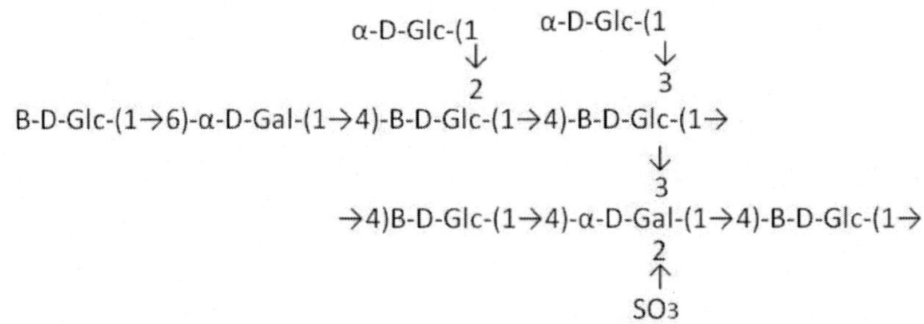

Figure 5. Sulfated exopolysaccharide unit of *Alteromonas macleodii*.

Several species of halophilic bacteria from the genus *Halomonas* synthesize sulfated exopolysaccharides. These species include *H. ventosae, H. anticariensis* [49], *H. eurihalina* [50], and *H. muara* [51]. Bejár and colleagues [50] studied the chemical composition of exopolysaccharides from 19 different strains belonging to *Halomonas eurihalina*. They showed that several strains of *Halomonas* produced SPs and that the sulfate content ranged from 11% to 24% [50]. In addition, it was suggested that the SP synthesis in *Halomonas* was not related to the hypersaline conditions of the environment in which this bacterial group survives, as also observed in the case of seagrass [6].

Table 2. Monosaccharide composition and properties of SPs from some microorganisms

Taxon	Species	Sugar residues	Properties
Cyanobacteria	*Arthrospira platensis*	Rha (50%), Rib, Man, Fru, Gal, Xyl, Glu, GlcA, GalA	Antiviral [14], Anti gastric ulcer [15], antiangiogenic [16]
	Aphanothece sacrum	Fuc	Antiviral [18]
	Aphanothece halophytica	Xyl, Glu, Gal	Antiviral [19]
Cryptophyceae	*Cryptomonas obovata*	Fuc (42%), Gal (26%) Rha (15%) small proportion of other monosaccharides	ND [25]
Archaeplastida	*Porphyridium* SP	Xyl, Glu, Gal, GlcA	Anti-inflammatory [26], antitumor [27], antioxidant [28], antiviral [29]
SAR	*Phaeodactylum tricornutum*	Man, GlcA	ND [33]
	Navicula directa	Fuc (26.6%), Xyl (25.0%), Gal (20.7%), Man (13.1%), Rha (8.7%)	Antiviral [34]
Alveolatas	*Cochlodinium polykrikoides*	Man, Gal, Glu	Anticoagulant [35], antiviral [36]
	Gyrodinium impudicum	Gal, UroA	Immunomodulatory [35]
Prokaryotes	*Rhizobium meliloti*	(LPS) Man, Gal, Fuc, Glu, Xyl, Ara	ND [41]
	Halomonas maura	Man (34%), Gal (14%) Glu (29%) and GluA (21%),	Metal chelating [54], Immunomodulatory [54]
	Mesorhizobium loti	(LPS) Man, Gal, Fuc, Glu, Xyl, Ara	ND [42]
	Sinorhizobium meliloti	(LPS) Man, Gal, Fuc, Glu, Xyl, Ara	ND [42]
	Azospirillum brasilense	(LPS) Man, Gal, Fuc, Glu, Xyl, Ara	ND [43]
	Mycobacterium tuberculosis	(glycolipids) Tre	ND [44]
	Pseudomonas sp.	Gal, Glc, Gal	Antitumor [45]

Ara – Arabinose; Fru – Fructose; Fuc – Fucose; Gal – Galactose; GalA – Galacturonic acid; GlcA – Glucuronic acid; Glc – Glucose; Man – Mannose; Rib – Ribose; Rha – Rhamnose; Tre – Trehalose; UroA – Uronic acid; Xyl – Xylose; ND – not determined.

Mauran is a sulfated exopolysaccharide produced by the halophilic bacterium *Halomonas maura*. This 4.7 × 10^6 Da SP is composed of mannose (34%), galactose (14%), glucose (29%), and glucuronic acid (21%), with a sulfate content that corresponds to 6.5% (w/w) [51]. However, its structural features have not been elucidated yet. Moreover, nanoparticles have been synthesized with mauran and chitosan using simple polyelectrolyte complexation of anionic and cationic polysaccharides. These nanoparticles showed potential for sustained drug delivery, cancer chemotherapy, and bioimaging [52]. Mauran has also been used to synthesize 120 nm-sized nanofibers. These structures were evaluated as a scaffold for testing migration and proliferation of several cell lines and differentiation of mesenchymal stem cells. The results suggested that mauran nanofibers have great potential for tissue engineering studies [53]. In addition, this polysaccharide also shows heavy metal chelating property and immunomodulating effect [54]. Further, mauran does not affect the mouse fibroblast (L929 cell) viability and cell antioxidant system [55].

3. Outlook

The information available in the literature suggests that a few species of plants and microorganisms synthesize SPs, a conclusion that would require further confirmation through the analysis of other species of plants and microorganisms. To facilitate this task, it would be useful to undertake *in silico* studies involving the genomes from different plant and microbial species in order to verify the presence of genes involved in the synthesis of SPs. Furthermore, conclusive structural characterization should be performed with the polysaccharides obtained so far to confirm their sulfation. In the case of microorganisms, it is important to determine the best growing conditions that favor the large-scale production of SPs. Finally, the molecular mechanisms that influence the synthesis of SPs in plants and microorganisms, as well as those that influence their biological or pharmacological properties, need to be better understood with the aim of exploiting the biotechnological potential of SPs.

References

[1] Hayakawa, Y.; Hayashi, T.; Hayashi, T.; Niiya K, Sakuragawa N. 1995. Selective activation of heparin cofactor II by a sulfated polysaccharide isolated from the leaves of *Artemisia princeps*. *Blood Coagulation & Fibrinolysis*. 6 (7), 643-649.

[2] Choosawad, D.; C.; Leggat, U.; Dechsukhum, C.; Phongdara, A.; Chotigeat, W. 2005. Anti-tumour activities of fucoidan from the aquatic plant *Utricularia aurea* lour. Songklanakarin *Journal of Science and Technology*, 27 (3), 799-807.

[3] Gerencer, M.; Turecek, P. L.; Kistner, O.; Mitterer, A.; Savidis-Dacho, H.; Barrett N.P. 2006. In vitro and in vivo anti-retroviral activity of the substance purified from the aqueous extract of *Chelidonium majus* L. *Antiviral Research*, 72 (2), 153-156.

[4] Dang, Z.; Feng, D.; Liu, X.; Yang, T.; Guo, L.; Liang, J.; Hu, F.; Cui, F.; Feng, S. 2013. Structure and antioxidant activity study of sulfated acetamido-polysaccharide from *Radix Hedysari*. *Fitoterapia*, 89, 20-32.

[5] Fan, J.; Wua, Z.; Zhaoa, T.; Suna, Y.; Yea, H.; Xua, R.; Zenga, X.; 2014. Characterization, antioxidant and hepatoprotective activities of polysaccharides from *Ilex latifolia* Thunb. *Carbohydrate Polymers*, 101, 990– 997.

[6] Aquino, R.S.; Landeira-Fernandez, A.M.; Valente, A.P.; Andrade, L.R.; Mourão, P.A. 2005. Occurrence of sulfated galactans in marine angiosperms: evolutionary implications. *Glycobiology*, 15, 11-20.

[7] Aquino, R.S.; Grativol, C.; Mourão, P.A. 2011. Rising from the sea: correlations between sulfated polysaccharides and salinity in plants. *PLoS ONE*, 6 (4), e18862.

[8] Silva, J. M. C.; Dantas-Santos, N.; Gomes, D. L.; Costa, L. S.; Cordeiro, S. L.; Costa, M. S. S. P.; Silva, N. B.; Freitas, M. L.; Scortecci, K. C.; Leite, E. L.; Rocha, H. A. O. 2012. Biological activities of the sulfated polysaccharide from the vascular plant *Halodule wrightii. Revista Brasileira de Farmacognosia*, 22, 94-101.

[9] Dantas-Santos, N.; Gomes, D. L.; Costa, L. S.; Cordeiro, S. L.; Costa, M. S. S. P.; Trindade, E. S.; Franco, C. R. C; Scortecci, K. C.; Leite, E. L.; Rocha, H. A. O. 2012. Freshwater Plants Synthesize Sulfated Polysaccharides: Heterogalactans from Water Hyacinth (*Eicchornia crassipes*). *International Journal of Molecular Sciences*, 13, 961-976.

[10] Nwodo, U. U.; Green, E.; Okoh, A. I. 2012. Bacterial exopolysaccharides: Functionality and prospects. *International Journal of Molecular Sciences*, 13 (11) 14002-14015.

[11] Matsunaga, T.; Sudo, H.; Takemasa, H.; Wachi, Y. 1996. Sulfated extracellular polysaccharide production by the halophilic cyanobacterium *Aphanocapsa halophytia* immobilized on light-diffusing optical fibers. *Applied Microbiology and Biotechnology*, 45, 24-27.

[12] Yim, J.H.; Kim, S.J.; Ahn, S.H.; Lee, H.K. 2003. Optimal conditions for the production of sulfated polysaccharide by marine microalga *Gyrodinium impudicum* strain KG03. *Biomolecular Engineering*, 20 (4-6), 273-280.

[13] Panoff, J-M.; Priem, B.; Morvan, H.; Joset, F. 1988. Sulphated exopolysaccharides produced by two unicellular strains of cyanobacteria, *Synechocystis* PCC 6803 and 6714. *Archives of Microbiology*, 150, 558- 563.

[14] Lee, J-B.; Hayashi, T. Hayashi, K.; Sankawa, U.; Maeda, M.; Nemoto, T.; Nakanishi, H. 1998. Further purification and structural analysis of calcium spirulan from *Spirulina platensi. Journal of Natural Products*, 61, 1101-1104.

[15] Loke, M. F.; Lui, S. Y.; Ng, B. L.; Gong, M.; Ho, B. 2007. Antiadhesive property of microalgal polysaccharide extracton the binding of *Helicobacter pylori* to gastricmucin. *FEMS Immunology and Medical Microbiology*, 50 (2), 231-238.

[16] Kaji, T.; Fujiwara, Y.; Hamada, C.; Yamamoto, C.; Shimada, S.; Lee, J.B.; Hayashi, T. 2002. Inhibition of cultured bovine aortic endothelial cell proliferation by sodium spirulan, a new sulfated polysaccharide isolated from *Spirulina platensis. Planta medicinal,* 68, (6), 505-509.

[17] Ngatu, N. R.; Okajima, M. K.; Yokogawa, M.; Hirota, R.; Eitoku, M.; Muzembo, B.A.; Dumavibhat, N.; Takaishi, M.; Sano, S.; Kaneko, T.; Tanaka, T.; Nakamura, H.; Suganuma, N. 2012. Anti-inflammatory effects of sacran, a novel polysaccharide from *Aphanothece sacrum*, on 2,4,6-trinitrochlorobenzene-induced allergic dermatitis *in vivo. Annals of* Allergy, *Asthma Immunology*,108 (2), 117-22.

[18] Ogura, F.; Hayashi, K.; Lee, J.B.; Kanekiyo, K.; Hayashi, T. 2010. Evaluation of an edible blue-green alga, *Aphanothece sacrum*, for its inhibitory effect on replication of herpes simplex virus type 2 and influenza virus type A. *Bioscience, Biotechnology, and Biochemistry,* 74 (8), 1687-90.

[19] Zheng, W.; Chen, C.; Cheng, Q.; Wang, Y.; Chu, C. 2006. Oral administration of exopolysaccharide from *Aphanothece halophytica* (Chroococcales) significantly inhibits influenza virus (H1N1)-induced pneumonia in mice. *International Immunopharmacology,* 6 (7), 1093-9.

[20] Philippis, R. D.; Vincenzini, M. 1998. Exocellular polysaccharides from cyanobacteria and their possible applications. *FEMS Microbiology Reviews*, 22, 151-175.

[21] Poupot, R.; Martinez-Romero, E.; Promé, J.C. 1993. Nodulation factors from Rhizobium tropici are sulfated or nonsulfated chitopentasaccharides containing an N-methyl-N-acylglucosaminyl terminus. *Biochemistry*, 32 (39) 10430-5.

[22] Cheng, J.; Huang, N.; Lur, H.; Kuo, C.; Lu, M 2009. Characterization and biological fuctions of sulfated polysaccharides from sulfated-salt treatment of Antrodia cinnamomea. *Process Biochemistry*, 44, 453-459.

[23] Cheng, J.; Chang, C.; Chao, C.; Lu, M 2012. Characterization of fungal sulfated polysaccharides and their synergistic anticancer effects with doxirubicin. *Carbohydrate Polymers*, 90, 134-139.

[24] Chang, C.; Lur, H.; Lu, M.; Cheng, J. 2013. Sulfated polysaccharides of Armillariella mellea and their anti-inflammatory activities via NF-ĸB suppresion. *Food Research International*, 54, 239-245.

[25] Giroldo, D.; Vieira, A. A. H. 2002. An extracellular sulfated fucose-rich polysaccharide produced by a tropical strain of *cryptomonas obovata* (cryptophyceae). *Journal of applied phycology*, 14, 185-191.

[26] Matsui, M. S.; Muizzuddin, N.; Arad, S.; Marenus, K. 2003. *Applied Biochemistry and Biotechnology,* 104 (1), 13-22.

[27] Gardeva, E.; Toshkova, R.; Minkova, K.; Gigova, L. 2009. Cancer protective action of polysaccharide, derived from red microalga *Porphyridium cruentum* - A biological background. Biotechnology *& Biotechnological*, 23(1), 783-787.

[28] Tannin-Spitz, T.; Bergman, M.; van-Moppes, D.; Grossman, S.; Arad S. M. 2005. Antioxidant activity of the polysaccharide of the red microalga *Porphyridium* sp. *Journal of Applied Phycology*, 17, 215-222.

[29] Huleihel, M.; Erukhimovitch, V.; Talyshinsky, M; Karpasas, M. 2002. Spectroscopic Characterization of Normal Primary and Malignant Cells Transformed by Retroviruses. *Applied Spectroscopy*, 56, (5), 640-545.

[30] Keidan, M.; Friedlander, M. Arad; S. M. 2009. Effect of Brefeldin A on cell-wall polysaccharide production in the red microalga *Porphyridium sp.* (Rhodophyta) through its effect on the Golgi apparatus. *Journal of* Applied Phycology, 21, 707-717.

[31] Simon-Bercovitch, B.; Bar-Zvi, D.; Arad, S. M. 1999. Cell-wall formation during the cell cycle of *Porphyridium* sp. (Rhodophyta). *Journal of Phycology*. 35 (1), 78-83.

[32] Arad, S. M.; Friedman, O. D.; Rotem, A. 1988. Effect of Nitrogen on Polysaccharide Production in a *Porphyridium* sp. *Applied and Environmental Microbiology*, 54 (10), 2411-2414.

[33] Ford, C. W.; Percival, E. 1965. The carbohydrates of *Phaeodactylum tricornutum*. Part I. Preliminary examination of the organism, and characterisation of low molecular weight material and of a glucan. *Journal of the Chemical Society*, 7035-7041.

[34] Lee, J-B.; Hayashi, K.; Hirata, M.; Kuroda, E.; Suzuki, E.; Kubo, Y.; Hayashi, T. 2006. Antiviral sulfated polysaccharide from Navicula directa, a diatom collected from deep-Sea water in toyama bay. *Biological Pharmaceutical Bulletin*, 29, (10), 2135-2139.

[35] Hasui, M.; Matsuda, M.; Okutani, K.; Shigeta, S. 1995. In vitro antiviral activities of sulfated polysaccharides from a marine microalga (*Cochlodinium polykrikoides*) against human immunodeficiency virus and other enveloped viruses. *International Journal of Biological Macromolecules,* 17 (5), 293-297.

[36] Yim, J.H.; Kim, S. J.; Ahn, S. H.; Lee, H. K. 2003. Optimal conditions for the production of sulfated polysaccharide by marine microalga *Gyrodinium impudicum* strain KG03. *Biomolecular Engineering,* 20, 273-280.

[37] Yim, J. H.; Son, E.; Pyo, S.; Lee, H. K.; 2005. Novel sulfated polysaccharide derived from red-tide microalga *Gyrodinium impudicum* strain KG03 with immunostimulating activity *in vivo*. *Marine Biotechnology*, 7, 331-338.

[38] Bae, S.Y.; Yim, J. H.; Lee, H. K.; Pyo, S. 2006. Activation of murine peritoneal macrophages by sulfated exopolysaccharide from marine microalga *Gyrodinium impudicum* (strain KG03): involvement of the NF-kappa B and JNK pathway. *International Immunopharmacology,* 6, 473-484.

[39] Yim, J. H.; Kim, S. J.; Ahn, S. H.; Lee, C. K.; Rhie, K. T.; Lee, H. K. 2004. Antiviral effects of sulfated exopolysaccharide from the marine microalga *Gyrodinium impudicum* strain KG03. *Marine Biotechnology*. 6, 17-25.

[40] Kim, M.; Yim, J. H.; Kim, S-Y.; Kim, H. S.; Lle, W. G.; Kim, S. J.; Kang, P-S.; Lee, C-K. 2012. In vitro inhibition of influenza A virus infection by marine microalga-derived sulfated polysaccharide p-KG03. *Antiviral Research*, 93, 253-259.

[41] Lerouge, P.; Roche P.; Faucher C.; Maillet, F.; Truchet, G.; Promé, J. C.; Dénarié, J. 1990. Symbiotic host-specificity of Rhizobium meliloti is determined by a sulphated and acylated glucosamine oligosaccharide signal. *Nature*, 344, 781-784.

[42] Townsend II, G.E.; Forsberg, L.S.; Keating, D.H. 2006. *Mesorhizobium loti* produces nodPQ-dependent sulphated cell surface polysaccharides. *Journal of* Bacteriology, 188 (24), 8560-8572.

[43] Vanbleu, E.; Choudhury, B.P.; Carlson, R.W.; Vanderleyden, J. 2005 The nodPQ genes in *Azospirillum brasilense* sp7 are involved in sulphation of lipopolysaccharides. *Environmental Microbiology,* 7(11), 1769-1774.

[44] Goren, M. B.; Grange, J. M.; Aber, V. R.; Allen, B. W.; Mitchison, D. A. 1982. Role of lipid content and hydrogen peroxide susceptibility in determining the guinea-pig virulence of *Mycobacterium tuberculosis*. *British Journal of Experimental* Pathology, 693-700.

[45] Matsuda, M.; Yamori, T; Naitoh, M.; Okutani, K. 2003. Structural Revision of Sulfated Polysaccharide B-1 Isolated from a Marine Pseudomonas Species and Its Cytotoxic Activity Against Human Cancer Cell Lines. *Marine Biotechnology,* 5, 13-19.

[46] Arvaniti, A.; Karamanos, N.K.; Dimitracopoulos, G.; Anastassiou, E. D. 1994. Isolation and characterization of a novel 20-kDa sulfated polysaccharide from the extracellular slime layer of *Staphylococcus epidermidis*. *Archives of Biochemistry and Biophysics*, 308 (2), 432-8.

[47] Raguénès G. H.; Peres, A.; Ruimy, R.; Pignet, p.; Christen, R. Loaec, M.; Rougeax, H.; Barbier, G.; Guezennec, J. G. 1997. *Alteromonas infernus* sp. nov., a new polysaccharide-producing bacterium isolated from a deep-sea hydrothermal vent. *Journal of Applied Microbiology*, 82(4), 422-430.

[48] Cambon-Bonavita, M.A.; Raguenes, G.; Jean, J.; Vincent, P.; Guezennec, J. 2002. A novel polymer produced by a bacterium isolated from a deep-sea hydrothermal vent polychaete annelid, *Journal of Applied Microbiology*. 93(2), 310–315.

[49] Mata, J. A.; Béjar, V.; Llamas, I.; Arias, S.; Bressollier, P.; Tallon, R.; Urdaci, M. C.; Quesada, E. 2006. Exopolysaccharides produced by the recently described halophilic bacteria *Halomonas ventosae* and *Halomonas anticariensis*. *Research in Microbiology*, 157, 827–835.

[50] Béjar, V.; Llamas, I.; Calvo, C.; Quesada, E. 1998. Characterization of exopolysaccharides produced by 19 halophilic strains of the species *Halomonas eurihalina*. *Journal of Biotechnology*, 61, 135–141.

[51] Arias, S.; del Moral, A.; Ferrer, M. R.; Tallon, R.; Quesada, E.; Béjar, V. 2003. Mauran, an exopolysaccharide produced by the halophilic bacterium *Halomonas maura*, with a novel composition and interesting properties for biotechnology. *Extremophiles*, 7, 319–326.

[52] Raveendran, S.; Poulose, A. C.; Yoshida, Y.; Maekawa, T.; Kumar, D.M. 2013. Bacterial exopolysaccharide based nanoparticles for sustained drug delivery, cancer chemotherapy and bioimaging. *Carbohydrate Polymers*. 91, 22–32.

[53] Raveendran, S.; Dhandayuthapani, B.; Nagaoka, Y.; Yoshida, Y.; Maekawa, T.; & Kumar, S. 2013. Biocompatible nanofibers based on extremophilic bacterial polysaccharide, mauran from *Halomonas maura*. *Carbohydrate Polymers*, 92, 1225–1233.

[54] Llamas, I.; Moral, A. D.; Martinez-Checa, F.; Arco, Y.; Arias, S.; Quesada, E. 2006. *Halomonas maura* is a physiologically versatile bacterium of both ecological and biotechnological interest. *Antonie van Leeuwenhoek*, 89, 395–403.

[55] Raveendran, S.; Palaninathan, V.; Chauhan, N.; Sakamoto, Y.; Yoshida, Y.; Maekawa, T.; Mohanan, P. V.; Kumar, D. S. 2013. *In vitro* evaluation of antioxidant defense mechanism and hemocompatibility of mauran. *Carbohydrate Polymers*, 98 (1), 108–115.

In: Sulfated Polysaccharides
Editors: M. Gama, H. Nader, H. de Oliveira Rocha
ISBN: 978-1-63482-975-5

Chapter 11

METHODS OF CHARACTERIZATION OF SULFATED OLIGO-POLYSACCHARIDES

Annamaria Naggi* and Giangiacomo Torri
Istituto di Ricerche Chimiche e Biochimiche “G. Ronzoni”, Milan, Italy

ABSTRACT

Sulfated polysaccharides (SPs) are widely distributed in many organisms where they regulate several biological properties. They belong to the class of block polymers or heteropolymers and respect to other macromolecules, present a higher level of complexity, not only for the primary structure, but also for their microheterogeneity, that is due to fine differences in structure, either of the constituent monosaccharides, their arrangement, linkage type, degree and position of substitution together with variation in molecular weight distribution. Due to the improvements in the sensitivity of analytical techniques, such microheterogeneity is likely to become an increasingly important factor to be considered.

In this chapter we will give an overview of established methodologies and new approaches useful to establish the structure of sulfated polysaccharides with particular attention to the application of nuclear magnetic resonance, also combined with the use of a recent statistical approach in detecting principal components, and mass spectrometry for the structural identification of oligomers.

1. INTRODUCTION

At the simplest level, polysaccharides consist of more than 10 monosaccharidic units. Their linear or branched combinations through glycosidic linkages with elimination of water form oligosaccharides (from 2 to 10 monomers) and polysaccharides (from 10 to potentially thousands of units). Within this framework there is a great number of distinct combinations including structural modifications provided by nature or by synthetic derivatizations, such as sulfations. In a polysaccharide chain each monosaccharide constituent adopts a particular

* Email: naggi@ronzoni.it.

cyclic structure (most commonly, a hexo-pyranose) and a peculiar anomeric configuration (α or β) at carbon 1. Based on the Fischer projection formulae, most the monosaccharide residues of polysaccharides belong to the D- series, while others (especially some algal components, or iduronates in glycosaminoglycans) belong to the L-series. In conformational representations, the α and β configuration at the anomeric C1 determines the position of its OH or substituent being respectively axial or equatorial with respect to the plane of the ring when the conformation is $^4C_1$ [1].

The structural contribution of the anomeric configuration is especially important in determining the shape of the polysaccharide chains, which also depends on the conformation of the component residues. These residues can adopt different conformations, including chairs, boats, or skewed forms and the only one energetically most favorable will predominate. In general, the most favored forms are those in which bulky groups (such as OH, CH_2OH or COOH) are sufficiently distant from each other, as in the optimal case of β-D-glucopyranose or β-D-glucuronic acid in the $^4C_1$ chair form, in which all groups lie equatorial with respect to the plane of the ring. Other regular arrangements of linkage geometries lead to overall structures which become more complex and more difficult to elucidate as the degree of sequence heterogeneity and branching increases. Irregularities in the sequence of a polysaccharide may sometimes have a great impact on the overall structure, as exemplified in other chapters of this book.

Compared to other classes of biological macromolecules (for example, proteins, that consist of a limited number of amino acids usually connected in a linear way through peptide bonds) polysaccharides present a higher level of complexity in their primary structure and offer a major challenge to structural investigations.

Polysaccharides that have some residues bearing sulfate substituents are widely distributed in many organisms. They include, i.a, carrageenans, fucoidans, and glycosaminoglycans (GAGs).

The presence of several anionic groups determines in these molecules, as in the case of GAGs, the capacity to interact with proteins and to modulate their biological properties. For this reason, numerous attempts have been made to mimic the structure and activity of GAGs through sulfation of a variety of polysaccharides [2, 3], chemical synthesis of sulfated oligosaccharides [4-6] and their mimetics [7]. Since some of the structural motifs in originally sulfated molecules or induced by introducing extra sulfate groups are responsible for a number of patho-physiological interactions, in order to establish structure-activity relationships it is important to characterize their structure.

In this chapter we will give an overview of established methodologies and new approaches useful to establish the structure of sulfated polysaccharides (SPs). More in detail, we will take into consideration aspects relating to the determination of the molecular weight and its distribution, the evaluation of the composition homogeneity with respect to both size and charge density, compositional analysis though selective chemical or enzymatic depolymerization, the use of mass spectrometry (MS) for the structural identification of oligomers, and nuclear magnetic resonance (NMR) also combined with the use of a recent statistical approach in detecting principal components. We will also briefly mention application of molecular modeling and NMR methods to determine the conformation of sulfated oligosaccharides in solution, both in the free state and in complexes with proteins.

2. Isolation and Homogeneity

Separation and purification of polysaccharides from natural sources can take advantage from their different solubility and binding properties compared to other biological species, such as proteins and lipid components. The majority of polysaccharides is water soluble and are not affected by treatments designed to denature proteins, they can be precipitated from water solutions by alcohols or other organic solvents and, in the case of SPs with hydrophobic quaternary compounds [8, 9]. Alternatively ionic exchange chromatography is employed [10, 11]. Formation of insoluble complexes with metal ions is exploited for isolation and purification of polysaccharides such as dermatan sulfate [12] and galactomannans [13].

The isolation, purification and characterization processes are limited by the intrinsic heterogeneity and molecular weight dispersion of polysaccharide samples. A symmetrical gel permeation or electrophoresis profile is normally considered a sufficient proof for polysaccharide homogeneity, and can be supported by an ion-exchange chromatography or an electrophoretic analysis. This type of analytical approach has been adopted by some pharmacopoeias as a criterion of purity as in the case of heparins that must be analyzed by a SAX HPLC method able to identify polysaccharides with different degree of sulfation such as dermatan sulfate (a natural impurity) and oversulfated chondroitin sulfate (OSCS), a toxic contaminant [14].

Capillary electrophoresis (CE), due to its high resolving power and sensitivity, has been applied in the analysis of intact GAGs and GAG-derived oligo- and di-saccharides [15], it was also proposed as an alternative method to monitor presence of contaminants in heparin preparations. CE has been shown useful for a reliable quantification of OSCS in heparin samples, but also allowed the determination of dermatan sulfate (DS), a common heparin impurity [16] CE with laser-induced fluorescence detection have been used for the separation and quantification of carrageenan polysaccharides [17] and capillary zone electrophoresis to optimize the extraction process of fucoidans [18].

Interaction with proteins is a fundamental property for the biological activity of SPs. These highly acidic macromolecules can bind non-specifically to any basic domain on a protein surface, but also to some very specific protein substructures having high affinity for particular carbohydrate sequences, the best-known example being the heparin pentasaccharide with high affinity for antithrombin III (AT) [19]. These properties can be exploited both for extraction purposes and/or to study their biological properties.

Affinity chromatography, in function of the different ionic strength required for eluting polysaccharides from immobilized proteins, can detect and quantify the presence of binding structures. Typical examples are the isolation of AT affinity species from heparin [20] and from LMW heparin [21] as shown in Figure 1.

The same approach was applied with other proteins such as heparin cofactor 2 [22] or growth factors and their receptors [23, 24].

Developments on a greatly reduced scale fractionation of GAG oligosaccharide mixtures were obtained with surface noncovalent affinity mass spectrometry (SNA-MS). Target proteins previously adsorbed onto a solid matrix were bound *in situ* with a mixture of oligosaccharides, whose components can be selectively eluted using increased ionic strength (i.e., NaCl), and their structures determined by matrix-assisted laser desorption ionization (MALDI) MS [25]. MALDI MS can also be used to discriminate between specific and nonspecific protein binding, by displacing the oligosaccharides aspecifically bound to a target protein with different concentrations of another protein, as in the case of LMWH oligomers and ATIII versus albumin [26].

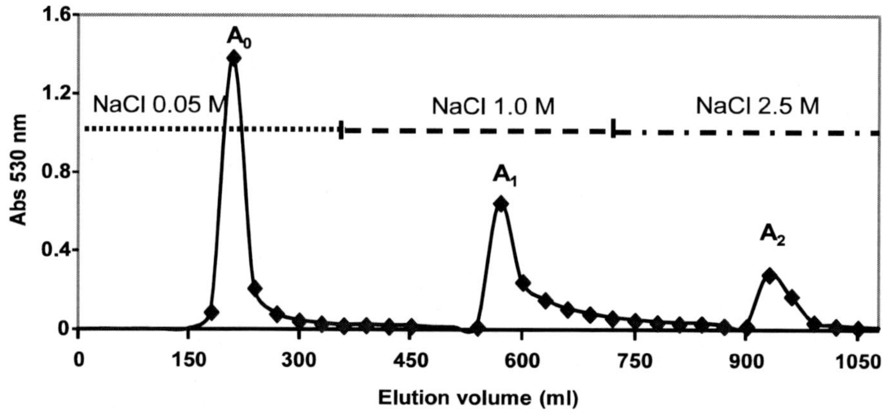

Figure 1. Affinity chromatography on AT-Sepharose of Lovenox, with a three-step salt gradient. A0 no affinity fraction, A1 and A2 affinity fractions at different interaction strength. The salt concentration used for elution is indicated by the broken line. Sample detection performed trough a colorimetric approach (carbazole).

Dynamic affinity capillary electrophoresis (ACE) was used for determining the binding constants of SPs to ATIII [27, 28], the dissociation constant for a heparin-binding domain of amyloid precursor protein (APP) to both heparins and heparan sulfate [29]. Other approaches for quantitative analysis of the interactions of SPs (mainly heparin and heparan sulfate) with proteins have been reviewed by Turnbull et al. [30].

3. Molecular Weight and Its Distribution

Biological activities of sulfated oligo- and poly-saccharides [31-36] are closely related to their molecular weights [37] Early on, the method most commonly used for the determination of the molecular weight and size distribution was the acrylamide gel electrophoresis [38, 39]. Currently the high-performance size-exclusion chromatography (HPSEC) is the method of choice [40]. Both these methods require appropriate reference standards, derived from the same component under study, whose molecular weight should be also determined with an alternative technique, because, as shown by Bertini et al. for similar sulfated GAGs, different backbones and sulfation degrees can generate different hydrodynamic volumes (Figure 2) [41].

The use of light scattering measurement avoids the long time consuming work of isolation and characterization of the reference standard of the same component to be analyzed. When combined with gel filtration chromatography and different detectors such as viscometer, refraction index and or UV/Vis spectrophotometer (HP SEC TDA), it allows the determination of Mw, Mn and their distribution, taking in consideration that, for this last, the results could depend on the column properties [42].

Particular attention should be paid on the choice of the static light scattering (LS) instrument, the low angle being the proper one for polymers (LALS). Analyzing low molecular weight oligomers, due to sensitivity problems, the current algorithm based on multiangle system (MALS) it seems not the proper one, while the right angle approach (RALS) seems to be the most appropriated one.

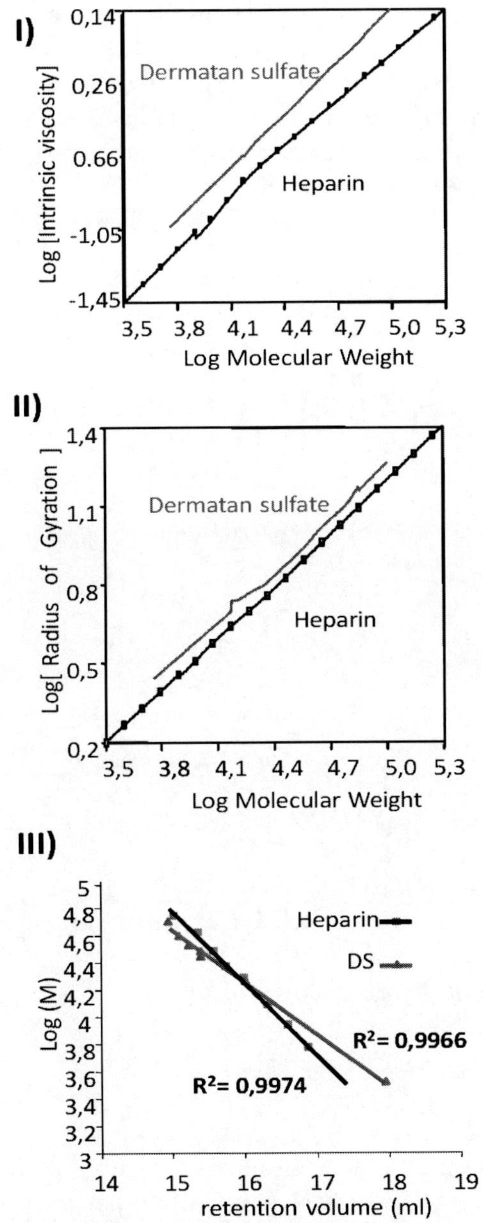

Figure 2. HP SEC TDA data obtained on two usually assumed similar sulfated polymers: dermatan sulfate and heparin. a) Comparison of Mark-Houwink plots (I) and relation between radius of gyration and molecular weight (II). b) (III) Different slopes of conventional calibration curves.

Being such HP SEC TDA technique considered, rightly or wrongly, too much sophisticated to be of wide use, the 'traditional' HPSEC determination of the molecular weight is still required as identification test for the low molecular weight heparins from both the Unites States and European Pharmacopoeia, though with different methods, and it is now proposed its introduction also in the monograph of unfractionated heparin [43, 44].

Polyacrylamide gel electrophoresis (PAGE) was also used for determination of molecular weights of heparin and LMW heparin preparations [45, 46], being particularly useful when the amount of polysaccharide sample available is particularly low [47].

Appropriate chromatographic stationary phases, such as Bio-Gel P6/P10 [21] or Superdex 30, [48] can give high resolution profiles of low-molecular weight polysaccharide mixtures and used for chain mapping characterization. An example of a separation profile of a LMWH on a GPC column is shown in Figure 3.

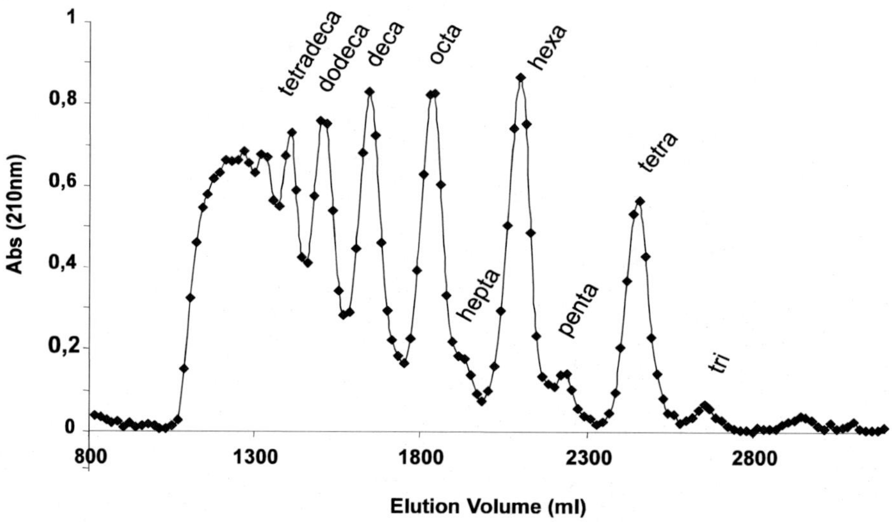

Figure 3. Size exclusion chromatography (SEC) elution profile on Biogel P6 of a LMWH (enoxaparin).

4. Structure Determination

4.1. Summary of Classic Approaches

In most cases, it is essential to approach the problem of primary structure sequence elucidation from a number of different angles in order to arrive at information concerning linkage position and configuration. A general, and it must be stressed, oversimplified classification of the arrangement of the constituent units may be described by the following terminology: linear or branched, homopolymer, co-polymers, block polymers, and heteropolymers. However this classification is inadequate when some fine differences among the components of a sample are considered. For example in GAGs there is usually a distribution of molecular weights, and very often small differences in structure, either of the constituent monosaccharides, their arrangement, linkage type, or the degree and position of substitution, which should be detected. These small but significant differences in structure are collectively termed microheterogeneity. With improvements in the sensitivity of the analytical techniques, microheterogeneity is likely to become more and more an increasingly important matter to be investigated.

In a few instances it is possible to sequence a polysaccharide by NMR alone but even then it is strongly advisable to verify the structure by at least another method. Usually, a large number of approaches are employed including chemical, enzymatic depolymerization and spectroscopic methods. The first practical hurdle facing the investigator is to obtain a sufficient quantity of suitably pure material on which to carry out a structural investigation. In many cases, this is the limiting factor and may determine which techniques are deemed to be the most suitable. Examples of such investigations are those concerning some proteoglycans from human aorta [49] or heparan sulfate from human brain [50].

The hydrolysis of a polysaccharide down to its constituent oligo- or mono-saccharides followed by their chromatographic separation and identification is the most old and still common method of carrying out a compositional analysis as for the elucidation of primary structures of several complex sulfated galactans and xylogalactans [51].

Typical acid hydrolysis conditions are characterized by the use of 1N acid solution at 100 C° for two hours, but in certain cases, stronger conditions are required. Glycosidic linkages involving uronic acids or non N-acetylated aminosugars are known to be more resistant to acid hydrolysis [52]. This kind of hydrolysis can give preliminary information about the sulfated polysaccharide backbone but no information about the degree and pattern of sulfation.

To identify and/or quantify the mono or oligosaccharides obtained, paper (PC), thin layer chromatography (TLC), gas-liquid (GLC), high performance liquid (HPLC), supercritical fluid [53] chromatographies and electrophoresis [54] can be employed. GLC and capillary GLC require a pre-derivatization to obtain volatile derivatives such as methyl ethers, trimethylsilyl ethers and alditol acetates. Electron -impact mass spectrometry (e.i. MS) can also be used.

It is possible to identify certain linkage types on the basis of their different degrees of chemical and enzymatic susceptibility and this may also provide information on the sequence. Beta elimination by alkaline treatment of uronic acid containing polysaccharides [55, 56] and deaminative degradation with nitrous acid, to cleave the glycosidic linkages of the 2-amino sugars, has been employed especially for structural studies on GAGs [57, 58]. This method has also been used to study the composition of keratan sulfate from bovine articular cartilage, allowing the preservation of the fucose residues [59].

Enzymatic hydrolysis preserving sulfate groups was extensively used to determine the disaccharide composition of GAGs by CE, SAX HPLC or ion-pair HPLC chromatography [60-63].

A number of procedures have been designed to selectively derivatize the glycan reducing terminal unit, with a group to improve the sensitivity of detection, and allow the characterization of heparin/HS saccharides obtained from small pieces of tissue or cell specimens [64].

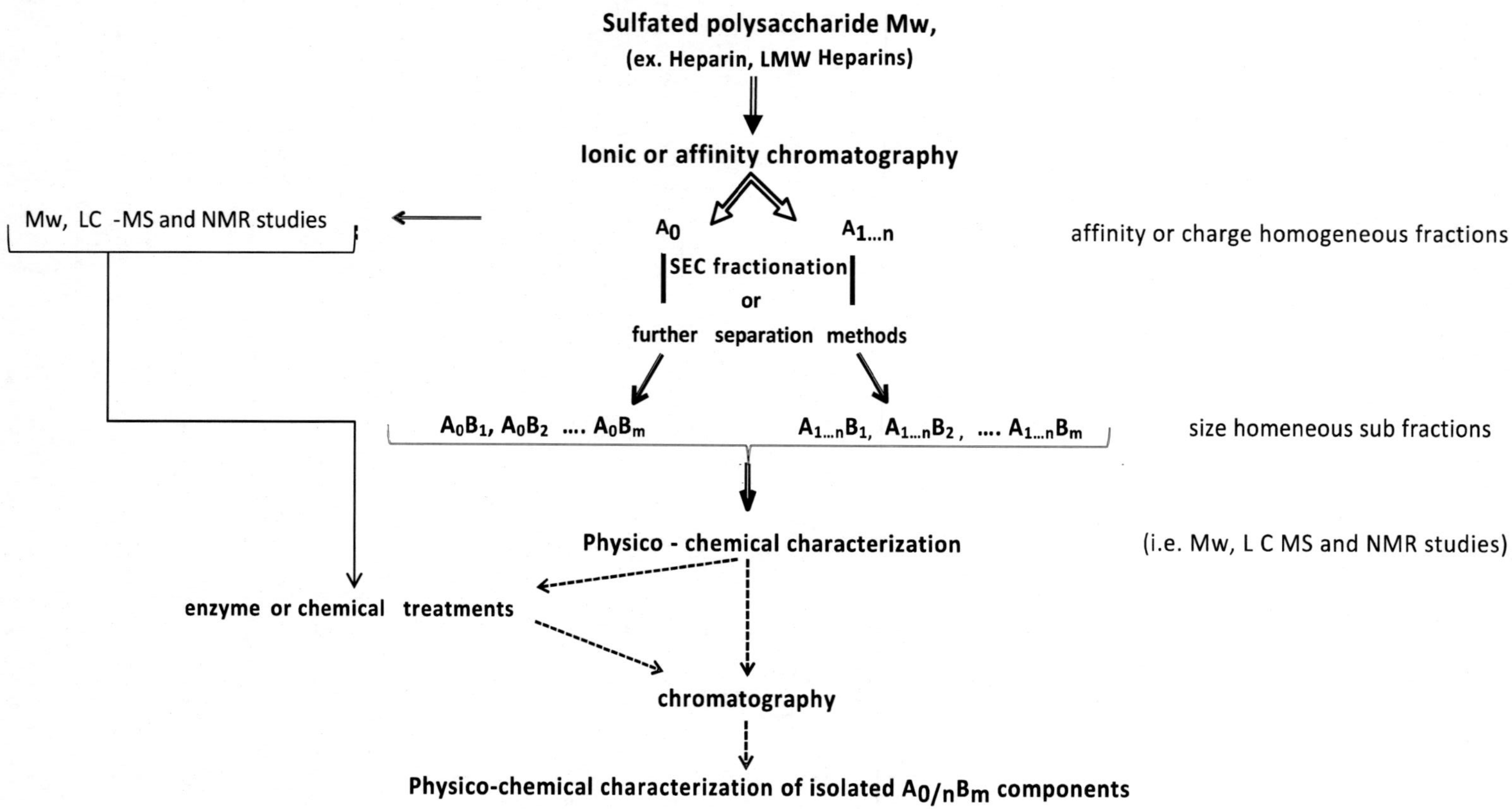

Figure 4. Scheme of a sequential structural determination of a sulfated polysaccharide.

Mixture of oligosaccharides obtained by enzymatic hydrolysis using specific endoglycosidases can be separated, characterized and further hydrolyzed in a systematic way with exoglycosidases giving sequence information starting from the non-reducing end unit. This approach has been used for sequencing heparin or heparan sulfate oligosaccharides [65-68]. The Figure 4 shows a scheme summarizing the typical structural characterization approach of a sulfated polysaccharide.

4.2. Nuclear Magnetic Resonance

The subject of NMR spectroscopy applied to polysaccharides has already been dealt with in several monographs [69-71] and our aim here is to give an update with specific references to multidimensional homo- and hetero- nuclear techniques, useful for facing NMR problems even in both the most common and also the most recent applications of NMR techniques for characterizing SPs in solution. NMR spectroscopy has become a discipline in its own right and, because of the sophistication of the experimental application, skilled spectroscopists are required. This is a particular need in the carbohydrate field where NMR spectroscopy has to be considered the 'queen' technique for the molecular characterization and study of sulfated oligo- and polysaccharides.

In general, the NMR spectra of a macromolecular system separated signals are given for each type of magnetically active nucleus (^{1}H, ^{13}C, ^{31}P, ^{15}N). In the case of the polysaccharides two main magnetically active nuclei frequently occur: the proton (^{1}H) and an isotope of carbon (^{13}C). ^{1}H and ^{13}C spectroscopy, used individually or in parallel, are useful in determining not only the structure of the polysaccharide but also in evaluating other parameters such as the degree of homogeneity or heterogeneity, the degree of structural change due to chemical or enzymatic derivatization or degradation. Other nuclei, like ^{31}P in the poly-deoxy ribonucleotides and deuterium (^{2}H), can be used for very specialized applications in polysaccharide NMR spectroscopy. On the other hand nitrogen is not as important as it is in the study of proteins but a recent publication suggests an interesting use of ^{15}N nucleus for the characterization of the microstructure of heparin and heparan sulfate using N-sulfoglucosamine as probe residue [72].

There are three fundamental experimental measures in NMR spectroscopy: chemical shift, signal intensity and coupling constant. Chemical shift is a measure of the degree of deshielding experienced by a nucleus, and is related to its chemical environment. Signal intensity is a measure of the number of times an excitation process has occurred, and can be related to the number of nuclei giving rise to a particular signal. The coupling constant is a measure of the interaction between nuclei, and is independent of the applied magnetic field.

The interaction of the various nuclear magnetic moments is a factor peculiar to NMR spectra and is the very basis of the actual evolution of the NMR discipline. This nuclear magnetic interaction leads to spin- spin interactions that generate coupling between nuclei of identical species, as seen in ^{1}H spectra, and nuclei of different species magnetically active. For example, in case of ^{13}C spectroscopy this nucleus, which is normally in contact with hydrogen atoms and other carbon atoms, will give rise to heteronuclear coupling (^{1}H-^{13}C) starting point for bidimensional heteronuclear correlation experiments (see later).

The homonuclear ^{13}C-^{13}C coupling being of little statistical significance, due to its low natural abundance (1.1%), in early times were not taken in account for habitual use [73].

If during the measurement the hydrogen nuclei are kept in the excited state, impeding relaxation, (the typical monodimensional (1 D) ^{13}C proton decoupled experiment) the nuclei cannot contribute to the spin-spin coupling and the result is a ^{13}C spectrum consisting of as many signals as are the magnetically different carbon atoms present in the system. On the other hand, such system simplification is not possible in ^{1}H spectroscopy as homonuclear interactions are involved and recourse must be made to selective saturation. This leads to either the techniques of selective excitation or to the traditional, selective homonuclear decoupling methods being this last of less common use.

Because of the low dispersion of carbohydrate NMR signals, the coexistence, in the same dimension of two independent variables, chemical shift and scalar coupling, is one of the causes of NMR spectrum interpretation difficulties and it led to the development of bidimensional techniques.

To characterize a glycan of unknown structure, NMR spectroscopy provides all the experimental strategies for structural and conformational determination. Furthermore, the probe cryogenic technology associated with high field magnets has widened the possibility to carry out analyses on sub-nano-molar scale, also in the carbohydrate field. Now it is possible to detect glycoconjugates in complex biological samples, identify contaminants in pharmaceutical products, perform studies on carbohydrate interaction with proteins and develop quantitative analysis of complex carbohydrates [36, 74-77]. More recently the monosaccharide composition of complex carbohydrates, determined by quantitative-HSQC through $J_{C\text{-}H}$ dependence for the anomeric carbons exclusively by NMR spectroscopy, was found to be comparable to those obtained from the GC–MS-carboxy reduction approaches [78].

In any case it is necessary to select, from the numerous mono- and bidimensional available techniques, the most suitable one for each experimental aim.

4.3. ^{13}C NMR

Compared with proton spectroscopy, ^{13}C spectroscopy has a greater signal dispersion 200 ppm compared to 10 ppm for ^{1}H, that becomes respectively about 60 to 2 ppm for the carbohydrate ring atoms due to the lower natural abundance of the nucleus and the lower value of its gyromagnetic constant leading to a decrease in sensitivity, by a factor of about 5600 with respect to hydrogen.

^{13}C spectroscopy owes its widespread use to the relative simplicity of its traces and, as already illustrated, the heteronuclear couplings can easily be removed by decoupling hydrogen hence each signal represents a specific nuclear situation. This is probably the reason why ^{13}C NMR still plays a significant role in the polysaccharide field but this advantageous simplification of the spectra has its price. The intensity and the area of the signals of each carbon nucleus become dependent on the number of hydrogens bonded to it (Overhauser effect, nOe), a factor that makes ^{13}C spectra non-quantitative [73].

It has been demonstrated that, in both sulfated and non-sulfated polysaccharide spectra, quantitative evaluations are always possible by comparing signals derived from atoms of the same nature e.g., those of primary alcohol groups.

The ^{13}C spectrum of a polysaccharide can be divided into different spectral regions characteristic of each carbohydrate functional group. The most informative is usually the region containing the anomeric signals, i.e., those of the ^{13}C nuclei involved in acetal or hemiacetal bonds, a region ranging from about 95 to 115 ppm. It must be kept in mind that the value of the chemical shift of the anomeric signal is a function of different factors; the ring type, the anomeric configuration, its implication (or not) in a glycoside bonding, the position of the glycosidic bond and, lastly, conformation or conformational equilibria effects.

The spectrum can first be considered through the analysis of the anomeric region; from the number of signals present one can establish the number of glycosidic units that are connected in the polymer chain and whether we have a homo- or hetero-polymer. In the case of hetero-polymers it will be possible, from the relative ratio of the signals, to establish whether a copolymer is involved or not, and identify the number of glycosidic units included in the repeating unit.

Given that the objective is the identification of the various residues, a more detailed analysis of the spectrum is needed to establish the configuration of their glycosidic bonds and their position. Thus, as the compositional data are readily available, a first assignment can be made. The literature reports important reviews concerning this area [70, 79].

The aldehyde and ketone carbonyl and carboxyl signals fall downfield, around 160-200 ppm, whilst those of methyls, methyl esters and aliphatic substituents can be found upfield, 10-35 ppm. The signals of secondary alcohol groups, usually prevailing numerically, fall in the central, and consequently the most crowded, part of the spectrum, 66-80 ppm, whilst the primary alcohol groups fall around 60-64 ppm.

The presence of substituents on alcoholic groups, such as sulfates (see Figure 5), causes a shift, its sign and magnitude being a function of the electronegativity of the substituent e.g., a 6-9 ppm shift downfield indicates the formation of ether bonds [80, 81] This kind of shift, usually can contribute to signal dispersions and, in case of structurally homogeneous compounds, simplify the spectrum interpretation. Identification of signals from primary, secondary, tertiary and quaternary nuclei could be done by employing the Distorsionless Enhancement by Polarization Transfer (DEPT) technique. The technique takes advantage of the modulation of the magnetic vectors of protons bonded to the ^{13}C nuclei and induces a better signal to noise ratio compared to the normal acquisition procedure.

It is dignus of note that is generally considered an unappropriated the use of DEPT technique for quantitative evaluation, because the intensity of signals is once more dependent on proton effects.

An interesting technique, based on the ^{1}H - ^{2}H isotopic effect, have been developed and found particularly useful for carbohydrates to distinguish alcohols from ethers, especially in the case of particularly complex heterogeneous structures. In the case of exchangeable alcoholic protons it can be observed that under conditions of strict analytical equivalence, using either depleted H_2O or high titre D_2O as solvent, the variation of the proportion of isotopes results in a small variation of the chemical shift of the ^{13}C signal of the carbon nucleus attached to the hydroxyl residue [82, 83].

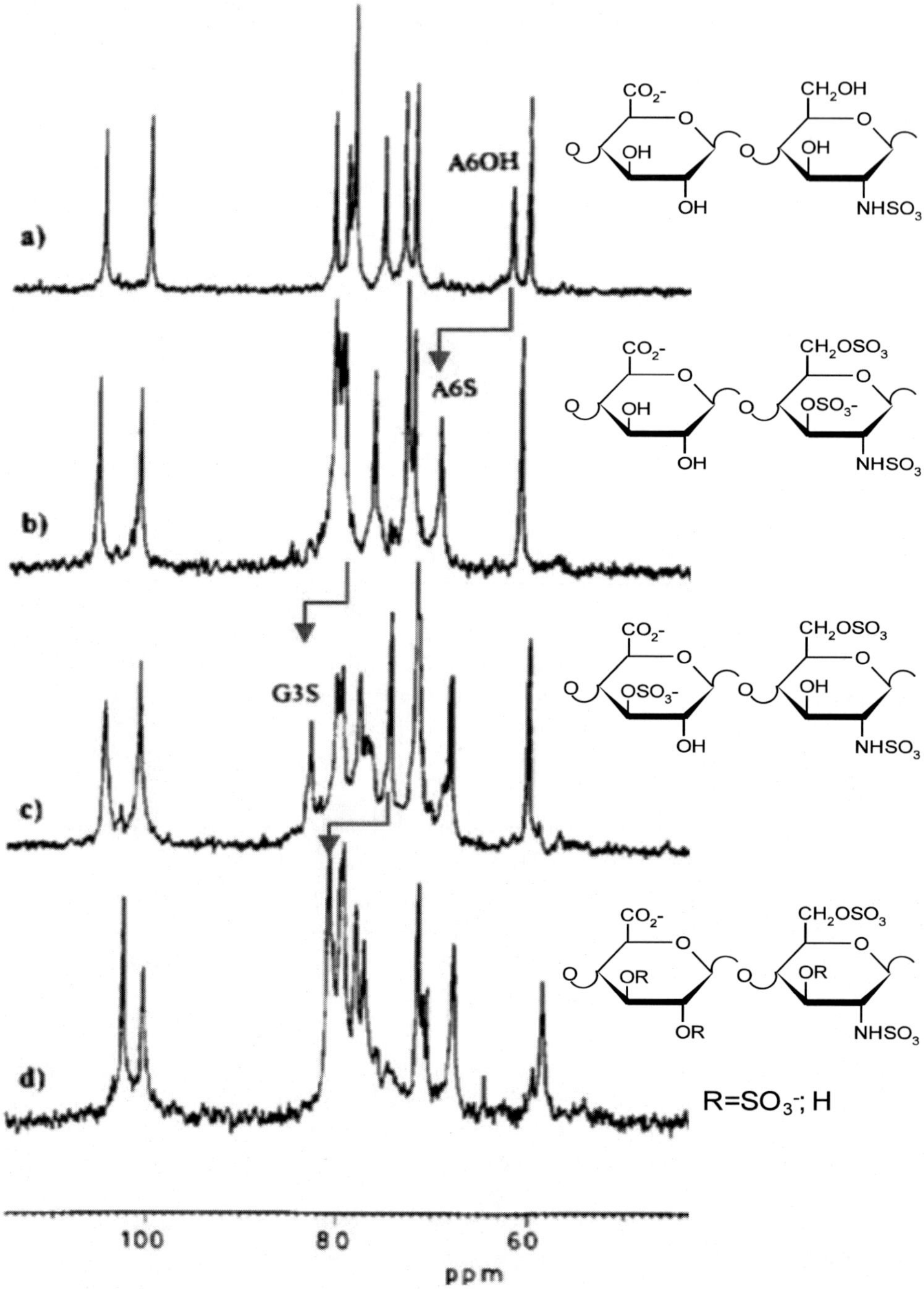

Figure 5. [13]C spectrum of a polymer constituted by α -NAcetyl D-glucosamine 1-4 β D-glucuronic acid disaccharide repeating unit (SAH, a, alias N-sulfated K5). (82) Shift effect induced by sulfation of different alcohol groups: (b) SAH-6S: 6 O-sulfated N-sulfated α D-glucosamine 1-4 β D-glucuronic acid; (c) SAH-6S,G3S: 6 O-sulfated N-sulfated α D glucosamine 1-4 3O- sulfated β D-glucuronic acid; (d) SAH-polyS: randomly -O--sulfated N-sulfated K5.

4.4. ^{1}H NMR

In general, NMR spectroscopy of hydrogen is the most widespread, being the most sensitive and usually the most informative. The proton spectrum is characterized not only by chemical shift (ppm) and by the coupling constant between neighboring hydrogens (^{3}J); in several cases it is also possible to observe long range coupling constants (^{4}J). Another characteristic, peculiar to the ^{1}H spectrum, is that through the integration of signals, a relative comparison of all the components of the spectrum can be made; accurate quantitative evaluations are also possible by the addition of internal references to the sample solution. Conformational studies often employ ^{1}H NMR, evidencing the relationship between the coupling constant and the dihedral angle of adjacent protons [83 - 84].

The ^{1}H NMR spectrum of a polysaccharide is characterized by strong signal overlapping due to similarities in the polymeric component structure and the multiplicities of each signal. As previously anticipated, the signals are compressed into a spectral region normally in the order of 1-2 ppm. For this reason, the so-called second order phenomena often take place, caused by the close proximity of the resonances making the reading of the spectrum even more difficult. In the most frequent case the use of a high magnetic field spectrometer could help in terms of sensitivity, allowing a decrease of solution concentration, i.e., viscosity, but not really in terms of direct increase of signal dispersion. In general, the nature of the chemical environment of the majority of polysaccharide nuclei is similar and there are only two nuclei which can be significantly differentiated; primarily the anomeric hydrogen and somewhat those of the primary alcohols. Obviously any variations in structure, such as sulfates ester, but also deoxy residues, double bonds, carboxyls, phosphates, amino groups and ketones, are useful for the spectra interpretation as they cause a chemical shift usually comprised between 0,2-0,7 ppm. Other factors can act as shift agents, among these the solvent effect, ionic strength and nature of ions present in solution.

Variations in chemical shift can also be utilized to study binding capacity between the polysaccharides themselves or to other molecules such as proteins, in this case, the limiting factor becomes the association constant of the complex, together with the complex stability, its solubility and the complexity of the resultant spectra.

A recent developed use of principal component analysis (PCA) technique, that is a statistical approach independent of any previously determined parameters [85-89], had opened new potentiality of ^{1}H NMR spectra of polysaccharides. The aim of PCA is to reduce the dimensionality of a dataset to find the most effective way to compare complex data allowing the objective comparison of many spectra when these are difficult to compare by visual inspections. PCA is specifically advised for proton spectra analysis of pharmaceutical macromolecules as a safeguard in detecting unusual structural modifications introduced during manufacturing processes and possible presences of contaminants of biological drugs, including active biological principles of different origin, that are all potential sources of adverse reactions [90, 91]. The use of PCA applied to proton NMR spectra succeeded recently in detecting contamination of some pharmaceutical heparin preparations, causative of hundreds cases of adverse events in many countries [92].

A general approach for constructing a suitable illustrative heparin library employing multivariate analysis was proposed and illustrated by Rudd and al. [93].

Further statistical elaborations based on PCA approach are available. It is the case of the 2D-Cos filtering that allows the detection of trace of contaminants even if they are of the same nature of the compound under analysis, i.e., heparins of different origins [94].

A drawback of such an approach is the need of an apposite distortionless spectrometer having high spectrum line basis linearity.

4.5. 2D NMR

To obviate to the signal overlapping in monodimensional spectroscopy (1D) of polysaccharides, the use of bidimensional spectroscopy (2D) has become common. It spreads the NMR signals onto a plane in such a way that the numerical coordinates of the signals depend on the experimental conditions to which the nuclear spin systems are subjected. The spectrum shows a series of peaks that develop in the third dimension (intensity) perpendicular to the plane defined by Cartesian axes (fl, f2).

Bidimensional spectra are usually represented by the "stacked plots" method (of spectral traces translated in both dimensions of the plane) that permits a three-dimensional visualization of the peaks, but poorly lends itself to a numerical evaluation, or by the most frequently used "contour plots" where the spectrum is represented by the peak projection on a plane by means of lines corresponding to sections through the "intensity" dimension of the signals.

Through the bidimensional technique it is possible to correlate two nuclei that are of either a similar (^{1}H-^{1}H,) or different nature (^{1}H-^{13}C) (homonuclear and heteronuclear correlations respectively, also termed homoscalar and heteroscalar correlations) in such a way as to identify signals involved in the reciprocal interaction, and transfer the available information from one nuclear spectrum to the other. The natural abundance of the various magnetically active nuclei and the relative sensitivity of each couple with suitable technical and instrument availability make homonuclear ^{1}H and heteronuclear ^{1}H-^{13}C correlation experiments very probative. Other experiments which need longer instrument time are possible, such as correlative approaches of the bidimensional technique which could also allow the separation of the two variables normally present on the same axis; the chemical shift and the coupling constant.

The homonuclear correlated spectrometry is the most widespread technique. The COSY spectrum (COrrelation SpectroscopY) takes advantage of a well-defined starting point e.g., an anomeric signal, and sequentially to assign the signals of the neighboring protons. In analyzing the sulfated homo-polysaccharides, because of the shift effect induced by the presence of sulfated groups, in some case this technique could be sufficient to permit a complete recognition of the structure. In case of sulfated complex heteropolymers, the structural analysis of their oligo- and poly-saccharidic units can be carried out by the TOCSY (TOtal Correlation SpectroscopY). The experiment, based on the propagation of the magnetization from one atom to its neighbor, allows the identifications of all the atoms of a carbohydrate ring. The correlation signals appear, in the contour plot map, lined to their parent signal, for example starting from the anomeric signals.

Having cited the most diffuse 1H 2D NMR experiments, others useful for specific aims are available but, due to their complexity and being their description out of the scope of this review, one can refer to dedicated texts or to NMR specialists.

4.6. Heteronuclear Correlated Spectra

Early on the bidimensional heteronuclear ^{13}C-^{1}H spectroscopy (HSQC), was considered as a very useful technique allowing maximum signal spreading in the ^{13}C NMR spectrum. The dispersion of ^{1}H spectrum signals along to the ^{13}C axis allows not only to identify overlapping signals but also approximation of the ^{1}H coupling constants to be read. Problems of sensitivity, and hence of availability of instrument time, viscosity and, not least, sample availability can make difficult to obtain a 2D spectrum. Recent developments in the 2D spectrometry and the availability of more and more sophisticated instruments have made possible to improve performance and application of this technique in a routine way up to be used for quantitative purposes.

The quantitative HSQC experiments needs to be set using average values of the expected $^{1}J_{C\text{-}H}$ couplings of the individual atoms. The ''tune'' coupling set in the experiment and the real coupling ''true'' values for each atom cause a variation of the integrated peak volumes proportional to the differences in their values [95].

In porcine mucosal heparin HSQC spectra it have been observed that, for glucosamine and uronic acid residues, differences up to 5- to 8-Hz among values, do not affect their calculated content. On the contrary, comparison among volumes of cross-peaks, corresponding to atoms with larger differences of $^{1}J_{C\text{-}H}$ values can generate significant errors in the compositional analysis [96, 97].

An interesting example of the application of HSQC technique is shown in Figure 6 where the spectrum of a semi-synthetic sulfated maltotriose *alfa – beta* C-C linked dimers (SMTCs) is compared with its proton 1D spectrum and that of its non-sulfated precursor [98, 99].

The typical shifts induced by the sulfates on the oligosaccharide signals is particularly evident from the comparison of the 1D spectra, while the HSQC spectrum lead to variations in both the coordinates of the bidimensional map, in such a way as to highlight both the direct effects on the peaks, due to sulfation, and those indirect effects, the so called neighboring alfa and beta shift effects, for instance arising from the insertion of charge into the structure. The full number of possibilities which are available including various correlative techniques, some of which may not be feasible in a given case, can also be integrated by studies of analogous model compounds or products of chemical and or enzymatic degradations.

Table 1 shows the complete proton and carbon assignments of the derivatives with different sulfation pattern of K5PS, a polymer constituted by α -N-Acetyl D-glucosamine 1-4 β D-glucuronic acid disaccharide repeating unit. Through a comparison between the chemical shifts of each derivative one can estimate the magnitude of the direct and indirect effects of sulfation as was already graphically shown in Figure 5 for the ^{13}C spectra.

The most advanced spectrometers are able to perform sophisticated experiments such as Heteronuclear Multiple Bond Correlation (HMBC) and others characterized mainly by the combination of HSQC with other techniques derived from the 1D and 2D homonuclear experiments. Experiments such as hetero DEPT, hetero COSY and hetero TOCSY are very informative for the characterization of macromolecular structures and are becoming of more and more large use.

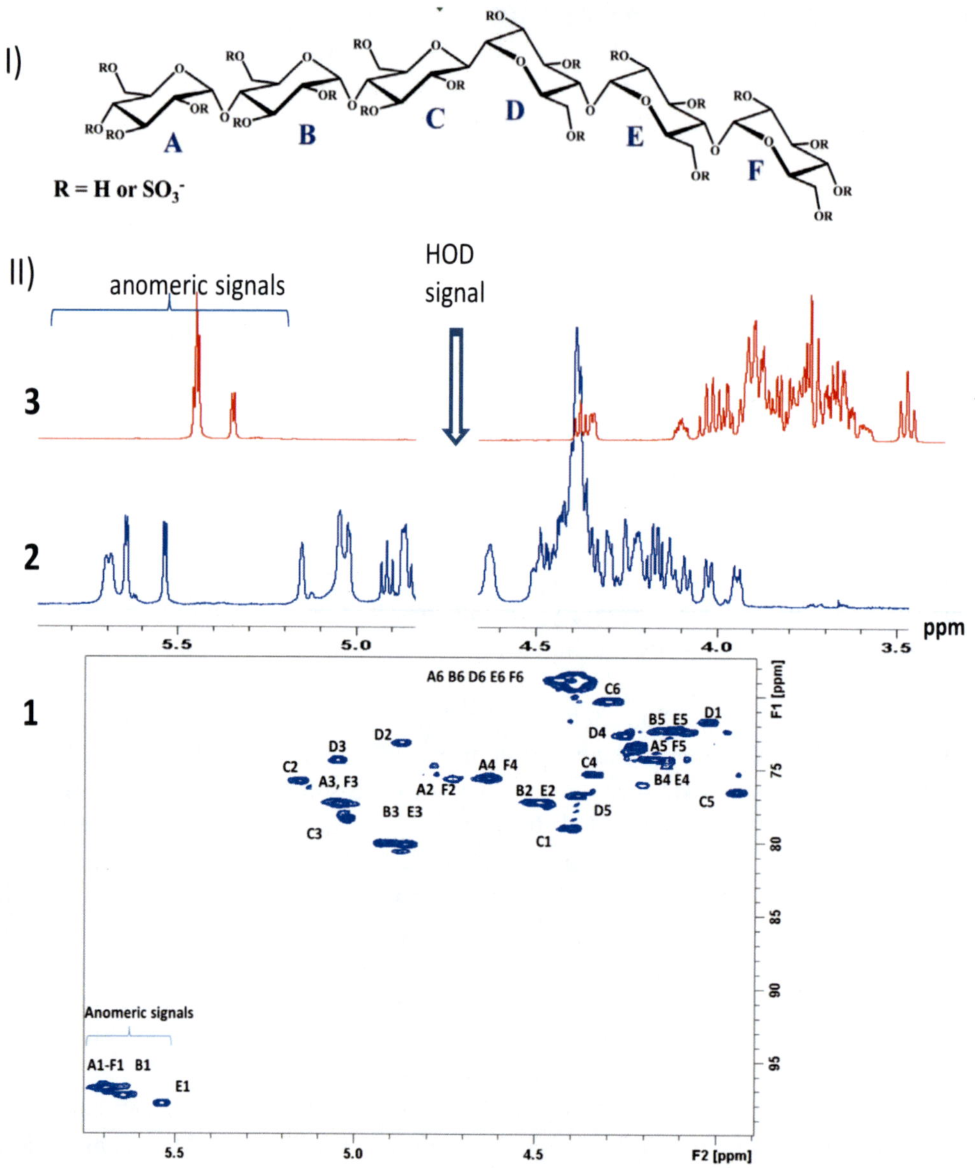

Figure 6. I) Structure of semi-synthetic maltotriose alfa – beta C-C linked dimers. Letters are indicating the different glucose rings; II) HSQC spectrum of a fully sulfated derivative oligosaccharide mimic (1) in comparison with its proton 1D spectrum (2) and that of its non-sulfated precursor (3). Letters with subscripts indicate both the carbon and proton ring atoms.

As shown before, HSQC spectra, simplifying the problem of overlapping signals, provide structural information about sample, whether it actually consist of different chemical entities or of heterogeneous polymers. To explore HSQC spectra, also in this case, statistical correlation spectroscopy have been used (HSQCcos, see Figure 7). Correlating a single component or signal, with every other component or signal, in a series of two-dimensional spectra, make possible to see which signals are each other associated. Such correlation spectra can be used to generate information concerning the composition of mixtures of molecules whose sequences are heterogeneous [100].

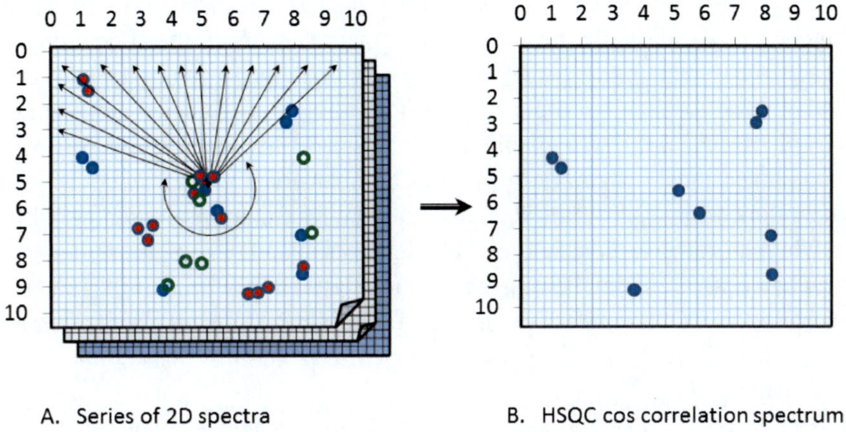

Figure 7. Schematic of the Process Behind Correlation Spectroscopy Applied to Two-Dimensional Data. (A) Single element or signal from a series of two-dimensional spectra (element h,h in this example) is correlated with every other element of the spectra. (B) The result is a correlation/HSQCcos spectrum, which reports all the signals that correlate highly with the signal that was initially interrogated.

Table 1. ^{1}H and ^{13}C chemical shifts of different derivatives of K5PS, polymer constituted by *α -N-Acetyl D-glucosamine 1-4 β D-glucuronic acid disaccharide repeating unit.* Derivatives: AH and SAH, N-deacetylated and N-sulfated glucosamine (A) respectively; SAH-6S, N, 6-O- sulfate A; SAH- N,6S,G3S, N, 6 -O sulfated A and 3-O-sulfated glucuronic acid (G); SAH-6S,G2S, N, 6 -O sulfated A and 2-O-sulfated G; SAH-polyS, randomly sulfated A and G

Prevalent sequence	K5-PS (1)		AH(2)		SAH (3)		SAH-6S (4)		SAH- 6S,G3S (5)		SAH-6S,G2S b (6)		SAH-polyS (8)	
Glucosamine														
HI/C1	5.36	99.6	5.66	98.9	5.58	99.9	5.56	100.8	5.41	100.7	5.44	100.9	5.44	100.2
H2/C2	3.89	56.1	3.41	57.1	3.27	60.5	3.27	61.1	3.26	60.1	n.d.	n.d.	3.48	58.4
H3/C3	3.86	73.	3.99	70.9	3.68	72.4	3.67	73.1	3.75	71.9	n.d.	n.d.	4.62	77.9
H4/C4	3.64	81.1	3.75	80.7	3.68	80.7	3.70	80.6	3.73	80.0	n.d.	n.d.	3.95	77.0
H5/C5	3.82	72.0	3.76	74.7	3.79	73.3	3.96	72.3	3.99	71.8	n.d.	n.d.	4.08	71.4
H6;H6'/C6	3.84;3.84	62.2	3.84;3.90	62.2	3.81-3.83	62.4	4.37;4.17	69.5	4.42;4.24	68.1	n.d.	n.d.	4.52;4.31	67.6
Glucuronic acid														
HI/C1	4.48	105.2	4.61	105.5	4.55	105.1	4.58	105.5	4.69	104.6	4.74	102.8	4.91	102.4
H2/C2	3.37	76.3	3.46	76.1	3.41	75.6	3.36	76.4	3.64	76.9	4.14	82.3	4.39	79.3
H3/C3	3.69	78.9	3.83	78.6	3.83	78.7	3.82	79.5	4.52	82.7	3.98	77.3	4.65	80.8
H4/C4	3.78	79.1	3.44	79.7	3.90	78.8	3.84	79.8	4.12	77.6	3.85	79.1	4.17	79.7
H5/C5	3.78	79.1	4.14	76.7	3.83	79.3	3.78	80.3	3.91	79.5	3.88	79.5	3.88	79.5
SO_3^-/COO^-	0		0		1		2.2		>2		2.5		2.3	

4.7. Conformational Information

Characterized the primary structure and all the signals of the ^{1}H and ^{13}C spectrum assigned as previously described, for a conformational study such an approach and route adopted cannot directly allow the measurement of the values of all the ^{3}J coupling constants of the system and the interatomic distances. It is known that the ^{3}J values depend on values of the dihedral angles formed by the planes where the proton nuclei are found but 2D spectroscopy does not lend itself to this aspect. Instead, this information can be supplied through monodimensional spectra derived from bidimensional experiments. The result is a selective monodimensional spectrum, recorded with the maximum resolution permitted by the system, showing the coupling constants even of signals situated in complex regions.

Other techniques can be applied to obtain conformational information, among them the ^{1}H 1D- selective COSY. The selective impulse on a signal allows the identification of the neighboring protons through the appearance of the corresponding signals in phase dispersion, and the measurement of the ^{3}J coupling constants.

As for the corresponding 2D spectra, this technique, when applied to polysaccharide spectra, can resolve only relatively simple local situations. In fact lack of magnetization transferring beyond the neighboring nucleus resulted in an overcrowded spectral region with signal overlap hindering further investigation. To bypass this hindrance it is more practical to make use of total ID correlation (ID-TOCSY) allowing the selective excitation of a nucleus to be transferred to all the protons of a monosaccharide ring and the measurement of J^{3}.

The nuclear Overhauser effect (nOe) is another useful NMR tool for conformational studies. It arises from relaxation processes facilitated by dipole-dipole interactions which occur through space, not through bonds as is the case in scalar coupling. The magnitude of the effect is related to the magnetic field component at the nucleus under observation and to the distance between the interacting dipoles, there being a rapid fall-off between the strength of the interaction and the distance between the nuclei.

In polysaccharides, this effect can be useful for a variety of reasons. Interactions between nuclei on either side of a glycosidic linkage and between nuclei within a given spin system can be observed. That allows to approach conformational questions about not only the glycosidic linkage but also the sugar ring. There are two types of nOe experiments, the first is the two dimensional one (NOESY) which can afford an improved resolution through signal spread and the simultaneous observation of multiple interactions, which is efficient especially in providing qualitative information. The second is the one dimensional experiment, by which it is possible to measure the signal integration more accurately, which is essential for approaching quantitative measurements. Peculiar advantage of nOe polysaccharide one dimensional spectra is that the anomeric region is usually well resolved allowing integrals to be accurately measured otherwise for this crowded region of the spectrum it can be problematic.

Conventional nOe experiments can sometimes fail to reveal interactions between suitably spaced nuclei because for certain medium sized molecules, the maximum possible nOe passes through a minimum. In such cases, it is strongly advisable to use an alternative type of experiment, namely the nOe in the rotating frame (ROESY). This experiment operates through a completely different mechanism to the conventional nOe experiment. This method does not show a minimum at the same point as for the nOe experiment, but this useful complementary technique requires the optimization of the spin-lock period, a procedure which is best to be carried out by an experienced operator.

There are many examples of use of nOe techniques in the carbohydrate characterization studies especially for the identification of glycosidic bonds and sequence of rings, it is also relevant its current use for interaction studies of sulfate oligosaccharide ligands to target proteins that can also be performed using the saturation transfer difference (STD) technique [101-105].

In most cases, actual conformations of poly-oligosaccharides in solution as predicted by theoretical conformational analysis, which in some cases is complicated by the co-existence of different equienergetic local conformations in a dynamic equilibrium, as in the case of iduronic acid-containing GAGs (2). On the other hand, the NMR data sometimes indicate not easily predictable conformations, as in the case of extensively sulfated chondroitin sulfate [106] and those and the heparan sulfate mimic C-C maltohexaose polysulfate (98), in which severe steric crowding of sulfate substituents cause conformational changes in glucopyranose rings as compared with the corresponding less sulfated glycans.

It must be noted that the complete 3D structure of polysaccharidic or oligosaccharidic sequences in solution has been determined only in a few favorable cases, especially for heparin oligosaccharides that could be prepared in a pure state, either synthetically or usually from LMWHs by careful separation methods, including affinity chromatography. Also, results from molecular modeling calculations should match those obtained experimentally by NMR. Several 3D structures have been obtained also for "active conformations" [102, 107, 108]

4.8. Mass Spectrometry Analysis

Mass spectrometry (MS) is widely used in the compositional characterization of oligo- and polysaccharides coupled with other techniques: gas chromatography-mass spectrometry (GC-MS), fast atom bombardament-mass spectrometry (FAB-MS), Matrix Assisted Laser Desorption ionization (MALDI), electrospray-mass spectrometry (ES-MS). As for SPs, the presence of sulfate groups was for some time a difficult problem to deal. The FAB behaviour of their O-acetylated and O-permetilated derivatives of sulfated oligosaccharides, in condition to preserve sulfate groups, are described for the first time by Dell [109,110] with promising prospectives for the characterization of a wide range of sulfated glycoconjugates, despite this several FAB spectra appear often complex especially in the presence of mixtures of oligomers. To measure the molecular ions of sulfated poly- oligosaccharides is quite difficult because of the chelation of metal cations and the trend to loss of sulfate groups. The development of approaches based on ES-MS and MALDI increased the application of MS to characterize sulfated oligosaccharides as reviewed by Zaia [111] for GAGs and by Ewart for marine SPs [112].

In the case of GAGs much work has been conducted in the development of methods for the analysis of fragments obtained by their enzymatic depolymerization coupling the MS with HPLC or CE. Method using ion trap instruments and standards have been reported [113, 114] Recently Wang et al. exploited for quantitative purposes, the MS response of heparin disaccharides separated by reverse phase ion-pairing ultra-high pressure liquid chromatography (RPIP-UHPLC) coupled with a quadrupole time-of-flight (Q-TOF) mass spectrometer [115]. The study has been devoted to the structural characterization of 20 lots of sodium heparin (active pharmaceutical ingredients, APIs) and 10 of low molecular weight heparin (LMWH) on the market.

More challenging is the approach to investigate sulfated oligosaccharide chain sequences both in natural and purified product sequences all being mixtures of different length chains. In case of pure oligosaccharides the analysis of MS/MS fragmentation pattern allows to define the sequence of different residues [116].

Analytical methods, capable of approaching complex mixtures of sulfated unit chains can be used to detect possible abnormalities such as increased or decreased levels of sulfation (see Figure 8), differences in the oligosaccharide size distributions or composition and are particularly appropriate to profiling LMW heparin preparations. For this specific aim two principal approaches can be applied: the first one based on-line size exclusion chromatography coupled with ESI, equipped with an on-line desalting system, proposed by Henriksen, that allowed the determination of more than 60 components in a LMWH sample [117]. The second one based on the reverse phase ion-pair HPLC was developed also by other groups [118-122] using gradients with different organic solvent (acetonitrile or methanol) and cationic surfactants, like protonated alkylamines. The advantage to use alkylamines is based on their volatility that make separation easily coupled with detection by MS providing direct information regarding oligosaccharide size, extent of sulfation and acetylation other than to identify modifications resulting from the depolymerization process such as the 1,6-anhydrosugars in enoxaparin. Moreover a single type of C18 column can provide an array of separation selectivities by simply changing the ion pair reagent in the mobile phase.

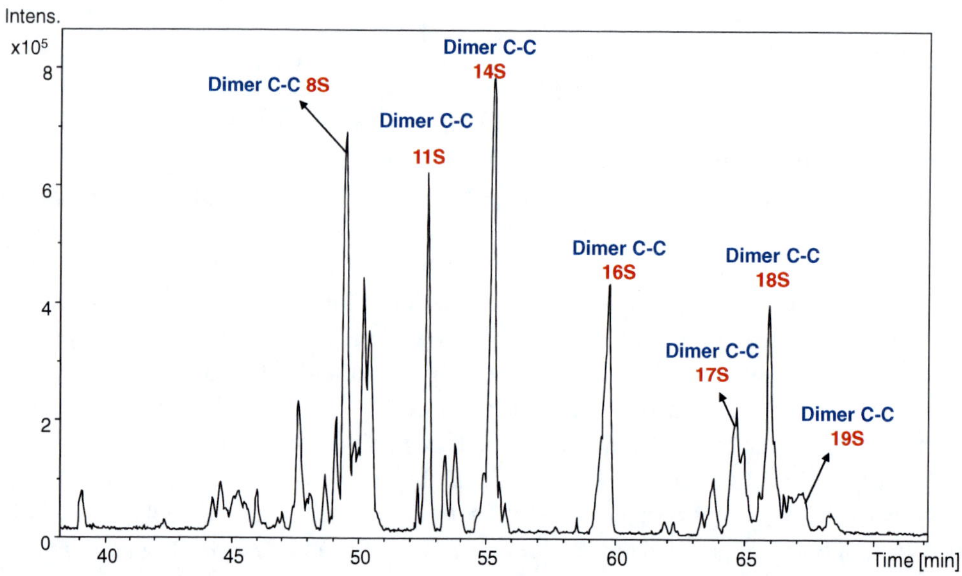

Figure 8. Chromatographic profile of a mixture of semi-synthetic maltotriose beta – beta C-C linked dimers having different degree of sulfation. The employed technique is the reverse phase ion pair ultra-performance liquid chromatography associated with a electrospray ion trap mass spectrometer (RP IP UPLC – ESI IT MS) (99).

This approach applied to periodate oxidized heparins and LMWHs, along with their enzymatic depolimerization products, also provides information on the structure of terminal residues and the length of regular sequences [123, 124]. Recently Linhardt reported the direct characterization of LMWHs by chip-based amide hydrophilic interaction chromatography (HILIC) using a HILIC-Fourier transform (FT)-ESI-MS platform to characterize and compare two currently marketed LMWH products using an approach requiring no special sample preparation steps [125]. This HILIC system relies on cross-linked diols rather than amide chemistry, affording highly resolved chromatographic separations using a relatively high percentage of acetonitrile in the mobile phase, resulting in stable and high efficiency ionization.

MS may be also used to evaluate the affinity of sulfated oligosaccharides to target proteins as was done by Dubin et al. for heparin octasaccharides [126] and by Viskov et al. for a heparin dodecasaccharide [105].

CONCLUSION

An impressive number of analytical tools and variants of each of them is now available for the structural characterization of sulfated poly-and oligosaccharides. The most suitable approach should be chosen on consideration of preliminary information on compositional analysis, presence, complexity and frequency of branching, and of sulfation pattern. Although, in favourable cases, such a preliminary insight can be provided by mono-dimensional NMR spectroscopy alone, the structural complexity associated with size- and sequence heterogeneity often requires orthogonal approaches, typically a combination of ad hoc separation methods and 2D NMR measurements. Coupling HPLC separation and MS spectrometric analysis of fragments obtained by enzymatic or chemical fragmentation of polysaccharides is also becoming an invaluable approach for their structural characterization.

REFERENCES

[1] Aspinall, G.O. 1970. *Polysaccharides*. Pergamon Press, N.Y.

[2] Casu, B. 1985.Structure and biological activity of heparin. *Adv. Carbohydr. Chem. Biochem. 43, 51-134.* Academic Press, New York.

[3] Alban, S. 1997. Carbohydrates with anticoagulant and antithrombotic properties. In: Witczak ZJ & Nieforth KA editors.. *Carbohydrates in Drug Design,* New York, Marcel Dekker, *209-276.*

[4] Basten, J., Jauran, G., Olde-Hanter, B., Petitou, M., van Boekel, C.A.A., 1992. Biologically active heparin-like fragments with a "non-glycosamino" glycan structure. Part 2 : a tetra-O-methylated pentasaccharide with high affinity for antithrombin III. *Bioorg. & Med. Chem. Lett.* 2, 901–990.

[5] Noti, C., Seeberger, P. H. 2005. Chemical Approaches to Define the Structure-Activity Relationship of Heparin-like Glycosaminoglycans. *Chemistry & Biology* 12, 731–756.

[6] Dulaney, S.B, Xuefei, H. 2012. Strategies In Synthesis of Heparin/Heparan Sulfate Oligosaccharides: 2000–Present. *In: Adv. Carbohydr. Chem. Biochem,* New York Academic Press. 67: 95–136.

[7] Casu, B., Naggi, A., Torri, G. 2010. Heparin-derived heparan sulfate mimics to modulate heparan sulfate-protein interaction in inflammation and cancer. *Matrix Biol.* 29, 442-452.

[8] Scott, E.J. 1965. Fractionation by precipitation with quaternary ammonium. In: J. N. BeMiller, M. L. Wolfrom Ed R. Whistler. *Methods Carbohydr. Chem.* New York, Academic Press, 22, 38-44..

[9] Rodriguez, M.L., Jann, B., Jann, K. 1988. Structure and serological characteristics of the capsular K4 antigen of Escherichia coli O5: K4: H4, a fructose-containing polysaccharide with a chondroitin backbone. *Eur. J. Biochem.* 177, 117-124.

[10] An, J., Oneil., M.A., Albersheim, P., Darvill, A.G. 1994. Isolation and structural characterization of β-D-glucosyluronic acid and 4- O-methyl β-D-glucosyluronic acid-containing oligosaccharides from the cell-wall pectic polysaccharide, rhamnogalacturonan I. *Carbohydr. Res.* 252, 235-243.

[11] Gruter, M., Billy, D.,Dewaard, P., Kuiper, J., Kamerling, J.P, Vliegenthart, J. 1994. Structural studies on a cell wall polysaccharide preparation of Lactococcus lactis subspecies Cremoris H414. *J. Carbohydr. Chem.*13, 363-382.

[12] Cifonelli, J.A., Ludowieg, J., Dorfinan, A. 1958. Chemistry of beta-Heparin (Chondroitin-sulfuric Acid-B). *J. Biol. Chem.* 233, 541.

[13] Iacomini, M., Schneider, C.L., Gorin, P.A.J. 1985. Comparative studies on the polysaccharides of Cladonia alpestris, Cladonia confusa, and Cladonia amaurocraea. *Carbohydr. Res.*142, 237-251.

[14] Trehy, M.L., Reepmeyer, J.C., Kolinski, R.E., Westenberger, B.J., Buhse, L.F. 2009. Analysis of heparin sodium by SAX/HPLC for contaminants and impurities. *J. Pharm. Biomed. Anal.*49, 670–673.

[15] Volpi, N., Maccari, F., Linhardt, R.J. 2008. Capillary electrophoresis of complex natural polysaccharides. *Electrophoresis* 29, 3095–3106.

[16] Somsen, G.W, Taka, V., Toraño, J.S, Jongen, P.M.J.M, de Jong, G.J. 2009. Determination of oversulfated chondroitin sulfate and dermatan sulfate impurities in heparin by capillary electrophoresis. J. *Chromatogr. A* 1216, 4107–4112.

[17] Roberts, M.A, Zhong, H.J., Prodolliet, J., Goodall, D.M. 1998. Separation of high-molecular-mass carrageenan polysaccharides by capillary electrophoresis with laser-induced fluorescence detection. *J. Chromatogr. A* 817, 353-366.

[18] Guo, H., Liu, F., Jia, G., Zhang, W., Wu, F. 2013. Extraction optimization and analysis of monosaccharide composition of fucoidan from Saccharina japonica by capillary zone electrophoresis. *J. Appl. Phycol.* 25, 1903-1098.

[19] Lindhal, U., Backstrom, U.G., Thunberg, L., Leder, I.G. 1980. Evidence for a 3-O-sulfated D-glucosamine residue in the antithrombin-binding sequence of heparin. *Proc. Nat. Acad. Sci. USA* 77, 6551-6555.

[20] Lindahl, U., Backstrom, G., Hook, M., Thunberg, L., Fransson, L.A., Linker, A. 1979. Structure of the antithrombin-binding site in heparin. *Biochemistry* 76, 3198-3202.

[21] Bisio, A., Vecchietti, D., Citterio, L., Guerrini, M., Raman, R., Bertini, S., Eisele, G., Naggi, A., Sasisekharan, R., Torri G. 2009, Structural features of low-molecular-

weight heparins affecting their affinity to antithrombin. *Thromb Haemost* 102: 865–873.

[22] Tollefsen, D.M. 1997. *Heparin cofactor II.* Chemistry and Biology of Serpins, In: Springer US, 35-44..

[23] Kreuger, J., Prydz, K., Pettersson, R.F., Lindahl, U., Salmivirta, M. 1999. Characterization of fibroblast growth factor 1 binding heparin sulfate domain. *Glycobiology* 9, 723-729.

[24] Loo, B.M., Kreuge, J., Jalkanen, M., Lindahl, U., Salmivirta, M. 2001. Binding of heparin/heparan sulfate to fibroblast growth factor receptor 4. *J. Biol. Chem,* 276, 16868-16876.

[25] Keiser, N., Venkataraman, G., Shriver, Z., Sasisekharan, R. 2001. Direct isolation and sequencing of specific protein-binding glycosaminoglycans. *Nat. Med.* 7, 123-128.

[26] Bisio, A., Urso, E., Naggi, A., Torri, G. 2014, unpublished data.

[27] Varenne, A., Gareil, P., Colliec-Jouault, S., Daniel, R. 2003. Capillary electrophoresis determination of the binding affinity of bioactive sulfated polysaccharides to proteins: study of the binding properties of fucoidan to antithrombin. *Anal. Biochem.* 315, 152-159.

[28] Gotti, R., Parma, B., Spelta, F., Liverani, L. 2013. Affinity capillary electrophoresis in binding study of antithrombin to heparin from different sources. *Talanta,*105, 366–371.

[29] McKeon, J., Holland, A. 2004. Determination of dissociation constants for a heparin-binding domain of amyloid precursor protein and heparins or heparan sulfate by affinity capillary electrophoresis. *Electrophoresis* 25, 1243–1248.

[30] Powell, A.K., Yates, E.A., David, G., Fernig, D.G, Turnbull J.E. 2004. Interactions of heparin/heparan sulfate with proteins: Appraisal of structural factors and experimental approaches. *Glycobiology* 14, 17R-30R.

[31] Baba, M., Snoeck, R., Pauwels, R., de Clercq E. 1998. Sulfated polysaccharides are potent and selective inhibitors of various enveloped viruses, including herpes simplex virus, cytomegalovirus, vesicular stomatitis virus, and human immunodeficiency virus. *Antimicrob. Agents Chemother.* 32, 1742-1745.

[32] Nishino, T., Aizu, Y., Nagumo T. 1991. The influence of sulfate content and molecular weight of a fucan sulfate from the brown seaweed Ecklonia kurome on its antithrombin activity. *Thromb. Res.* 64, 723–73.

[33] Qi, H., Zhao, T., Zhang, Q., Li, Z., Zhao, Z., Xing, R. 200. 5Antioxidant activity of different molecular weight sulfated polysaccharides from Ulva pertusa Kjellm. *J. Appl. Phycol.* 17, 527–534.

[34] Waltons, K.W. 1954. Investigation of the toxicity of a series of dextran sulphates of varying molecular weight. *British J. Pharmacol. Chemother.*9, 1–14.

[35] Kitajima, H., Takum, S., Morimoto, M. 2000. Histological Analysis of Murine Colitis Induced by Dextran Sulfate Sodium of Different Molecular Weights. *Experimental Animals* 49, 9-15.

[36] Guerrini, M., Beccati, D., Shriver, Z., Naggi, A., Viswanathan, K., Bisio, A., Capila, I., Lansing, J.C, Guglieri, S., Fraser, B., Al-Hakim, A., Gunay, S.N, Zhang, Z., Robinson, L., Buhse L., Nasr, M., Woodcock, J., Langer, R., Venkataraman, G., Linhardt, R., Casu, B., Torri, G., Sasisekharan, R. 2008 Oversulfated chondroitin sulfate is a contaminant in heparin associated with adverse clinical events. *Nat. Biotechnol.* 26, 669-675.

[37] Cipriani, T.R., Gracher, A.H.P, de Souz, L.M, Fonseca, R.J.C., Belmiro, C.L.R., Gorin, P.A.J., Sassaki, G.L., Iacomini, M. 2009. Influence of molecular weight of chemically sulfated citrus pectin fractions on their antithrombotic and bleeding effects. *Thromb. and Haemost.* 101, 795-990.

[38] Gumaraes, M., Mourao, P.A.S. 1997 Urinary excretion of sulfated polysaccharides administered to Wistar rats suggests a renal permselectivity to this polymers based on molecular size. *Biochim. Biophys. Acta* 1335: 161-172.

[39] Hilbor, J.C, Anastassiadis, P.A. 1969. Acrylamide gel electrophoresis of acidic mucopolysaccharides. *Anal. Biochem.* 31, 51-55.

[40] Chaidedgumjorn, A., Suzuki, A., Toyoda, H., Toida, T. 2002. Conductivity detection for molecular mass estimation of per- O-sulfonated glycosaminoglycans separated by high-performance size-exclusion chromatography. *J. Chromatogr. A* 959, 95-102.

[41] Bertini, S., Bisio, A., Torri, G., Bensi, D., Terbojevich, M. 2005. Molecular Weight Determination of Heparin and Dermatan Sulfate by Size Exclusion Chromatography with a Triple Detector Array. *Biomacromolecules* 6, 168-173.

[42] Rochas, C. 1989 Average molecular weight and molecular weight distribution of agarose and agarose-type polysaccharides. *Carbohydr. Pol.* 10, 289–298.

[43] *In-Process Revision: Heparin Sodium.* USP Monography. 37-NF32.

[44] Mulloy, B. 2012. Structure and physicochemical characterisation of heparin. In: *Heparin-A Century of Progress* Springer, Berlin-Heidelberg, 77-98.

[45] Cowman, M.K., Slahetka, M.F., Hittner, D.M., Kim, J., Forino, M., Gadelrab, G. 1984. Polyacrylamide-gel electrophoresis and alcian blue staining of sulfated glycosaminoglycan oligosaccharides. *Biochem. J.* 221, 707-716.

[46] Edens, R.E., Al-Hakim, A., Weiler, J.M., Rethwisch, D.G., Fareed, J., Linhardt, R.J. 1992. Gradient polyacrylamide gel electrophoresis for determination of molecular weights of heparin preparations and low molecular weight heparin derivatives. *J. Pharm. Sci,* 81, 823-827.

[47] Vieira, R.P., Mouriio, P.A.S. 1988. Occurrence of a unique fucose-branched chondroitin sulfate in the body wall of a Sea Cucumber. *J. Biol. Chem.* 263, 18176-18183.

[48] Safaiyan, F, Kolset, S.O, Prydz, K, Gottfridsson, E, Lindhal, U, Salmivirta, M. 1999. Selective effects of sodium chlorate treatment on the sulphation of heparan sulfate. *J.Biol. Chem.*, 274, 36267-36273.

[49] Formaro, M., Senes, A., Soccolini, F., Coinur, R., Cherchi, G.M.A. 1994. Reversed phase HPLC method for the simultaneous determination of all monosaccharides contained in galactosaminoglycan isomers from human aorta proteoglycans. *Carbohydr. Res.* 255, 27-39.

[50] Lindahl, B., Eriksson, L., Lindahl, U. 1995. Structure of heparan sulphate from human brain, with special regard to Alzheimer's disease. *Biochem. J.* 306. 177-184.

[51] Usov, A.I. 1998. Structural analysis of red seaweed galactans of agar and carrageenan groups. *Food Hydrocolloid*, 12, 301–308.

[52] Englyst, H.N., Quigley, E.M., Hudson, G.J. 1994. Determination of dietary fibre as non-starch polysaccharides with gas–liquid chromatographic, high-performance liquid chromatographic or spectrophotometric measurement of constituent sugars. *Analyst.* 119, 1511-1518.

[53] Leoy, J., Lemoine, J., Ricart, G., Michalski, J.C., Montreuil, J., Fomet, B. 1990. Separation of oligosaccharides by capillary supercritical fluid chromatography and analysis by direct coupling to high-resolution mass spectrometer: application to analysis of oligomannosidic N-glycans. *Anal. Biochem.*184, 235-243.

[54] Kakehi, K., Honda, S. 1986. Profiling of carbohydrates, glycoproteins and glycolipids. *J.Chromatogr.* 379, 27-55.

[55] Lindberg, B., Lonngren, J. 1976. Specific degradation of polysaccharides containing uronic acid residues.In: BeMiller E. R. & Whistler J. N. *Methods in Carbohydrate Chemistry VII.* Academic Press, New York, *142-148.*

[56] Kiyohara, H., Hirano, M., Wen, X.G., Matsumoto, T., Sun, X.B, Yamada, H. 1994 Characterisation of an anti-ulcer pectic polysaccharide from leaves of Panax ginseng C.A. Meyer. *Carbohydr. Res.* 263, 89-101.

[57] Foster, A.B., Martlew, E.F., Stacey, M. 1953. The anticoagulant activity of heparin. *Chem Ind. (London)* 34, 825-826.

[58] Cifonelli, A.J. 1976. *Methods Carbohydr. Chem. VII.* In: Whistler R. & BeMiller J.N. Academic Press, New York, 139-141.

[59] Brown, G.H., Huckerby, T.N., Morris, G., Nieduszynski, A. 199. 2Degradation of articular cartilage keratan sulphates using hydrazinolysis and nitrous acid. Environment of fucose residues. *Biochem. J.* 286, 235-241.

[60] Ampofo, S.A., Wang, H.M., Linhardt, R.J. 1991 Disaccharide compositional analysis of heparin and heparan sulfate using capillary zone electrophoresis. *Anal. Biochem.* 199, 249–255.

[61] Murata, K., Murata, A., Yoshida, K. 1995. High-performance liquid chromatographic identification of eight constitutional disaccharides from heparan sulfate isomers digested with heparitinases. *J. Chromatog. B: Biomed. Sci.Applic.* 670, 3-10.

[62] Karamanos, N.K., Vanky, P.. Tzanakakis, G.N., Tsegenidis, T., Hjerpe, A. 1997. Ion-pair high-performance liquid chromatography for determining disaccharide composition in heparin and heparan sulfate. *J. Chromatogr. A.* 765, 169-179.

[63] Toyoda, H., Yamamoto, H., Ogino, N., Toida, T., Imanari, T. 1999. Rapid and sensitive analysis of disaccharide composition in heparin and heparan sulfate by reversed-phase ion-pair chromatography on a 2 μm porous silica gel column. *J. Chromatogr. A*, 830, 197-201.

[64] Skidmo, M., Atrih, A., Yates, E.A, Turnbull, J.E. 2009. Labelling heparan sulphate saccharides with chromophore, fluorescence and Mass Tags for HPLC and MS separations. *Methods Mol. Biol.* 534, 157-169.

[65] Getting, P., Home, A. 1992. One- and two-dimensional 13C-n.m.r. characterization of two series of oligosaccharides derived from porcine intestinal mucosal heparin by degradation with heparinase. *Carbohydr. Res.* 223, 81-98.

[66] Larnkjaer, A., Hansen, S.H, Ostergaard, P.B. 1995. *I*solation and characterization of hexasaccharides derived from heparin. Analysis by HPLC and elucidation of structure by 1H NMR. *Carbohydr. Res.* 266, 37-52.

[67] Tsuda, H., Yamada, S., Yamane, Y., Yoshida, K., Hopwood, J.J., Sugahara, K. 1996. Structural variants with apparent biosynthetic precursor-product relationships for the Antithrombin III-binding site. *J. Biol. Chem.*271, 10495-10502.

[68] Vivè, R.R., Pye, D.A., Salmivirta, M., Hopwood, J.J., Lindahl, U., Gallagher, J. T. 1999. Sequence analysis of heparan sulphate and heparin oligosaccharides. *Biochem. J.* 339, 767–773.

[69] Perlin A.S., Casu B. 1982. Spectroscopic methods. In: Aspinall G.O. *The polysaccharides* I, New York, Academic Press, 133-193.

[70] Gorin P.A.J. 1981. 13C NMR spectroscopy of polysaccharides. In: Tipson D. & Horton R. S. *Adv.Carbohydr. Chem.* New York, Academic Press, *38,* 13-104..

[71] Hricovıni M., Nieto P.M, Torri G. 2002. NMR of Sulfated Oligo- and Polysaccharides. In: Jimenez-Barbero J. & Peters T. *NMR spectroscopy of glycoconjugates,* Weiheim, Wiley-VCH, 189–229..

[72] Langeslay, D.J., Beeche, C.N., Naggi, A., Guerrini, M., Torri, G., Larive, C.K. 2013. Characterizing the Microstructure of Heparin and Heparan Sulfate Using N-Sulfoglucosamine 1H and 15N NMR Chemical Shift Analysis. *Anal. Chem.* 85, 1247−1255.

[73] Wehrli F.W., WirthlinT. 1978. *Interpretation of carbon-13 NMR spectra.* Arrowsmith J.W. Ltd, Bristol, UK,.

[74] Guerrini, M., Beccati, D., Shriver, Z., Naggi, A., Viswanathan, K., Bisio, A., Capila, I., Lansing, J.C, Guglieri, S., Fraser, B., Al-Hakim, A.,Gunay, L.S., Zhang, Z., Robinson, L., Buhse, L., Nasr, M., Woodcock, J., Langer, R., Venkataraman, G., Linhardt, R.J., Casu, B., Torri, G., Sasisekharan, R. 2008. Oversulfated chondroitin sulfate is a contaminant in heparin associated with adverse clinical events. *Nat Biotechnol.*, 26: 669-675.

[75] Mayer, M., Meyer, B. 1999. Characterization of ligand binding by saturation transfer difference NMR spectroscopy. *Angew. Chem. Int. Ed.* 38, 1784–1788.

[76] Sassaki, G.L., Riter, D.S., Santana, A.P., Guerrini, M., Lima, M.A., Cosentino, C., Souza, L.M., Cipriani, T., Rudd, T.R., Nader, H.B., Yates, E.A., Gorin, P.A.J., Torri, G., Iacomini, M. 2011. A robust method to quantify low molecular weight contaminants in heparin: Detection of tris(2-n-butoxyethyl) phosphate. *Analyst*, 136, 2330–2338.

[77] Torri, G., Guerrini, M. 2008. Quantitative 2D NMR analysis of glycosaminoglycans. I. In: Wawer, B. Diehl, U. Holzgrabe. *NMR spectroscopy in pharmaceutical analysis.* Elsevier, New York, 407–428.

[78] Sassaki, G., Guerrini, M., Serrato, V.R., Paixão, A.S., Carlotto, J., Simas-Tosin, F., Cipriani, T.R., Iacomini, M., Torri, G., Gorin, P.A.J. 2014. Monosaccharide composition of glycans based on Q-HSQC NMR. *Carbohydr. Pol.* 104, 34–41.

[79] Bock K., Pedersen C., Pedersen H. 1984. Carbon-13 NMR data for oligosaccharides.In: R.S Tipson.& H. Horton. *Adv. Carbohydr. Chem.* Academic Press, *42, 193-225.*

[80] Yates E.A., Santini F., Guerrini M., Naggi A., Torri G., Casu B. 1996. 1H and 13C NMR spectral assignments of the major sequences of twelve systematically modified heparin derivatives. *Carbohydr. Res.* 294, 15-27.

[81] Yates, E.A., Santini, F., De Cristofano, B., Payre, N., Cosentino, C., Guerrini, M., Naggi, A., Torri, G. Hricovini, M. 2000. Effect of substitution pattern on 1H, 13C NMR chemical shifts and 1J (CH) coupling constants in heparin derivatives. *Carbohydr. Res.* 329, 239-247.

[82] Casu, B., Grazioli, G., Razi, N., Guerrini, G., Naggi, A., Torri, G., Oreste, P., Tursi, F., Zoppetti, G., Lindahl, U. 1994. Heparin-like compounds prepared by chemical

modification of capsular polysaccharide from E. coli KS. *Carbohydr. Res.* 263, 271-284.

[83] Derome, A.E. 1987. *Modem NMR Techniques for Chemistry Research.* Pergamon Press,.

[84] Altona, C. 1996. Vicinal Coupling Constants and Conformation of Biomolecules.In: D. M. Grant & R. Morris. *Encyclopedia of NMR,* New York, Wiley, 4909-4923,

[85] Ruiz-Calero, V., Saurina, J., Galceran, M.T., Hernandez-Cassou, S., Puignou, L. 2000. Potentiality of proton nuclear magnetic resonance and multivariate calibrationmethods for the determination of dermatan sulfate contamination in heparin samples. *Analyst.* 125, 933–938.

[86] Ruiz-Calero, V., Saurina, J., Galceran, M.T., Hernandez-Cassou, S., Puignou, L. 2002. Estimation of the composition of heparin mixtures from various origins using proton nuclear magnetic resonance and multivariate calibration methods. *Anal Bioanal Chem.* 373, 259-265.

[87] Ruiz-Calero., Saurina, J., Hernandez-Cassou, S.,Galceran, M.T., Puignou, L. 2002. Proton nuclear magnetic resonance characterisation of glycosaminoglycans using chemometric techniques. *Analyst.* 127, 407–415.

[88] Duarte, I.F., Barros, A., Almeida, C., Spraul, M., Gil, A.M. 2004. Multivariate analysis of NMR and FTIR data as a potential tool for the quality control of beer. *J Agric Food Chem.* 52, 1031–1038.

[89] Almeida, C., Duarte, I.F., Barros, A., Rodrigues, J., Spraul, M., Gil A.M. 2006. Composition of beer by 1H NMR spectroscopy: Effects of brewing site and date of production. *J Agric Food Chem*, 54, 700–706.

[90] Rudd, T.R., Skidmore, M.A., Guimond, S.E., Cosentino, C., Torri, G., Fernig, D.G., Lauder, R.M., Guerrini, M., Yates, E.A. 2009. Glycosaminoglycan origin and structure revealed by multivariate analysis of NMR and CD spectra. *Glycobiol.* 19, 52-67.

[91] Zang, Q., Keire, D.A., Wood, R.D., Buhse, L.F., Moore, C.M., Nasr, M., Al-Hakim, A., Trehy, M.L., Welsh, W.J. 2011. Determination of galactosamine impurities in heparin samples by multivariate regression analysis of their 1H NMR spectra. *Anal. and Bioanal. Chem.* 399, 635-649.

[92] Zang, Q., Keire, D.A., Wood, R.D., Buhse, L.F., Moore, C.M., Nasr, M., Al-Hakim, A., Trehy, M.L., Welsh, W.J. 2011. Class modeling analysis of heparin 1H NMR spectral data using the soft independent modeling of class analogy and unequal class modeling techniques. *Anal. Chem*, 83, 1030-1039.

[93] Rudd, T.R., Gaudesi, D., Skidmore, M.A., Ferro, M., Guerrini, M., Mulloy, B., Torri, G. Yates, E.A. 2011. Construction and use of a library of bona fide heparins employing 1H NMR and multivariate analysis. *Analyst*, 136, 1380-1389.

[94] Rudd, T.R., Macchi, E., Gardini, C., Muzi, L., Guerrini, M., Yates, E.A., Torri, G. 2012. How to find a needle (or anything else) in a haystack: two-dimensional correlation spectroscopy-filtering with iterative random sampling applied to pharmaceutical heparin. *Anal. chem.* 84, 6841-6847.

[95] Heikkinen, S., Toikka, M.M., Karhunen, P.T., Kilpela, I.A. 2003. Quantitative 2D HSQC (Q-HSQC) via suppression of J-dependence of polarization transfer in NMR spectroscopy: application to wood lignin. *J. Am. Chem. Soc.* 125, 4362–4367.

[96] Guerrini, M., Naggi, A., Guglieri, S., Santarsiero, R., Torri, G. 2005. Complex glycosaminoglycans: profiling substitution patterns by two dimensional NMR spectroscopy. *Anal. Biochem.* 337, 35-47.

[97] Yates, E.A., Santini, F., De Cristofano, B., Payre, N., Cosentino, C., Guerrini, M., Naggi, A., Torri, G., Hricovini, M. 2000. Effect of substitution pattern on 1H, 13C NMR chemical shifts and $^{1}J_{C-H}$ coupling constants in heparin derivatives. *Carbohydr. Res.* 329, 239-247.

[98] Coletti, A., Elli, S., Macchi, E., Galzerano, P., Zamani, L., Guerrini, M., Torri, G., Vismara, E. 2014. Conformational changes of 1-4-glucopyranosyl residues of a sulfated C-C linked hexasaccharide. *Carbohydr. Res.* 389, 134-140.

[99] Vismara, E., Coletti, A., Valerio, A., Naggi, A., Urso, E., Torri, G. 2012. Anti-metastatic semi-synthetic sulfated maltotriose C-C linked dimers. Synthesis and characterisation. *Molecules* 17, 9912-9930.

[100] Rudd, T., Macchi, E., Muzi, L., Ferro, M., Gaudesi, D., Torri, G., Casu, B., Guerrini, M., Yates, E.A. 2013. Unravelling structural information from complex mixtures utilizing Correlation Spectroscopy applied to HSQC spectra. *Anal. Chem.* 85, 7487–7493.

[101] Guerrini, M., Guglieri, S., Beccati, D., Torri, G., Viskov, C., Mourier, P. 2006. Conformational transitions induced in heparin octasaccharides by binding with antithrombin III. *Biochem. J.* 399, 191–198.

[102] Guerrini, M., Hricovíni, M., Torri, G. 2007. Interaction of heparins with fibroblast growth factors: conformational aspects. *Curr. Pharmac. Des.* 13, 2045-2056.

[103] Guerrini, M., Elli, S., Gaudesi, D., Torri, G., Casu, B., Mourier, P., Herman, F., Boudier, C., Lorenz, M., Viskov, C. 2010 Effects on molecular conformation and anticoagulant activities of 1,6-anhydrosugars at the reducing terminal of antithrombin-binding octasaccharides isolated from low-molecular-weight heparin enoxaparin. *J. Med. Chem.* 53, 8030–8040.

[104] Guerrini, M., Guglieri, S., Casu, B., Torri, G., Mourier, P., Boudier, C., Viskov, C. 2008. Antithrombin-binding octasaccharides and role of extensions of the active pentasaccharide sequence in the specificity and strength of interaction. Evidence for very very high affinity induced by an unusual glucuronic acid residue. *J. Biol. Chem.* 283, 26662–26675.

[105] Viskov, C., Elli, S., Urso, E., Gaudesi, D., Mourier P., Herman, F., Boudier, C., Casu, B., Torri, G., Guerrini, M. 2013. Heparin Dodecasaccharide Containing Two Antithrombin-binding Pentasaccharides Structural Features and Biological Properties. *J. Biol. Chem.* 288, 25895-907.

[106] Muruyama, T., Toida, T., Imanari, T., Yu, G., Linhardt, R.J. 1998. Conformational changes and anticoagulant activity of chondroitin sulfate following its O-sulfonation. *Carbohydr. Res.* 306, 35-43.

[107] Angulo, J., Hricovíni, M., Gairi, M., Guerrini, M., de Paz, J.L., Ojeda, R., Martin-Lomas, M., Nieto P.M. 2005. Dynamic properties of biologically active synthetic heparin-like hexasaccharides. *Glycobiology* 15, 1008-1015.

[108] Guerrini G., Viscov C. 2014. *Antithrombin-binding oligosaccharides: structural diversity in a unique function. Glycocon. J.* (in press).

[109] K hoo, K.H., Morris, H.R., McDowell, R.A., Dell, A., Maccarana, M., Lindahl, U. 1983. FABMS/derivatisation strategies for the analysis of heparin-derived oligosaccharides. *Carbohydr. Res.*, 244, 205–223.

[110] Dell, A., Rogers, M.E, Thomas-Oates, J.E. 1988. Fast-atom-bombardment mass-spectrometric strategies for sequencing sulphated oligosaccharides. *Carbohydr. Res.* 179, 7–19.

[111] Zaia J. 2005. Principles of mass spectrometry of glycosaminoglycans. *J. Biomacromol Mass Spectrom.* 1, 3-36.

[112] Jiao, G., Yu, G., Zhang, J., Ewart, H.S. 2011. Chemical Structures and Bioactivities of Sulfated Polysaccharides from Marine Algae. *Marine Drugs* 9, 196-223.

[113] Saad, O.M., Leary, J.A. 2003. Compositional analysis and quantification of heparin and heparan sulfate by electrospray ionization ion trap mass spectrometry. *Anal Chem.* 75, 2985-2995.

[114] Thanawiroon, C., Rice, K.G., Toida, T., Linhardt, R.J., 2004. Liquid chromatography/mass spectrometry sequencing approach for highly sulfated heparin-derived oligosaccharides. *J. Biol. Chem.* 279 (4), 2608–2615.

[115] Wang, B., Buhse, L.., Al-Hakim, A., Boyne, I.M.T., Keire, D.A. 2012. Characterization of currently marketed heparin products: analysis of heparin digests by RPIP-UHPLC–QTOF-MS. *J. Pharmac. Biomed. Analysis* 67–68, 42–50.

[116] Liu, R., Miao Chen, Y. X., Weïwer, M., Zhou, X., Bridges, A.S., DeAngelis, P. L., Zhang, Q., Linhardt, R.J., Liu, J. 2010. Chemoenzymatic Design of Heparan Sulfate Oligosaccharides *J. Biol. Chem.* 285:34240-34249.

[117] Henriksen, J., Ringborg, L.H., Roepstorrf, P. 2004. On-line size-exclusion chromatography/mass spectrometry of low molecular mass heparin. *J. Mass Spectrom.* 39, 1305–1312.

[118] Doneanu, C.E., Chen, W., Gebler, J.C., 2009. Analysis of oligosaccharides derived from heparin by ion-pair reversed-phase chromatography/mass spectrometry. *Anal. Chem.* 81, 3485-3499.

[119] Henriksen, J., Roepstorff, P., Ringborg, L.H. 2006. Ion-pairing reversed-phased chromatography/ mass spectrometry of heparin. *Carbohydr. Res.* 341, 382–387.

[120] Kuberan, B., Lech, M., Zhang, L., Wu, Z.L., Beeler, D.L., Rosenberg, R.D. 2002. Analysis of heparan sulfate oligosaccharides with ion pair-reverse phase capillary high performance liquid chromatography-microelectrospray ionization time-of-flight mass spectrometry. *J. Am. Chem. Soc.* 124, 8707–8718.

[121] Patel, R.P., Narkowicz, C., Jacobson, G.A. 2009. Effective reversed-phase ion pair high-performance liquid chromatography method for the separation and characterization of intact low-molecular-weight heparins. *Anal. Biochem.* 387 (1), 113–121.

[122] Langeslay, D.J., Urso, E., Gardini, C., Naggi, A., Torri, G., Larive, C.K. 2013. Reversed-phase ion-pair ultra-high-performance-liquid chromatography–mass spectrometry for fingerprinting low-molecular-weight heparins. *J. Chromat. A* 1292, 201-210.

[123] Alekseeva, A., Casu, B., Torri, G., Pierro, S., Naggi, A. 2013. Profiling glycol-split heparins by high-performance liquid chromatography/mass spectrometry analysis of their heparinase-generated oligosaccharides. *Anal. Biochem.* 434, 112-122.

[124] Alekseeva, A., Casu, B., Cassinelli, G., Guerrini, M., Torri, G., Naggi, A. 2014. Structural features of glycol-split low-molecular-weight heparins and their heparin lyase generated fragments. *Anal. and Bioanal. Chem.* 406, 249-265.

[125] Li, L., Zhang, F., Zaia, J., Linhardt R.J. 2012. Top-down approach for the direct characterization of low molecular weight heparins using LC-FT-MS. *Anal Chem.* 84, 8822–8829.

[126] Abzalimov, R.R., Dubin, P.L., Kaltash, I.A. 2007. Glycosaminoglycans as Naturally Occurring Combinatorial Libraries: Developing a Mass Spectrometry-Based Strategy for Characterization of Anti-Thrombin Interaction with Low Molecular Weight Heparin and Heparin Oligomers. *Anal. Chem.* 79, 6055-6063.

In: Sulfated Polysaccharides ISBN: 978-1-63482-975-5
Editors: M. Gama, H.B. Nader, H.A. de Oliveira Rocha © 2015 Nova Science Publishers, Inc.

Chapter 12

SULFATED POLYSACCHARIDES AND NANOTECHNOLOGY

Vera Carvalho*[*] *and Francisco Miguel Gama
Centre of Biological Engineering, University of Minho,
Braga, Portugal

ABSTRACT

Sulfated polysaccharides (SPs) are a complex group of biomolecules, many of which present relevant biological activities, such as antibiotic, anti-oxidant, anti-mutant, anti-coagulant, antiviral, antitumor, anti-inflammatory, immunomodulatory. Additionally, being biocompatible, biodegradable and suitable for processing into membranes, hydrogels and nanoparticulate systems, they have great potential for biomedical applications. Thus, SPs can be used for e.g., drug delivery systems, not just performing as carriers but also contributing to the bioactivity of the developed formulations. This chapter reviews the more recent research on the nanotechnological applications of SPs.

1. INTRODUCTION

Polysaccharides with high sulfate content are usually called sulfated polysaccharides (SPs). SPs include a complex group of macromolecules with a wide range of important biological properties [1, 2]. They are widespread in nature and can be been found in animals, plants, and microorganisms. Due to their physic-chemical properties and bioactivities, SPs have attracted great attention in nanotechnology [3], namely for the production of nanoparticles (NPs), liposomes, emulsions, polymeric micelles, hydrogels, polymeric scaffolds, etc. [4-7]. A wide variety of bioactivities have been recognized in SPs, as summarized in Table 1: antibiotic, anti-oxidant, anti-mutant, anti-coagulant, antiviral, antitumor, anti-inflammatory, immunomodulatory [1-3, 8-11].

[*] Corresponding author: Vera Carvalho. Centre of Biological Engineering, University of Minho, Campus de Gualtar, 4710-057 Braga, Portugal. E-mail: verascarvalho@gmail.com.

The bioactivity is dependent on composition and structural features such as molecular weight, degree of sulfation and position of the sulfate group [2, 3].

2. Nanotechnological Applications

Comprehensive descriptions of the nanotechnological applications of the most commonly used SPs will be provided in the next sections, highlighting applications of the different kinds, namely for drug delivery and tissue engineering.

2.1. Carrageenan

Carrageenan is a naturally occurring anionic sulfated linear polysaccharide obtained mainly from some red seaweed of the *Rhodophyceae* family [3, 35, 36]. There are several types of carrageenan with slightly different chemical structures and properties.

Based on their sulfation patterns and galactose distribution, carrageenans can be divided into three commercial and industrially important groups: kappa (κ), iota (i), and lambda (λ). These molecules differ on their gelling properties at room temperature as a result of their degree of sulfation, therefore κ- and i-carrageenans have strong gelling properties while λ-carrageenan is usually applied as a thickening agent [3, 35].

Carrageenans have been used for the production of bionanocomposites, nanostructured materials obtained by the combination of biopolymers and nanosized inorganic solids. The use of biopolymers can provide biocompatibility, biodegradability, and a non-toxic framework to the resulting nanocomposite, while the inorganic NPs are useful for multi-task applications that include imaging, sensing and diagnostic, and also drug delivery.

2.1.1. Carrageenan or Carrageenan-Based Nanogels (or Nanoparticles) (NPs)

Temperature responsive nanogels or nanoparticles (NPs) have promising characteristics for the development of smart drug delivery systems. External stimuli, such as pH and/or temperature, can change the gel network arrangement and dimensions, allowing the controlled release of encapsulated therapeutics. Responsive nanogels can "sense" and respond directly to pathophysiological disorders, thus configuring auspicious materials for the development of more efficient disease treatments based on smart drug delivery systems. In 2011, Daniel-da-Silva and collaborators described thermo-sensitive κ-carrageenan nanogels [37]. These carrageenans undergo a coil to helix conformation transition followed by helix aggregation that can be induced by cooling or monovalent cations like potassium. The κ-carrageenan gels are very interesting for temperature activated drug delivery applications, not only for their thermo-sensitive nature but also due to the carrageenan structure that includes various functional groups allowing additional derivatization and bioconjugation. κ-Carrageenan NPs have been prepared using an aqueous microemulsion of κ-carrageenan with KCl as the cross-linker, dispersed in a organic phase of n-heptane in the presence of the surfactant cetyltrimethyl ammonium bromide (CTAB). The microemulsions were optically clear, becoming slightly turbid as the temperature decreased due to the gelation of the κ-carrageenan.

Table 1. Various types of SP and their biological activities

Biological and pharmaceutical activities	Type of SP
Anti-viral	Exopolysaccharide [12] Fucoidan [2, 11, 13-16] Carrageenan [11, 17] Galactan [14] Heparin/heparan sulfate [11, 18-20] Dextran sulfate [21, 22]
Anti-tumor	Porphyran [2] Fucoidan [1, 2, 14, 15] Galactan [14] Heparin/heparan sulfate [23]
Anti-angiogenic	Fucoidan [14, 15] Galactan [14];
Anti-coagulant	Pullulan sulfate [24] Carrageenan [2, 17] Fucoidan [1, 2, 11, 13-15, 25] Dermatan sulfate [26] Galactan [14] Chitin sulfate [27] Chitosan sulfate [27] Chondroitin sulfate [28] Heparin/heparan sulfate [23]
Anti-oxidant	Fucoidan [2, 11, 17, 29] Laminaran [2] Alginic acid [2] Carrageenan [11, 17] Ulvan [11];
Anti-inflammatory	Fucoidan [13-15] Galactan [14] Chondroitin sulfate [30]
Anti-thrombotic	Fucoidan [14] Galactan [14] Chondroitin sulfate [28]
Anti-proliferative	Fucoidan [1, 13, 16] Heparin/heparan sulfate [23]
Immunomodelatory	Carrageenan [2] Fucoidan [1, 11, 16, 31] Heparin/heparan sulfate [23, 32]
Microbicidal	Chitosan [27] Carrageenan [20, 33]
Anti-fungal	Chitosan [27] Ulvan [34];

Stable microemulsions were obtained after cooling, no phase separation being observed. The dried NPs had spherical morphology, an average diameter of 74±15 nm and 31±10 nm for NPs obtained with 1.0 and 4.0 wt.% κ-carrageenan, respectively, and a zeta potential at physiological pH of *ca.* -20 mV.

It was observed that the κ-carrageenan NPs were thermo-sensitive at physiological temperature (between 37 and 45°C) and suffered reversible volume transition in response to thermal stimuli: when temperature increased from 25 to 45°C, a portion of the gel network suffered a gel-to-sol transition upon breaking of physical junctions, leading to a swelling effect and consequent particle size increase. This behavior was more pronounced in the more concentrated material, which results in more physical junctions available in the gel and also to osmotic effect. Methylene blue (MB) was loaded in the κ-carrageenan nanogels using a post-synthesis impregnation method and *in vitro* release studies were conducted. MB is a water-soluble cationic dye, here used as a model drug, mainly because it permits instant visual detection. It was found that the encapsulation efficiency and loading capacity values increased two fold by increasing carrageenan content from 1.0 to 4.0 wt.%, probably owing to a higher density of negatively charged sulfate groups available to interact with positively charged MB. Both nanogels demonstrated continuous MB release over time, up to 3-4 hours. It has been shown also that the thermo-sensitive nature of the κ-carrageenan NPs can modulate the *in vitro* release rate of an encapsulated model drug.

2.1.2. Carrageenan-Functionalized Magnetic NPs

Recently, Salgueiro et al. [38] reported the preparation of κ-carrageenan-coated magnetite (Fe_3O_4) NPs, which were tested as adsorbents for the magnetically assisted removal of MB, a cationic dye present in wastes from the textile industry. Dyes are present in the effluents of various industries such as textile, paper, polymers, and cosmetic, their discharge representing a serious problem with very harmful impact on the environment. MB is usually used for dying textiles and wood and although it is less hazardous than other dyes, it has several damaging effects. The methods commonly used to remove MB and other cationic dyes from industrial effluents are chemical precipitation, photochemical degradation and adsorption.

This is the mostly used method due to its simplicity, low cost and effectiveness. The search for environmentally friendly and low-cost dye adsorbents turned the attention to naturally obtained polysaccharides. As mentioned above, κ-carrageenan forms hydrogels stabilized by alkali-metal cations, and has the best gelation properties within the carrageenan family. The ability to bind cationic dyes comes from the sulfonate anions present in the polysaccharide structure. Despite promising properties, there are still limitations associated to the separation of the dispersed polysaccharide adsorbents from the treated effluents, which restricts their use in packed-bed sorption columns. Magnetic NPs, made of e.g., iron oxide, besides having a large surface area, have magnetic properties that allow a rapid and easy separation employing an external magnetic field. Magnetic Fe_3O_4 NPs were obtained by the coprecipitation method in aqueous medium. The presence of κ-carrageenan in the magnetic particles was confirmed by Fourier transform infrared spectroscopy (FTIR). The characterization of uncoated and carrageenan-coated magnetic NPs revealed that after coating the magnetization saturation decreased to roughly half the initial value (67.6 emu/g to 32.4 emu/g). This effect was reported for magnetic NPs coated with other polysaccharides and was attributed to the presence of diamagnetic material covering the NPs. Despite this, the coated magnetic NPs still allowed for their magnetically assisted separation.

It was stated that the amount of MB adsorbed increased with the pH due to lower competition between H^+ and MB and also lower protonation of sulfonated groups. The MB uptake by the coated NPs was studied using variable initial concentration of the dye. In all cases tested, fast adsorption took place and equilibrium was reached in approximately 5 min. At 23ºC and pH 9, the maximum MB absorption capacity of the carrageenan-coated magnetic NPs was found to be 185.3 mg/g, superior to the absorption capacity demonstrated by other magnetic sorbents tested. It was also showed that the NPs could be recycled using 1.0 M KCl; however, additional work is necessary (and ongoing) to improve the stability of the polymer shell. Iron oxide NPs have also been used in an increasing number of medical applications as drug targeting, bio-separation processes, cancer thermotherapy, and as contrast agent in magnetic resonance imaging (MRI) [39]. NPs smaller than 15 nm in diameter are normally required for superparamagnetic behavior to occur, avoiding particle agglomeration.

Routine chemical synthesis of Fe_3O_4 (magnetite) and γ-Fe_2O_3 (maghemite) powders includes the coprecipitation of a mixture of aqueous ferrous and ferric salts. Some strategies have been developed to inhibit iron oxide NPs agglomeration with simultaneously functionalization using a non-toxic biocompatible coat.

These include the coating of magnetic NPs using polymers or loading metal salt precursors into the polymer and inducing *in situ* synthesis of the magnetic NPs. Different studies have shown that some polysaccharides, as dextran, cyclodextrin and starch-based polymers, can interact with iron complexes, inhibiting the precipitation of iron hydroxide promoting the NPs synthesis under controlled conditions [40-42].

Daniel-da-Silva et al. conducted a study to elucidate the role of carrageenan type (κ, i and λ), its concentration, and nature of alkali-metal cation on the formation and characteristics of magnetite (Fe_3O_4) NPs [43, 44]. Magnetite NPs were produced by an *in situ* chemical coprecipitation method. FTIR results demonstrated that the synthesis of magnetite NPs in the presence of carrageenans involves the coordination of ferric (and ferrous) ions to the sulfate group of the polymer enhancing the physical cross-linking of the polymer chains and formation of a microgel at low pH. The presence of carrageenan had a significant effect on the magnetite particle size: NPs obtained in the presence of carrageenan showed lower average sizes with increasing polymer concentration. Moreover, the nature of the carrageenan played an important role, with i-carrageenan NPs showing higher average sizes compared to λ- and κ-carrageenans. Micrographs of κ-carrageenans gels containing iron oxide NPs showed that in contact with air, the initial black magnetite turns dark brown after approximately 1 day and orange-red after 1 week, in the air-exposed surface, that expands to the bottom of the reaction vessel over time. In fact, it is well known that magnetite solutions oxidizes very slowly to maghemite over several months and in the dry state this process can take years [45]. The presence of carrageenan seemed to accelerate the magnetite oxidation process and the actual rate of oxidation depended on the nature of the carrageenan used and its concentration.

For a biopolymer concentration of 20 g/L, the stability of the magnetite followed the trend i- > κ- > λ-carrageenan and for κ- and λ-carrageenan, the chemical stability decreasing with increasing polymer concentration. The i-carrageenan magnetite NPs demonstrated the lowest oxidation rate (as a consequence, highest stability), which could be directly related with the highest average size found for this system. Predictably, gel strength increased with biopolymer concentration, however this also leaded to a decreased stability of the magnetic NPs obtained.

It was also showed that in the presence of κ-carrageenan an enhancement of the dynamic moduli and increase of the gelling temperature occurred, due to the presence of NPs, and that this effect was less significant for i- and λ-carrageenan systems [44].

As already mentioned, carrageenan gelation takes place through chain association into double helices followed by the formation of helices aggregates, with a fundamental role of metal ions in gel stabilization. In the case of κ-carrageenan, potassium ions have a crucial function as helix-stabilizers enhancing aggregation and gelation. To confirm the importance of K^+ ions in gel consolidation, nanosized Fe_3O_4 was added to a 5 gL^{-1} κ-carrageenan solution without KOH. This led to a composite with inferior viscoelastic properties compared to the biopolymer matrix alone. The results showed that both the presence of Fe_3O_4 and high K^+ concentration contributed to the formation of a more stable helical aggregate and, as a consequence, a stronger gel. Overall, the results point to κ-carrageenan nanocomposites as the best candidate for future applications.

2.1.3. Carrageenan Nanocomposites

As seen in the previous study, besides inhibiting the spontaneous agglomeration of the iron oxide NPs, carrageenan enables the formation of smaller NPs and allows the chemical stability of the magnetite towards oxidation to be manipulated by choosing the carrageenan type and concentration [43]. Owing to the gelling properties of carrageenans, specifically κ-carrageenan, the obtained nanocomposites are expected to form thermoreversible gels that have the potential to be used for the controlled release of magnetic particles (or loaded drugs) by thermal and/or magnetic stimuli. Exploiting this hypothesis, Daniel-da-Silva reported the preparation of surface functionalized κ-carrageenan/iron oxide nanospheres by a reverse microemulsion method with thermally induced gelation of the polysaccharide [46]. The magnetic nanospheres were obtained by *in situ* chemical coprecipitation in the presence of the polysaccharide [43] and the resulting nanocomposite was utilized as the aqueous phase in the reverse microemulsion procedure. Then, the microemulsions were obtained by sonication of a quaternary system composed of *n*-heptane (organic phase), CTAB (surfactant), 1-butanol (co-surfactant) and the aqueous phase prepared previously allowing to change the size of the resulting NPs by modulating the droplet size of the dispersed phase, that depends on the ratio $[H_2O]/[CTAB]$ (w_0). The microemulsions were formed at 60ºC, followed by a cooling step to room temperature to induce thermal gelation of the polymer. The nanocomposites obtained consisted of biopolymer coated magnetite NPs and exhibited superparamagnetic behavior at room temperature. In addition, magnetization reached values close to saturation at relatively low fields (approximately 85% of saturation for H = 15 KOe, a field usually used in MRI) that makes these nanocomposites also interesting for imaging applications.

Carboxymethylation of κ-carrageenan allowed for further nanocomposite bioconjugation. The carboxyl groups were effectively attached to the surface of the NPs with substitution predominantly at the O-2 and O-6 positions of the sugar rings. To be applied in clinical setting, a nanocarrier should demonstrate thermo-sensitivity at temperatures slightly higher than 37ºC to about 44ºC. Differential scanning calorimetry (DSC) was used to evaluate the gel-sol transitions of the κ-carrageenan nanocomposites. The experiments showed that the obtained gelled NPs were highly thermo-sensitive at a physiological acceptable range of temperatures (37 to 45ºC), both before and after surface carboxymethylation, and that this behavior has extremely sensitive to the presence of ions, namely K^+ that increased the melting temperature from 41 to 45ºC.

To further investigate the potential of these nanocomposites for controlled biological applications, a bovine IgG antibody was conjugated to the surface of the magnetic NPs *via* a carbodiimide-mediated reaction. The so obtained optimized nanodelivery system could be used for targeted applications.

The incorporation of magnetic NPs, such as magnetite NPs, in carrageenan hydrogels resulted in thermoreversible hydrogel nanocomposites responsive to external magnetic stimuli, which can be used for magnetically and/or thermally activated drug release [44, 46]. To further characterize these nanocomposites, the swelling properties and the kinetics of the release of the model drug MB from the κ-carrageenan/magnetite hydrogel nanocomposite were also studied [47]. The nanocomposites were obtained by homogeneous dispersion of Fe_3O_4 NPs (obtained by chemical coprecipitation [43]) in a κ-carrageenan aqueous solution, followed by gelation of the mixture and by freeze-drying. The incorporation of Fe_3O_4 NPs increased the swelling of the hydrogels, and it was also observed that the elastic modulus increased with the Fe_3O_4 NPs content [44].

The influence of the matrix composition (κ-carrageenan and K^+ ions concentration) and of the loaded Fe_3O_4 NPs on the MB release profile was evaluated. All the nanocomposites showed a sustainable MB release as depicted in Figure 1a. A fast release was observed in the first two 2 h (burst release) but then it proceeded at lower rate until a steady concentration was achieved. The MB release rate was lower for increasing concentrations of κ-carrageenan and K^+ ions, the later having a more pronounced effect. The addition of Fe_3O_4 NPs (K40_MGN100) slowed the initial release that then progressed in a sigmoidal profile (Figure 1b). In order to prevent the premature dissolution of the hydrogel, release experiments were also made in phosphate buffered saline (PBS) with high K^+ ions concentration (1M). The direct consequence was the prolonged MB release, as seen in Figure 1c, and furthermore it was shown that at high K^+ concentration the MB release is correlated with gel strength (stronger gels limit MB release).

The incorporation of gold (Au) NPs into polymer hydrogels produces materials with light responsiveness for specific bio-applications such as sensing platforms [48], light activated bioadhesives [49] and drug delivery nanocarriers [50]. The optical properties of the nanocomposites arise from the surface plasmon resonance (SPR) effects of the Au NPs, the SPR band being sensitive to various parameters such as the size and shape. The Au nanospheres show the SPR effect in the visible region of the electromagnetic spectrum, however for some therapeutic applications it is appropriate to shift the optical absorption into the near-infrared where absorption of body tissue is smallest. This shift can be obtained using anisotropic Au nanorods. The effect of such NPs in the microstructure, thermal mechanical and release properties of thermosensitive κ-carrageenan hydrogels was investigated [51].

Nanocomposites were obtained by homogenous dispersion of both Au nanospheres and Au nanorods in a κ-carrageenan aqueous solution, followed by gelation of the mixture using K^+ ions. It was observed that all the nanocomposites exhibited higher elastic modulus, as compared to κ-carrageenan hydrogels. DSC thermograms presented a broad endothermic peak corresponding to the melting (gel-to-sol transition) of the hydrogel.

In the case of the nanocomposites, the peak shifted to higher temperatures confirming that the Au particles promoted the reinforcement of the κ-carrageenan hydrogel structure. The MB release kinetics varied according to the nature of the Au particles used as dispersed phase. In the case Au nanospheres, the MB release tends to be controlled by polymer relaxation and erosion, while for Au nanorods the MB release was diffusion controlled.

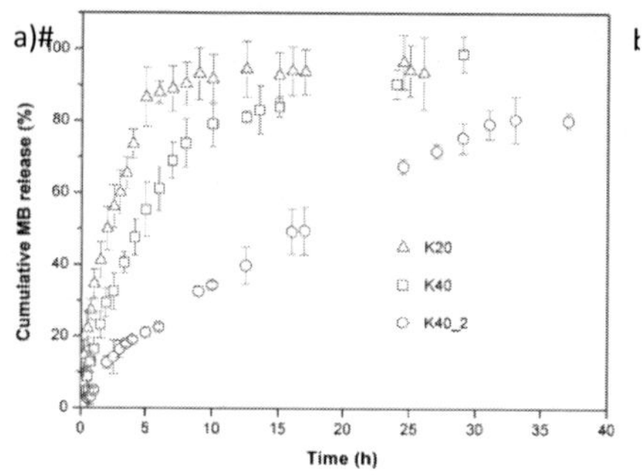

a

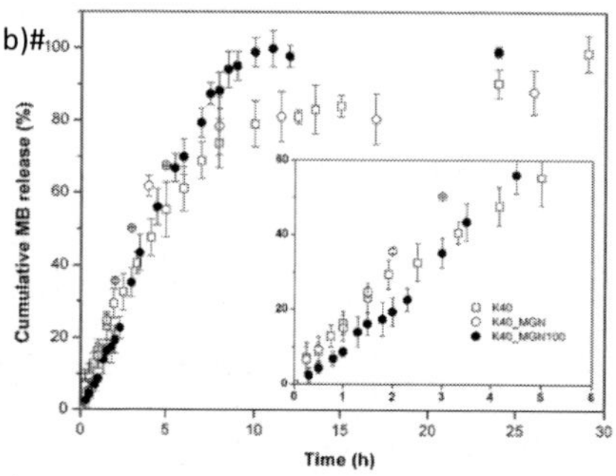

b

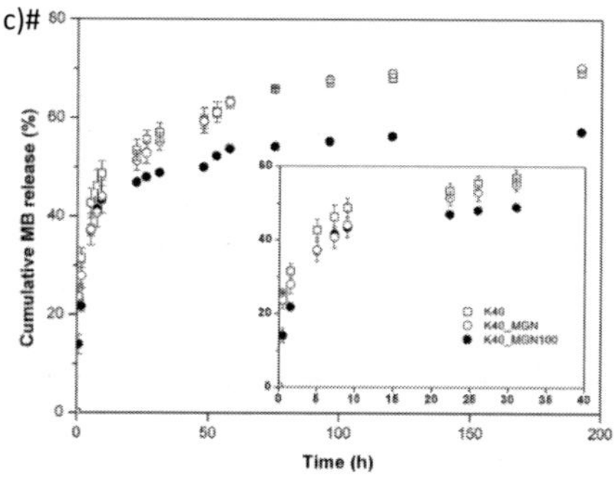

c

Adapted from [47], with permission.

Figure 1. *In vitro* MB release profiles from a) blank hydrogels obtained with varying concentrations of κ-carrageenan and K^+ ions (K20 and K40 with 1M KCl, K40_2 with 2M KCl); and from blank hydrogels (K40) and hydrogel nanocomposites with different loads of Fe_3O_4 NPs [K40_MGN with 0.025% (w/w) $m_{NPs}/m_{polymer}$ and K40_MGN100 with 2.50% (w/w) $m_{NPs}/m_{polymer}$] in b) PBS, and c) PBS with K^+ 1M. Inset: detail of the release profiles for the first 6h (b) and 40h (b).

Bearing in mind that the introduction of both Au nanospheres and nanorods originated stronger hydrogels, it was reckoned that MB release followed a polymer relaxation mechanism independent of the shape of the Au particles. So, it was suggested that the heterogeneous gel microstructure resulting from the incorporation of anisotropic Au NPs increased the tortuosity of MB path, rendering the MB release diffusion controlled.

The combination of NPs and pH sensitive hydrogels should give rise to new functional composites, e.g., to be used as drug carrier for gastro-intestinal tract release [52]. Two κ-carrageenan nanocomposite hydrogels, with silver (Ag) and magnetite (Fe_3O_4) nanofillers, were produced and its effect on the gastro-intestinal tract release was analyzed using MB as a model drug. To obtain the magnetite nanocomposite hydrogels, *in situ* magnetic NPs formation inside the hydrogel matrices was achieved by loading hydrogels with iron ions (Fe(II) and Fe(III)) up to their maximum capacity and by then immersing the loaded hydrogel in a sodium hydroxide solution to precede the reaction.

The presence of the magnetite NPs was indicated by the black color of the hydrogels after 4h. The formation of NPs in the hydrogel network resulted in a more compact structure that could affect MB release from the nanocomposites hydrogels. Under simulated digestive tract conditions, magnetite nanocomposites presented better performance by allowing for 30.43% of MB to be released in stomach medium and 13.47% in the intestine.

To further study the nanocomposites, genipin was used as a cross-linking agent in Ag and Fe_3O_4 nanocomposite hydrogels, which were then loaded with metal ions for *in situ* NPs formation. The authors concluded that the incorporation of Ag NPs in κ-carrageenan hydrogels improved the MB release in simulated gastro-intestinal tract compared to blank κ-carrageenan and magnetite NPs κ-carrageenan hydrogels.

Another work reported the synthesis and characterization of i-carrageenan nanocomposite with Ag NPs [53], aiming the development of the bactericidal properties. The synthesis of the biodegradable Ag nanocomposite hydrogels was achieved by a green process using acrylamide (AM) with i-carrageenan, and consisted in three steps, as shown in Figure 2. The reduction of the silver ions (Ag^+) to silver particles (Ag^0) occurred using cold leaf extracts of *Azadirachta indica* (neem leaf) solution (hence the green process). Transmission electron microscopy (TEM) confirmed that the Ag^0 NPs were dispersed in the hydrogels and had spherical shape with average size of 3±2 nm. The Ag^0 nanocomposite hydrogels had higher swelling ratio compared to plain P(IC-AM) hydrogels mainly due to the formation of Ag^0 NPs within the hydrogel that expanded the gel network and enhanced higher water molecules uptake capacity, an effect that was observed to be proportional to the i-carrageenan concentration. The antibacterial properties of the biodegradable nanocomposites were studied by assessing their ability to inhibit *Bacillus* and *E. coli* growth on agar discs.

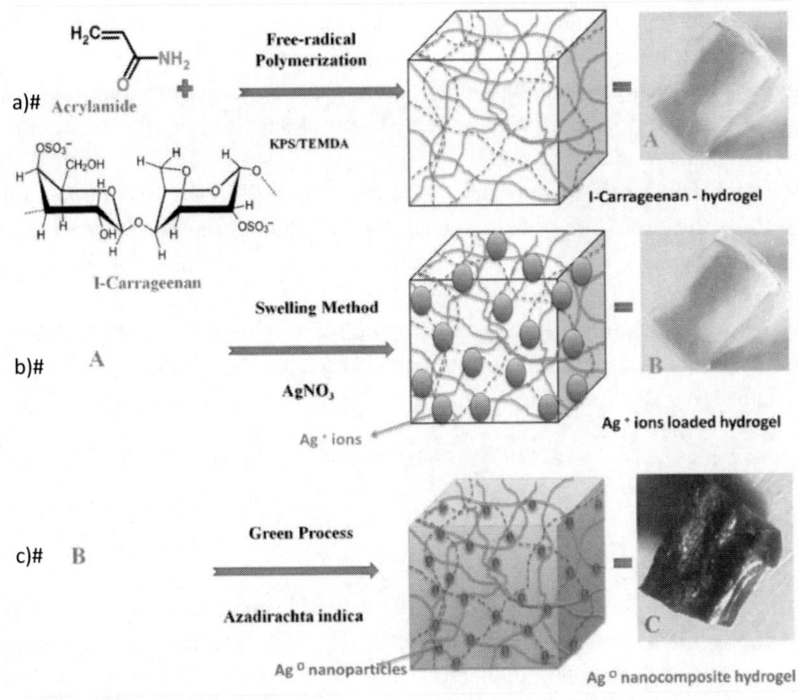

Adapted from [53], with permission.

Figure 2. Illustrative scheme of the preparation of biodegradable P(IC-AM) Ag nanocomposite: a) the fabrication of i-carrageenan hydrogel *via* free-radical reaction with AM [P(IC-AM)]; b) preparation of Ag^+ ions-containing hydrogels *via* swelling method; and c) synthesis of the Ag nanocomposite P(IC-AM) hydrogels *via* green process (the Ag NPs were obtained by reducing $AgNO_3$ using *A. indica* in the P(IC-AM) hydrogels).

After 48 h incubation, the inactivation of bacterial growth zones was noticeable in the Petri dishes when the Ag^0 nanocomposites were tested. The plain P(IC-AM) hydrogels showed no inhibition ability. Therefore, it was concluded that the inorganic biodegradable nanocomposites described could be efficiently used for biomedical antibacterial applications.

2.1.4. Carrageenan-Drug Complexes

Early in 2007, Dai et al., presented a nanosizing method for drug/carrageenan complexes, demonstrating that it enabled the solubility and dissolution rate of a poorly water-soluble compound [54]. Low oral bioavailability of poorly water-soluble drugs is a problem and several approaches have been used to improve bioavailability by increasing the drug dissolution rate and solubility.

Nanosizing techniques had already been used to enhance dissolution rate, thus improving oral bioavailability of such drugs [55, 56]. In Dai's work, λ-carrageenan was complexed with a synthetic poorly water-soluble compound, and then nanosized by wet-milling process. The compound/carrageenan amorphous complex formation was confirmed by X-ray diffraction (XRD) and DSC. The formation of an amorphous structure was assigned to the random distribution of the carrageenan chains, which may have created irregular structures in the complex. The complexation of the poorly water-soluble compound with the polysaccharide increased its aqueous solubility from 1-2 to 30 μg/mL by converting the compound's crystalline structure to an amorphous one. It is documented that compounds have higher solubility in an amorphous state than in a crystalline structure [57].

Predictably, the nanosuspension of the compound/carrageenan complex presented the highest aqueous solubility (39 μg/mL) since it combined the benefits of complexing with hydrophilic polymers and nanosizing.

Besides increasing the compound's aqueous solubility, the complex showed faster dissolution than the free compound and the nanosized compound/carrageenan complex exhibited the fastest dissolution rate of all samples tested, as a result of its smaller particle size and improved dispersion in the polymer network.

The results mentioned demonstrated that by converting the synthetized compound into an amorphous structure it was possible to increase both aqueous solubility and dissolution rate and that nanosizing further improved these properties, thereby possibly enhancing bioavailability.

2.2. Heparin

Heparin (HP) is a highly sulfated glycosaminoglycan that is generally used as an injectable anticoagulant. It is a linear polysaccharide of repeating units of uronic acid and glucosamine residues. The predominance of sulfate and carboxylate groups provides heparin the higher negative charge density among the known biomolecules [58].

Although being used mainly in medicine as an anticoagulant, its real physiological role in the body is still unclear.

It is normally stored in the secretory granules of mast cells and released only into the vasculature at sites of tissue injury, so a main defensive role of HP against invading microorganisms is proposed [32, 59].

2.2.1. Nanoparticles (NPs) for Heparin (HP) Controlled Delivery

HP and low-molecular-weight HP (LMWHP) have been used for decades to prevent venous thrombosis and pulmonary embolism in patients undergoing surgery [60, 61]. Due to their low oral bioavailability (probably owing to its size and negative charge, which provokes ionic repulsion from the negatively charged mucus and epithelial tissue), they have to be administered by parenteral route that involves careful patient monitoring. Additionally, to induce an anticoagulant effect, high doses are required, which can lead to bleeding complications. As a consequence, methods to increase HP and LMWHP bioavailability and half-life have been explored, namely the use of NPs as carriers for controlled delivery.

Early in 2002, Jiao et al. [62] reported the encapsulation of HP in polymeric nanoparticles (NPs) with biodegradable poly-ε-caprolactone (PCL), poly(lactic-*co*-glycolic acid) (PLGA) and the non-biodegradable positively charged polymers Eudragit RS and Eudragit RL (used alone or in combination) and accessed the *in vitro* and *in vivo* activity of the released HP after a single oral administration of HP-loaded NPs in rabbits. The HP encapsulation occurred by the double-emulsion and solvent-evaporation process, which is usually used for water-soluble drugs. The double-emulsion technique involves the formation of the secondary emulsion followed by the precipitation of the polymer and as a consequence the formation of the NPs holding a core region with a water phase where the HP is dissolved.

The mean NPs diameter depends primarily on the mixing velocity during the formation of the emulsions: more vigorous mixing result in smaller NPs. The encapsulation efficiency and control over the drug release depend on the nature of the polymer. In fact, due to the polyanionic nature of HP and the opposite polycationic nature of Eudragit, the highest encapsulation efficiencies were observed with Eudragit RL or Eudragit RS NPs (97 and 59%, respectively). Nevertheless, HP release from these formulations was very low, possibly due to the strong ionic interactions. The higher HP release was observed with PLGA NPs, alone or in combination with Eudragit, as compared to PCL. This fact could be explained by the high hydrophobicity of PCL that diminished the wettability of the NPs and as a consequence the diffusion of HP into the dissolution medium.

In order to have a therapeutic effect, significant amounts of the incorporated HP must be released from the polymeric NPs. For the *in vivo* study - oral administration in rabbits - the formulations with both suitable HP entrapment efficiency and high HP release were selected: Eudragit RL/PCL, Eudragit RS/PLGA and Eudragit /RSRL/PLGA. The best bioavailability was observed for Eudragit RL/PCL NPs with about 20% of the encapsulated HP being biologically active. Not surprisingly, oral bioavailability was lower than that achieved through the intravenous route; nevertheless, these results, especially those obtained with RL/ PCL NPs, open up further research possibilities for the *in vivo* oral administration of encapsulated HP in polymeric NPs.

Hoffart et al. [63] also used Eudragit RS in combination with poly(ε-caprolactone) to develop NPs for oral LMWHP (tinzaparin) administration. Tinzaparin, a complex mixture of different molecular weight species with an anti-Xa/anti-IIa ratio close to 2 was selected for this study. Because factor Xa generation occurs several steps earlier in the coagulation cascade than thrombin generation, inhibition of Xa has a profound effect on the later steps in coagulation. The results after oral administration in rabbits of two different tinzaparin doses (NP200 and NP600) were reported. The tinzaparin-loaded NPs were obtained by the emulsification and solvent evaporation process and resulted in NPs with mean diameter of 408±16 nm, which displayed an incorporation efficiency of 31.0±2.3%.

This low efficiency could be explained by the loss of non-incorporated tinzaparin into the external aqueous phase due to the hydrophilic nature of the drug that diffuses into the external aqueous phase before the precipitation of the polymers, during the double-emulsion process. The encapsulation of tinzaparin is thought to occur both in the core of the NPs and through surface adsorption by electrostatic interaction. In order to have therapeutic and prophylactic effects, biologically significant amounts of the incorporated tinzaparin had to be released from the polymeric NPs, furthermore these must protect tinzaparin from losing its biological activity during the encapsulation process that involves the exposure to organic solvents and shear stress during sonication. Tinzaparin release from the NPs was characterized by an initial burst release followed by a plateau, reaching 40% after 24 h. The initial burst phase was probably due to the release of tinzaparin adsorbed onto the cationic groups of Eudragit RS; the incomplete release could be justified by the strong electrostatic interactions between the drug and the polycationic polymer. After a single oral dose of tinzaparin-loaded NPs (both NP200 and NP600), anti-factor Xa activity was detected confirming that the drug was kept active, protected by the polymeric network in the GIT. The highest anti-factor Xa response was reported when NP600 was administered (1.68±0.14 anti-Xa U/mL 5 h after dosing) (Figure 3). As seen in Figure 3, despite the different highest concentrations reached, the absorption profiles were identical for the two formulations tested. Concerning bioavailability, both formulations demonstrated similar values with no significant difference (51.3±7.9% and 59±4.5% for NP200 and NP600, respectively).

It was postulated that the tinzaparin absorption could be a consequence of a strong and intimate interaction of the negatively charged mucus (specifically the sialic acid residues) and the positively charged NPs. Subsequently, Lamprecht and co-workers [64] tried to further elucidate the mechanisms of LMWHP absorption after being released from NPs. LMWHP-loaded NPs were obtained from different polymers (Eudragit RS, PLGA, and their 1:1 mixture) and their mucoadhesive properties were analyzed.

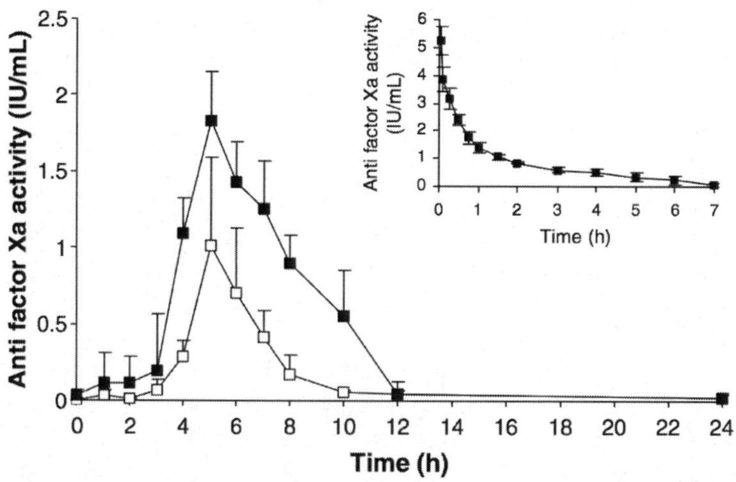

Reprinted from [63], with permission.

Figure 3. Anti-factor Xa activity over timer after a single oral administration in rabbits of LMWHP-loaded NPs. [600 anti-Xa U/kg (NP600), full square; 200 anti-Xa U/kg (NP200); open square] Inset: mean anti-factor Xa activity over time after i.v. administration in rabbits of LMWHP in solution (200 anti-Xa U/kg). Data are mean ± S.D, $n \geq 6$.

The NPs were produced by a method based on an oil/water emulsification-solvent evaporation method, as described above, where a fixed amount of polymer (200 mg) was used. All NP formulations presented a mean diameter size of about 100 nm. The interaction between Eudragit RS and the intestinal mucus was quantified using a resonant mirror system and a mucin-coated support. The results for the blank NPs indicated a strong interaction between Eudragit RS/PLGA NPs and mucin and that PLGA NPs showed very weak adhesion, certainly by merely sticking to the surface. With LMWHP-loaded NPs the responses were smaller, indicating the loaded NPs had fewer binding sites, possibly due to the negative zeta potential of the loaded NPs and the negative charge of the mucin surface. In studies with uncoated cells, all NPs formulations demonstrated similar extent of binding; however, in mucin-coated cell monolayers, the differences among the various NPs became more evident. Cell culture experiments with mucin-coated Caco-2 cells confirmed the results obtained with the resonant mirror system: the Eudragit RS and Eudragit RS/PLGA NPs bound tighter to mucin than PLGA NPs. Besides binding of NPs, the adhesion and uptake of LMWHP into the cell layer was evaluated overtime.

Generally, NP formulations translated into higher drug transport across the cell layer comparing to free drug solution. In the presence of mucin, binding and transcellular transport of LMWHP was higher for RS and RS/PLGA NPs than for PLGA NPs, which suggested a distinct effect of these NPs on HP transport. A two-step process preceding the uptake of the drug into the cells was proposed: initially the electrostatic interaction of the NPs with the mucus triggers the non-specific adhesion of the NP to the epithelial barrier; then the drug on the NP surface is released and replaced by mucin.

The conclusion is that the mucoadhesive NP properties are responsible for the extended and intensified contact of the NPs with the epithelial barrier therefore increasing drug absorption. Recently, new LMWHP-loaded NPs of a new synthetic block copolymers, poly (methyl methacrylate-*b*-trimethylaminoethyl methacrylate) (PMMA-*b*-PMAETMA) were reported [65]. The cationic acrylic block polymers with well-defined microstructure, charge and low polydispersity, were prepared by reversible addition-fragmentation chain-transfer (RAFT) polymerization and used to encapsulate bemiparin (ultra-LMWHP with an average molecular weight of 3600 g mol^{-1}).

PLGA and Eudragit RS NPs were used for comparison. Figure 4 depicts the block polymers NPs formation and bemiparin encapsulation by a two-step emulsion process. PLGA does not bind bemiparin, its encapsulation takes place in this case *via* packing during the emulsification process. Due to bemiparin negative charge, the positively charged Eudragit RS can bind the drug. The encapsulation of bemiparin into the block polymers NPs occurred when the amphiphilic block copolymers self-assembled into micelles.

The zeta potential indicated the NP stability in the suspension: NPs prepared with cationic systems without bemiparin had strongly positive zeta potential values in water owing to the ammonium charge group. The PLGA NPs exhibited a small negative zeta potential value probably owed to the carboxylic end groups of the polymer chain. When bemiparin was encapsulated, the zeta potential values became more negative, likely due to the presence of negatively charged bemiparin on the NP surface.

The NPs had mean size between 100 and 200 nm (depending on the polymer used) and a predominantly spherical morphology. The encapsulation efficiency was higher in the PMMA-*b*-PMAETMA and Eudragit RS NPs comparing to PLGA NPs, mainly due to the ionic bounds formed between the drug and the polymers.

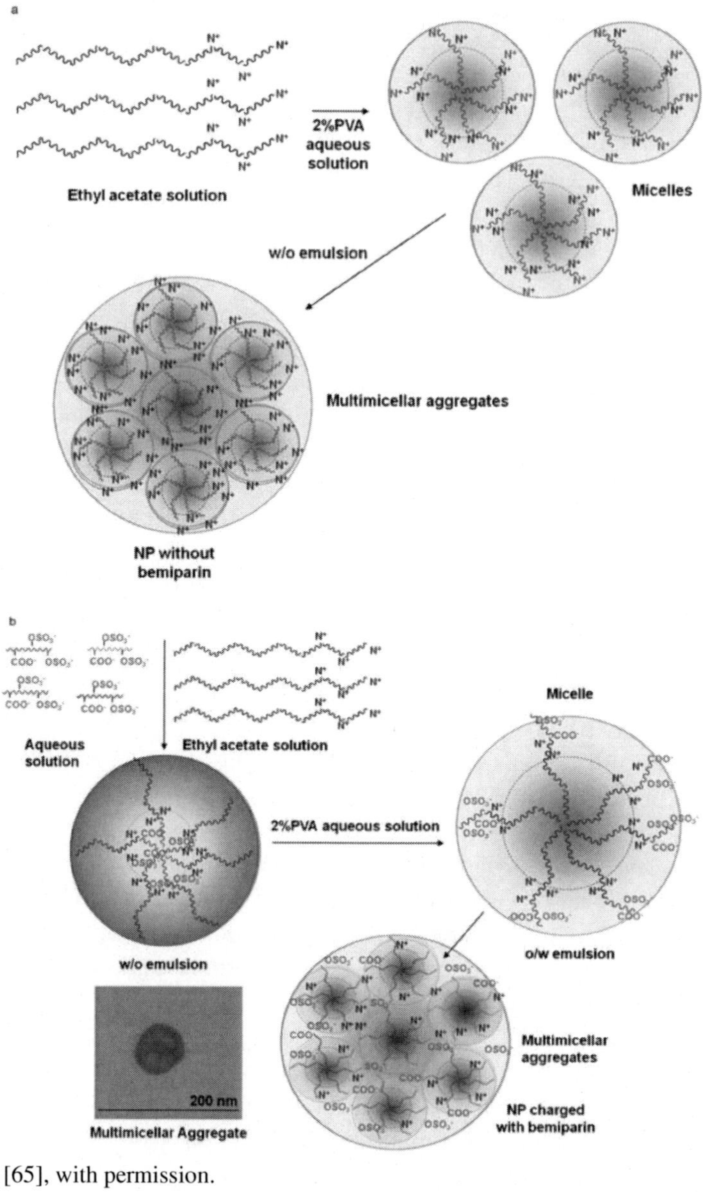

Reprinted from [65], with permission.

Figure 4. Explanatory scheme of NPs formation obtained by an emulsion technique using the block copolymers prepared *via* RAFT. (a) Preparation of non-loaded NPs and (b) preparation of loaded NPs with bemiparin. A TEM image of the NP system 86/14 PMMA-*b*-PMAETMA is also shown.

However the total amount of loaded bemiparin was identical for all the NP systems tested. The release profile at pH 7.4 was influenced by the chemical structure of the polymers over the 70 days analysis period: Eudragit RS NPs released the least amount of bemiparin; with PLGA NPs nearly 100% of bemiparin was released mainly because drug release from PLGA was achieved by diffusion of bemiparin across the flexible chains and/or erosion mechanisms of the PLGA biodegradable polymer; NPs prepared with PMMA-*b*-PMAETMA copolymers exhibited higher bemiparin release than Eudragit RS NPs probably due to the microstructure of the copolymers. These block copolymers formed a self-assembled system with a core of PMMA blocks and a shell of cationic PMAETMA complexed with bemiparin; this outer domain is very hydrophilic and certainly allowed better water diffusion compared to Eudragit RS.

Besides its anticoagulant activity, bemiparin has as a noteworthy regulation role of growth factor activity, including those of Fibroblast Growth Factors (FGFs) and Vascular Endothelial Growth Factor (VEGFs). The Baf32 cells were used to measure the activity of the ternary complexes formed with bemiparin, FGF2 and FGF receptor c1.

The formation of active ternary complexes induces cell proliferation. It was shown that increasing concentrations of encapsulated bemiparin in PLGA NPs resulted in increased cell proliferation, the highest dose of bemiparin encapsulated in PLGA NPs being equally active as free bemiparin. This could be attributed to the 40% release of bemiparin from PLGA NPs within 72h.

2.2.2. Heparin (HP)-Conjugates and HP-Based Nanoparticles (NPs)

HP-based/modified NPs and HP-conjugates have been widely studied as therapeutic delivery vehicles in several biomedical fields. Some HP-conjugates for cancer therapy have already been described such as HP-paclitaxel [66] and HP-doxorubicin (DOX) conjugates [67]. In order to develop a targeted drug delivery system for chemotherapy, NPs composed of folate-conjugated HP-based copolymers were synthetized [68]. DOX was effectively incorporated into HP and folate-poly(ethylene glycol)-conjugated HP (FPHP) NPs by a simple dialysis method. By self-assembling, the HP and FPHP NPs easily incorporated DOX by hydrophobic interactions between DOX and the polymer chains.

NPs smaller than 200 nm in diameter usually accumulate in the tumor tissue via the enhanced permeability and retention (EPR) effect while larger drug carriers are promptly scavenged by the mononuclear phagocytic system. Both kinds of DOX-loaded NPs had less than 200 nm, which may thus be advantageous. The *in vitro* release of DOX from the HP and FPHP NPs was studied in two different buffered solutions (pH 7.4 and 5.0). In all the conditions tested both formulations presented a typical biphasic release profile: an initially rapid, then slow and sustained over 2 weeks. However, the DOX release rates from both HP and FPHP NPs were accelerated when the pH was lowered from 7.4 to 5.0. So, when the injected NPs remain under normal physiological conditions (pH 7.4) DOX will expectedly be retained inside the NP for longer circulation times. As pH drops, as in endosomes/lysosomes or in the tumor tissue, a more accelerated release should happen. Thus, the pH-dependent DOX release is particularly useful to achieve tumor-targeted DOX delivery. The DOX uptake by KB cells was accessed by confocal laser scanning microscopy (CLSM), and it was shown that the intracellular uptake of DOX-loaded FPHP NPs was higher, as compared to the DOX-loaded HP NPs, presumably due to the folate targeting ligand. While free-DOX mainly accumulated in the nucleus whereas the majority of the DOX delivered by the NPs was observed still in the cytoplasm. However, DOX-loaded NPs (both HP and FPHP) were first taken into the cells by an endocytosis pathway, then localized in acidic endocytic compartments (endosomes and lysosomes) where the controlled release of DOX began due to lower pH. Ultimately, the released DOX would diffuse into the nucleus where it could exert its activity towards DNA. Altogether the results showed that DOX-loaded FPHP NPs had enhanced cellular uptake, increased targeting capacity and improved cytotoxicity on KB cells comparing to the DOX-loaded HP NPs.

Therapeutic proteins are becoming available for the treatment of several health problems. However, reduced stability and short circulation time poses major problems to the efficiency of these biopharmaceuticals. Heparinoids can specifically interact with several functional proteins such as growth factors, cytokines, extracellular matrix components and adhesion molecules [69], thus being very useful as a therapeutic agent in various pathological conditions. In 2010, self-assembled NPs composed of a HP-Pluronic conjugate were prepared for the intracellular delivery of ribonuclease A (RNase A) [70]. HP NPs have shown remarkable characteristics in terms of drug loading, intracellular fate, loading simplicity, elevated uptake efficiency and reduced size after drug loading. The HP-Pluronic NPs were obtained by self-assembly through hydrophobic interactions when the temperature was raised from 4°C to 37°C. The NPs were physically cross-linked by the hydrophobic interaction between the poly(propylene oxide) chains of Pluronic. Additionally, hydrogen bonds and electrostatic interactions supported the cross-linking. These NPs consisted of a gel-like structure that swelled and had variable size, high affinity for proteins, increased stability and negatively charged surface. The RNase A incorporation into the HP-Pluronic NPs occurred due to the electrostatic interaction during the process of NPs formation. Because the pI value of RNase A is approximately 9, it bears electrostatic affinity towards HP, which is negatively charged. Indeed, it was verified that the NPs had high affinity for RNase A (loading efficiency of about 80-99%). The RNase A release from the NPs was influenced by pH: at pH 7.4, and after 24 h, about 100% of the loaded RNase A was released, whereas only about 30% was released at pH 5.0. The cellular uptake of the RNase A-loaded HP-Pluronic NPs was examined using fluorescence activated cell sorter (FACS) and CLSM. The HP-Pluronic NPs demonstrated to have a critical role in cellular uptake as almost no cellular uptake was seen when free RNase A was added. The CLSM images showed that released RNase A was spread to most regions of the cytosol and even to the nuclei, indicating that the membrane-penetrating activity of HP is wider including endosomal, lysosomal or even nuclear membranes. The authors confirmed that the delivery of RNase A using the described HP-based NPs improved cytotoxicity in a RNase A concentration-dependent way. They hypothesized that due to the nuclear localization observed and high cellular uptake, the HP-Pluronic NPs facilitated apoptotic cell death through the release of RNase A not only to the cytosol but also in the nuclei. It was concluded that the developed NPs could not only be applied to the intracellular delivery of several proteins but could also be tailored to improve the efficiency of the delivery, making them suitable for cancer chemotherapeutics.

Recently, in 2013, a similar study using HP-Pluronic NPs was reported for combined chemotherapy with paclitaxel and deoxyribonuclease (DNase) [71]. The goal of combination chemotherapy is to combine drugs that act by different mechanisms, decreasing the chances of resisting cancer cells.

The aims of this study were to determine the efficiency of the multidrug nanocarrier developed for combination chemotherapy *in vitro*, and also to see if combined drugs with distinct bioactivities and physicochemical properties showed synergistic cytotoxicity when incorporated in the nanocarrier. Paclitaxel is an anticancer drug used in conventional chemotherapy that possesses activity against a broad range of tumor types by interfering with the normal microtubule breakdown and therefore the cytoskeleton functioning.

DNase is a nuclease that catalyzes the hydrolytic cleavage of phosphodiester linkages in the DNA backbone. Identically to what was described in the previous study [70], the HP-Pluronic NPs self-assembled when the temperature was raised to 37°C and the hydrophobic paclitaxel and the enzyme DNase were successfully encapsulated in the NPs network. DNase had high affinity for the highly sulfated HP resulting in a high loading efficiency even when paclitaxel was simultaneously loaded. As a consequence of drug encapsulation, the hydrodynamic properties of the NPs changed, both drugs increased its size in a concentration-dependent manner; paclitaxel loading increased the NPs surface charge, whereas DNase decreased it. As seen in Choi et al. [70], the HP-Pluronic NPs, besides being localized in the cytoplasmatic region, were also detected into the nuclei.

To test cytotoxicity of the formulations, HeLa cells viability was determined. In single-drug treatments a similar pattern and level of cytotoxicity was detected indicating the enhanced uptake efficiency of the incorporated drugs. This indicated that each individual drug was internalized into the cells and then released at the proper site of action. But more importantly, the superior effect of the combined therapy was demonstrated (Figure 5).

Based on the high specific binding of HP to several growth factors, various HP-based systems have been developed for the controlled release of those proteins [72, 73]. The binding of growth factors to HP stabilizes the proteins, inhibiting their denaturation or proteolysis and simultaneously enhances their binding to cellular receptors. Chung et al. presented a HP-functionalized PLGA NP for VEGF controlled release [72].

The HP-PLGA NPs were prepared by solvent diffusion method - slowly adding the organic phase with PLGA into the aqueous solution containing Pluronic F-127 and HP. The obtained monodispersed NPs had up to 4.7 wt.% HP in total, with most of the HP molecules essentially distributed at the surface layer. The localization of Pluronic molecules in the PLGA core was probably owed to the hydrophobic interaction between poly(propylene oxide) chains of Pluronic and PLGA.

Due to the favorable electrostatic interaction with HP, lysozyme was selected as a model protein to evaluate the release properties of the different NP formulations. The NPs with HP content of 2.4 and 4.7 wt.% showed linear and complete release profiles, with no initial burst release, up to 10 and 19 days, respectively. This release profile reflected the strong binding of the net positively charged lysozyme with the negatively charged surface layer of the NPs.

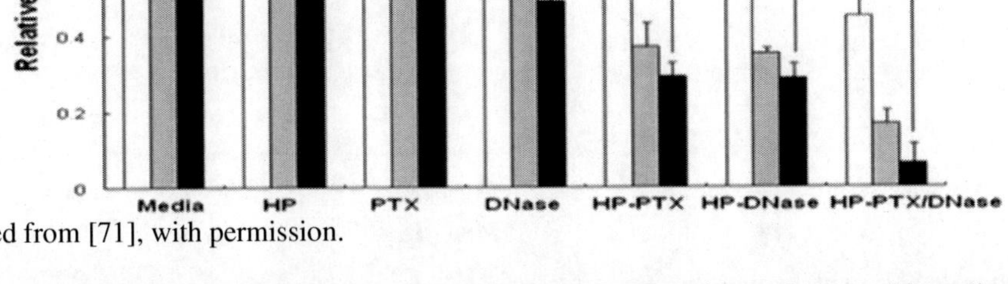

Reprinted from [71], with permission.

Figure 5. Cytotoxic effects of each free drug and HP-Pluronic NPs without drug or with paclitaxel and/ or DNase in HeLa cells for 24h. White, grey and black bars indicate treated drug doses 100, 250 and 500 μg/mL, respectively.

It was also demonstrated that the NPs effectively protected the protein from denaturation during loading and release. Based on this assay, the NPs with 4.7 wt.% HP content were used to evaluate the VEGF release. A similar, but prolonged, release profile was found: linear - with no initial burst - release up to 37 days, with roughly 85% of VEGF released (Figure 6). This prolonged release pattern could be the result of a more specific interaction between VEGF and HP than simple charge interaction. Based on these results, the developed HP-based NPs could be used in tissue engineering as a growth factor delivery component or as sustained release systems for HP-binding growth factors.

HP-functionalized chitosan (CS)/poly(γ-glutamic acid) (γ-PGA) NPs have also been prepared for the controlled release of FGF and HP [74], with great potential for enhancing ischemic tissue regeneration and simultaneously preventing blood vessel rethrombosis.

In this study, polyelectrolyte self-assembled NPs were produced by adding γ-PGA or γ-PGA/HP aqueous solutions to a CS aqueous solution under magnetic stirring, at room temperature. It was observed that the negatively charged HP produced a significant effect on the mean particle size (increase) and zeta potential (reduced) of the NPs. It is known that the ischemic and inflamed areas in the body have lower pH (<6.7) than surrounding tissues and blood (7.4). So, a "smart" and controlled release system should allow the sustained release of FGF under those conditions and then, after repairing the tissue, it should rapidly disintegrate. Therefore, the HP-CS/γ-PGA NPs were shown to be pH-sensitive: at pH 6.0 the NPs remained stable, however at pH 7.4 they began to swell and outer layer of the HP-CS/γ-PGA NPs became increasingly bubbled, suggesting that the NPs started to disintegrate.

So, these "smart" NPs are likely to release FGF or HP, depending on the pH variation. Due to the amount of positively charged CS used in the formulation, the loading efficiency of negatively charged HP was very high. The HP release from the NPs was significantly affected by the NPs stability in response to pH: at pH 6.0 the HP release was inhibited, becoming significant at pH 7.4 due to the NPs disintegration.

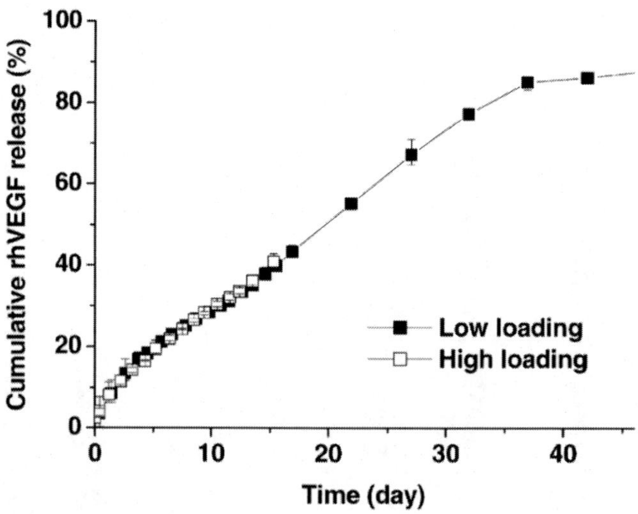

Reprinted from [72], with permission.

Figure 6. Cumulative VEGF release from the HP-PLGA NP with 4.7 wt.% HP content for two different VEGF loading amounts (n=3).

Accordingly, no significant anti-factor Xa activity was detected at pH 6.0, indicating negligible anticoagulant activity of the NPs due to poor HP release. Instead, at pH 7.4, the anti-factor Xa activity in blood plasma was maintained at 0.09-0.22 IU/mL, which is an effective concentration for the treatment of deep vein thrombosis and pulmonary embolism. At pH 7.4, more than 80% of loaded FGF was released within 24 h mainly because the NPs failed to ensure an adequate retention of loaded FGF, due to disintegration. On the contrary, at pH 6.0 the cumulative amount of FGF released was considerably reduced to about 52% when using HP-CS/γ-PGA NPs with 1.0 mg/mL loaded FGF. This result was attributed not only to the stability of the heparinized NPs at pH 6.0 but also to the high affinity of FGF to HP. This study also showed that the FGF released from the HP-CS/γ-PGA NPs retained its mitogenic activity and that the incorporated HP mediated the activation of FGF receptors, resulting in enhanced proliferation of human fibroblasts. The results additionally suggest that the FGF released from the HP-CS/γ-PGA NPs could actually stimulate tube formation of HUVEC that are involved in the angiogenesis process. As concluding remarks, the authors projected a schematic application and function for the multi-functional growth factor delivery system, depicted in Figure 7. In ischemic and inflamed areas (pH <6.7), the HP-CS/γ-PGA NPs could sustain the release of FGF enhancing both the proliferation of fibroblasts and the angiogenic tube formation by HUVEC. Then, in repaired tissue (pH = 7.4), the NPs would disintegrate leading to a continuous release of HP that would prevent harmful clots from forming in blood cells.

2.2.3. Heparin (HP)-Functionalized Metallic Nanoparticles (NPs)

Metallic NPs are easy to synthetize and have affinity for many biological molecules, which makes them attractive for the biomedical field (as already mentioned). Carbohydrates have already been incorporated into nanomaterials both as reducing and stabilizing agents [75]. HP was also used to modify the surface of metallic NPs, namely iron oxide and Au NPs, in order to avoid self-aggregation, enhance stability, and increase cellular uptake of the NPs [58] simultaneously providing targeted detection [76, 77] and inducing cellular apoptosis of cancer cells [78]. Kemp and collaborators [79, 80] described a clean method to prepare Au and Ag NPs with specific biological functions.

Stabilized Au and Ag NPs were obtained using HP derivatized with a diaminopyridine group (DAPHP) at the reducing end, synthesized through reductive amination [79].

In vitro studies indicated that the DAPHP NPs inhibited coagulation over the duration of the study (> 60 min). *In vivo* studies in a rat carrageenan model of paw edema showed the potential anti-inflammatory effect of the described NPs. The Au- and Ag-DAPHP NPs inhibited significantly the paw edema, with no systemic effects on hemostasis as compared to free HP. Additionally, the anti-angiogenesis properties of the developed NPs were investigated [80] in FGF-2-induced angiogenesis in the chick chorioallantoic membrane model. All the formulations tested (HP, DAHP, Au-glucose, Ag-glucose, Au-DAPHP and Ag-DAPHP NPs) demonstrated significant inhibition of FGF-mediated angiogenesis. However, higher inhibition was verified with Au-DAPHP or Ag-DAPHP NPs.

Similar results were obtained in the FGF-2 mouse matrigel model. The anticoagulant [79] and anti-angiogenic [80] properties of the reported nanocomposites could make them useful in several biological and biomedical applications such as psoriasis and other dermal accelerated pathological angiogenesis.

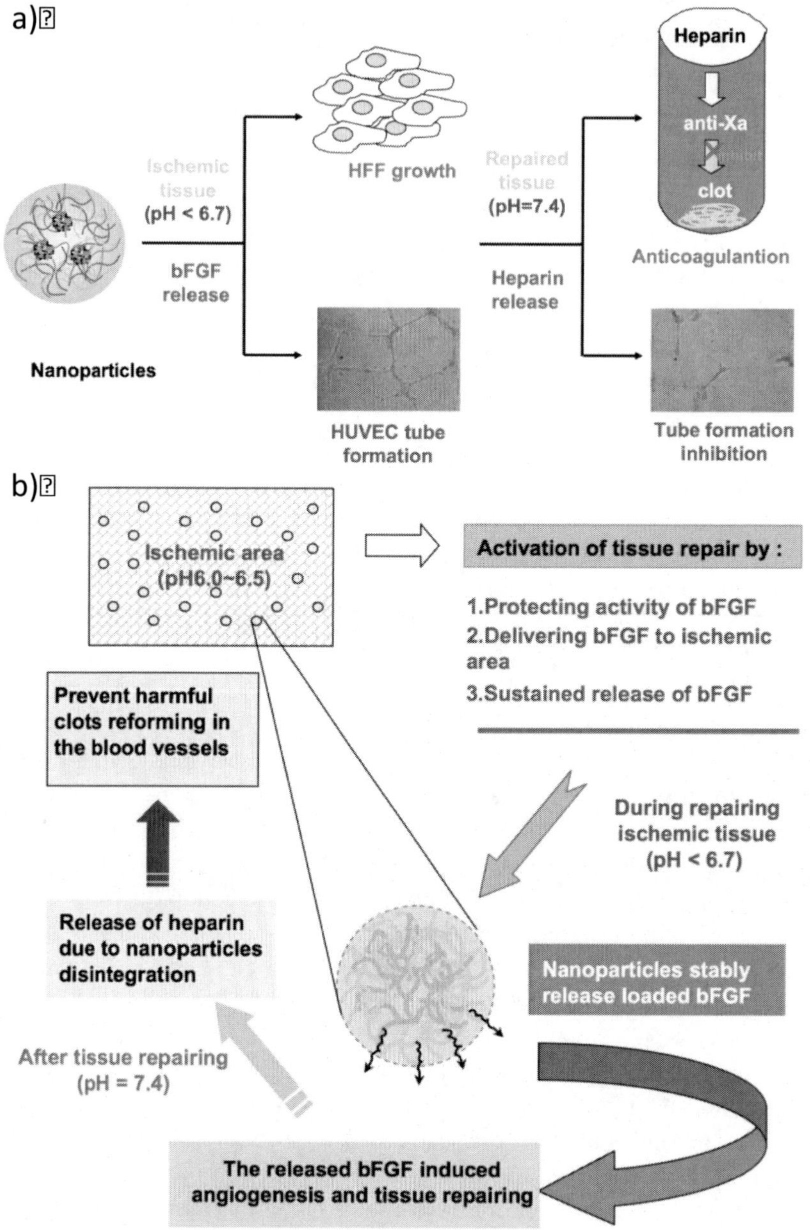

Adapted from [74], with permission.

Figure 7. (a) Schematic illustration of applications and function of the HP-CS/γ-PGA NPs for multi-functional FGF and HP delivery. (b) Brief schematic description of the mechanism for multi-functional delivery of FGF and HP.

2.3. Chitin/Chitosan – Associated with Sulfated Polysaccharides

Chitin is homopolymer of N-acetyl-D-glucosamine units that repeat to form a long chain in β-1,4 linkage. It is the major component of the cell walls of fungi, the exoskeleton of arthropods and insects, the radulas of mollusks, etc. Chitosan (poly[β-(1-4)-2-amino-2-deoxy-D-glucopyranose]) (CS) is a water soluble polycationic co-polymer of N-glucosamine and N-acetyl-glucosamine, that occurs naturally or is obtained by the partial N-deacetylation of chitin [81]. CS is biodegradable, non-toxic to animals, soluble in acidic solutions, available in various forms and much more tractable than chitin. As a result, CS presents characteristics that make it attractable for numerous biotechnology and biomedical applications.

A comprehensive review about CS nanoparticles (NPs) and their biomedical applications was published in 2010, by Pereira et al. [82]. The mentioned publication reviewed the different techniques to achieve CS-based NPs, their properties, and their applications in the biomedical field, namely as a drug delivery vehicle for biopharmaceuticals. In this review, the association of chitosan with SPs in the production of NPs will be reviewed.

2.3.1. Chitosan (CS)-Based Nanoparticles

In 2010 Grenha and colleagues reported the preparation and characterization of protein-loaded NPs achieved by ionic complexation of CS and carrageenan [83]. The NPs were formed by means of an electrostatic interaction of CS with carrageenan anions, in which the positively charged amino groups of CS interact with the negatively charged sulfate groups of carrageenan. The developed NPs obtained using a CS:carrageenan ratio of 3.5:1 had sizes between 350-650 nm and positive zeta potentials of +50-+60 mV. Ovalbumin was used as a model protein to evaluate the potential of the NPs as controlled delivery systems. Ovalbumin incorporation led to a significant reduction of size and zeta potential. This was possibly due to an effect of reticulation of the particles by the protein. The highest loading efficiency of about 17% was obtained for an initial amount of 30% ovalbumin.

The release profile of ovalbumin from the NPs revealed a sustained release of the protein for a period of 3 weeks. This behavior suggested that the protein was actually encapsulated in the solid matrix of the NPs and not at the surface of the particle, which would lead to an initial burst release. The ovalbumin probably established a strong interaction with the NPs mainly due their opposite charges. This release profile is ideal for drug delivery applications since it allowed sustained and prolonged protein release without the need of repeated administrations.

Later, Rodrigues et al. [84] prepared CS/carrageenan/tripolyphosphate (TPP) NPs and accessed the effect of the cross-linker TPP using the CS/carrageenan NPs obtained as mentioned previously [83]. The NPs were obtained by a very mild polyelectrolyte complexation/ionic gelation method, as described by Grenha et al. [83]. It was proved that the incorporation of TPP in the matrix of the NPs had a significant cross-linking effect resulting in changes in NP size, zeta potential and production yield. For the highest amount of TPP used, particle size decreased to 176-208 nm, zeta potential also diminished from +80 to about +50 mV, and the production yield increased 36%. The cross-linking effect explained the reduction in size as it induces the condensation of the polymeric chains resulting in smaller NPs.

The formulation containing TPP also evidenced improved and enhanced stability, showing no significant size variation for 250 days. The authors hypothesized that TPP acts like a stabilizer due to its cross-linking effect. Considering CS/carrageenan/TPP NPs properties such as the small size and high positive surface charge, a possible application in mucosal drug delivery could be envisaged.

In 2003, a NP delivery system employing the oppositely charged polymers CS and dextran sulfate (DS), and an arginine-rich hexapeptide (ARH) was conceived, aiming the development of protein/peptide delivery [85]. ARH peptide was proven to block the growth and metastasis of Vascular Endothelial Growth Factor (VEGF)-secreting HM7 human colon carcinoma in nude mice. The NPs were formed by self-assembling through phase separation and the process occurred at room temperature with no need for a stabilizer, heat or organic solvents, as it is desirable for the encapsulation of therapeutic proteins. The ARH peptide was incorporated into the NPs simply by mixing the CS and DS solutions containing the dissolved ARH peptide, at room temperature, *via* ionic interactions between the positively charged ARH peptide (pI> 9.0) and the negatively charged NPs. In general, the peptide was released in a biphasic way with an initial rapid release followed by a continuous slow release over several days.

Tissue regeneration is a very complex biological process that comprises chemotaxis, cell division, neovascularization and synthesis of extracellular matrix proteins. Stem cells are now extensively studied as powerful therapeutic tools for several applications due to their distinctive pluripotency and regenerative properties. However, cell therapy is often unsuccessful, since the majority of cells die after a few weeks following transplantation. So, chemotactic mobilization of stem cells from bone marrow is arising as an alternative strategy. This approach may reduce the exposure of the cells to a harsh environment and permit them to slowly remodel the insured site. The fundamental step in the mobilization strategy is to use an effective chemokine and a suitable controlled delivery system. Stromal cell-derived factor 1 (SDF-1) is a key chemokine involved in stem cell mobilization also playing a role in the biological and physiological functions of mesenchymal stem cells (MSC). Nevertheless, the use of SDF-1 is still limited mainly by its short half-life and ready degradation by enzymes. Huang and Liu investigated the use of CS-based NPs as a controlled delivery system for SDF-1 [86]. Three types of CS-based NPs were prepared by ionic gelation: CS/fucoidan NPs, CS-TPP NPs and CS/TPP/fucoidan NPs (TPP was used as a cross-linker, as mentioned previously). The SDF-1 loaded NPs were obtained by ionotropic cross-linking. The physico-chemical properties of the developed NPs were analyzed by FTIR, TEM and dynamic light scattering (DLS). The particle sizes of the CS-fucoidan, CS/TPP and CS/TPP/fucoidan NPs were 173-513, 182-299 and 173-403 nm, respectively, and the zeta potential varied from +3 to +33 mV. Negatively charged fucoidan and TPP can form an ionic complex with SDF-1 due to an electrostatic interaction. However, due to a stronger electrostatic interaction between fucoidan and SDF-1, the encapsulation efficiency of the CS-fucoidan NPs (77-99%) exceeded that of CS/TPP (80-89%). During the preparation of CS/TPP NPs, the TPP initially cross-links with SDF-1 and then with CS. So, TPP was a substantial positive influence on the amount of SDF-1 that can be incorporated in CS/TPP NPs.

However, the encapsulation efficiency of the CS/TPP/fucoidan NPs was lower (60-68%) than the other CS-based NPs perhaps because binding of the main groups of SDF-1 to the TPP might compete with that of fucoidan. The authors hypothesized that TPP occupied the majority of SDF-1 active sites due to its smaller molecular weight and higher diffusion rate, and this interfered with the specific interaction between SDF-1 and fucoidan, leading to lower encapsulation efficiency. SDF-1 release experiments displayed a biphasic profile, characterized by an initial 12 h burst followed by a slow release overtime.

It was also noticed that the CS/TPP/fucoidan NPs was most effective, thus further studies were made using only this system.

The CS/TPP/fucoidan NPs protected SDF-1 against proteolysis up to 12 h and also offered protection from inactivation by heat in the range of 37 to 77°C. As a heparinoid, it was postulated that fucoidan, like heparin, acts protecting SDF-1 against degradation. The mitogenic activity of the released SDF-1 was evaluated by determining its capacity to stimulate the proliferation of MSC in the absence of fetal bovine serum (FBS). Interestingly, the SDF-1 released from the CS/TPP/fucoidan NPs improved the MSC proliferation significantly more than SDF-1 alone. This result could be explained by the stabilization and protection offered by the NPs, and to the effect of fucoidan in regulating the dimerization and functionality of SDF-1 [87]. The ability of the released SDF-1 to induce chemotaxis was confirmed *in vitro* using a transwell system; further, the migration across the membrane closely followed the kinetics of the release of SDF-1 from the CS/TPP/fucoidan NPs. The developed system not only offered enzyme resistance and heat-protection, but also could be effective in promoting stem cell mobilization and repairing injured tissues.

Tissue regeneration in peripheral nerves or in the spinal cord after a transection injury is a complex process. Several strategies have emerged to increase the likelihood of axonal regeneration and functional recovery such as directing growth factors to injured areas. Huang and Yang also evaluated the feasibility of using CS-fucoidan NPs as potential carriers for nerve tissue-engineering applications [88].

The authors attempted to regulate the differentiation of PC12 cells by controlling basic Fibroblast Growth Factor (bFGF) release from the CS-fucoidan NPs. A major limitation of using bFGF is its degradation by enzymes and consequently its short half-life. This study demonstrated that the protein was effectively protected from heat and enzyme degradation by the NPs and that this effect was positively related to the amount of fucoidan. The bFGF release from the NPS was sustained up to 4 days with no burst release. The NPs had no cytotoxic effect on PC12 survival or growth, up to a concentration of 125 ng/mL. The PC12 cell line was also used to study how bFGF (free and encapsulated) affect neurons. The effect of bFGF release on PC12 cells was determined by analyzing the neurite length during a 3 day culture. The bFGF-loaded NPs preserved its ability to promote PC12 neurite growth, and furthermore it significantly decreased the amount of growth factor necessary for neurite extension. This finding was correlated to the presence of fucoidan, which, as a heparinoid, has the ability to bind bFGF and sustaining its bioactivity [89].

Pulmonary delivery systems using particulate drug carrier are very attractive for the treatment of several airway disorders like cystic fibrosis, asthma, pulmonary infections, pulmonary hypertension, and lung cancers. Ideal pulmonary drug delivery systems should provide direct and local delivery to the lungs allowing for both selective pulmonary effect and reduction of side effects. NPs have emerged as valuable carrier for efficient drug delivery to the lung epithelium without mociciliary clearance and phagocytic mechanisms.

Recently, novel CS/fucoidan NPs were developed and evaluated as potential carriers for antibiotics delivery to the lungs [90]. The NPs were obtained using a simple polyelectrolyte self-assembly method and characterized physicochemical by FTIR, TEM, and DLS. The size, zeta potential and stability depended on the weight ratio of CS to fucoidan. The NPs were not cytotoxic to A549 cells in concentrations ranging from 0.375 to 3 mg/mL and exhibited highly potent antioxidant effects by scavenging 1,1-diphenyl-2-picrylhydrazyl (DPPH), reducing the concentration of intracellular oxygen species (ROS) and superoxide anion in stimulated macrophages. The results suggested that the scavenging effect was mainly derived from fucoidan and that both CS and fucoidan contributed to the antioxidant effect of the developed NPs due to their individual properties (Table 1). In order to evaluate the CS-fucoidan potential as drug delivery systems, gentamicin was used as a model antibiotic in *in vitro* assays. The encapsulation efficiency of gentamicin in the NPs was 94%. The gentamicin release presented a biphasic profile: initial 10 h burst release (~90% gentamicin released), mainly due to the hydrophilic nature of the antibiotic used that can easily diffuse into the release medium; followed by a slow release rate (~99% after 72h) caused by polymer degradation. Given the suitability of regular administration to the airways, the released profile obtained may be of interest for clinical applications.

In tissue engineering, a porous scaffold is usually necessary to act as template for cell proliferation, differentiation and growth. Simultaneously, scaffolds can also act as controlled release systems that enable the delivery of growth factors ideally with rates matching the physiological need of the regenerating tissue. Delivering bioactive agents directly from the polymer matrix that is used for cell transplanting is the ideal situation. Nonetheless, the incorporation of bioactive molecules in scaffolds without loss or change of bioactivity is still a major challenge.

Aiming the development of scaffolds capable of accelerating regeneration, a CS/heparin (HP) NP-immobilized decellularized bovine jugular vein scaffold was created [91].

The main goal of this study was to obtain a sustained local release of VEGF in order to improve the vascularization of decellularized bovine jugular vein scaffold. The CS/HP NPs were obtained by self-assembly and their size varied from 67 to 132 nm. Interestingly, their zeta potential values were either positive or negative, depending on the relative ratio of the surface molecules, CS or HP. Although the authors hypothesized that NPs with higher ratio of HP to CS could bind more VEGF, it was verified that NPs with zeta potential of about +30 mV were more efficient. The NPs were immobilized to the scaffolds through chemical modification with 1-(3-Dimethylaminopropyl)-3-ethylcarbodiimide hydrochloride (EDC).

The scaffolds presented interconnected pores with an average diameter of 120 nm, which allowed for the NPs to penetrate deep into the matrix. Besides reinforcing the NPs structure, ECD was found to significantly reduce the initial burst release of VEGF, which besides extending the clearance time also reduced the toxicity, which may be associated to high concentrations. The system allowed the controlled release over long periods, e.g., for concentration of 50 ng/mL VEGFa release of ~37.12% in 30 days was observed. The recruitment of functional cells like endothelial and fibroblasts has a crucial role in regeneration whereas the extracellular matrix can provide support for growth and reproduction of the cells. *In vitro* studies showed that VEGF-loaded NPs clearly increased the proliferation of endothelial cells. The following *in vivo* studies proved that massive new capillary formation occurred in the subcutaneous implanted scaffolds immobilized with VEGF-loaded NPs at 4 weeks and 8 weeks after implantation. Remarkably, not only the scaffolds immobilized with VEGF-loaded NPs enhanced large vessel growth, but also the majority of the vessels were small capillaries.

Additionally, it was observed substantial cellular infiltration in the modified implants. The infiltrating cells were mainly fibroblasts and only a few macrophages. This fibroblast infiltration was complemented with significant synthesis of new collagen fibers and glycosaminoglycans suggesting that scaffolds immobilized with VEGF-loaded NPs could accelerate cell infiltration and tissue remodeling. In general, the described system presented high therapeutic potential due to its biocompatibility and promotion of accelerated regeneration.

In another work, a complex hybrid release system based on a three-dimensional (3D) foam of poly(D,L-lactic acid) (PDLLA) with incorporated CS/chondroitin sulfate NPs was described [92]. Bovine serum albumin (BSA) was used as model protein to evaluate the entrapment and *in vitro* release kinetics to study the potential application of these foams for dual protein release: the protein dispersed in the PDLLA foam should be released faster and the protein encapsulated in the CS/chondroitin sulfate NPs should be delivered in a slower way. The CS/chondroitin sulfate NPs were obtained by self-assembling and presented spherical shape and size ranging from 150 to 200 nm. The PDLLA foams displayed interconnected pores and a homogeneous distribution of the NPs over the 3D support.

Due to their hydrophobicity, the PDLLA foams took more than 6 months to degrade. By incorporating hydrophilic NPs, both the water uptake and the degradation rate of the foams increased. The incorporation of empty NPs into the PDLLA foams resulted in a more controlled protein release kinetics as compared to PDLLA foams alone: the PDLLA-only foams showed a burst release effect and most of the entrapped BSA was released in 1 day. The incorporation of the NPs had a positive effect on both the initial and long term release, maybe due to the electrostatic interaction between the positively charged NPs and the negatively charged BSA (pI = 5.3), therefore delaying the BSA diffusion out of the scaffold. The BSA release profile of foams with incorporated BSA-loaded NPs was also accessed. The NPs showed a controlled release profile, delivering the BSA at a constant rate, and their incorporation in the PLLDA foams slightly reduced the initial burst release, although it did not have significant effect in a long-term release.

CS and HP were also used to improve the relatively low tumor-targeting efficiency of bare PLGA NPs [93]. The surface functionalized PLGA NPs were obtained by the modified nanoprecipitation of PLGA, where precipitated PLGA particles were surface-modified with Pluronic conjugates with CS or HP. The obtained NPs had about 100 to 150 nm, and the surface functionalization with CS or HP was confirmed by positively or negatively increased zeta potential, respectively. In order to show the stability of the NPs in physiological environments, the NPs were incubated overnight in 100% FBS and then characterized in deionized water. In all cases tested, no noticeable changes in the size distribution were detected. Each kind of NPs obtained, P-PLGA (Pluronic), HP-PLGA (HP-Pluronic conjugate) and CP-PLGA (CS-Pluronic conjugate) were tested in 3T3 fibroblasts and SCC7 tumor cells to evaluate acute cytotoxic effect at the maximum dose used for the *in vivo* imaging (0.125 mg/mL). In both types of cells, viability was over 95% compared to the control without any NPs, indicating that neither NPs presented cytotoxicity. Cellular NPs uptake is a critical factor in the tumor targeting system since after diffusing into the tumor across leaky vessels free NPs not taken up by the tumors can simply diffuse back into the main bloodstream without accumulation. *In vivo* studies were performed whereby optical fluorescence images were captured after i.v. administration of all the NPs at a dose of 5 mg/kg mouse weight, in a athymic mice model of subcutaneous SCC7 tumor. The results showed that the functionalization of PLGA NPs was beneficial for the *in vivo* tumor accumulation of NPs, especially CP-PLGA NPs, which is an advantage for tumor targeting.

Helicobacter pylori are an important human pathogen that causes inflammatory responses thus constituting risk factors for gastric cancer and peptic ulcer. In 2009, Lin and colleagues used CS and HP, due to their mucoadhesive and anticoagulant properties, to prepare pH-responsive CS/HP NPs for gastric mucosal adhesion and for infiltration into the mucus layer at the sites of *H. pylori* infection [94]. The NPs were prepared using a simple ionic gelation method under aqueous-based conditions. The results showed that the obtained CS/HP NPs were stable at pH 1.2 to 2.5, allowing the protection of the encapsulated drug from destruction by gastric acids. *In vivo* assays demonstrated that the NPs localized to the spaces in the gastric villi of the mouse model and also infiltrated the cell-cell junctions and interacted locally *H. pylori* at the infection sites. However, encapsulation and controlled release of low molecular weight drugs (as antibiotics) in these NPs was difficult because the drugs could rapidly leak and quickly be released from the nanocarriers. To further enhance the CS/HP NPs properties, recently Lin et al., described multifunctional genipin-cross-linked fucose-CS/HP NPs to incorporate the antibiotic amoxicillin in order to target and directly make contact with the infected region on the gastric epithelium [95]. The fucose-CS/HP NPs were obtained by the ionic gelation of positively charged fucose-CS conjugate and negatively charged HP. The mean particle size ranged between 150 nm and 210 nm, with positive zeta potentials. The cell surface receptors recognized by *H. pylori* have fucose residues, so the authors assumed that the affinity of the fucose-conjugated NPs toward lectin receptors present on the bacteria surface could allow for the direct targeting.

Indeed, they demonstrated that the fucose-CS conjugate used could directly adhere and actively contact with the bacteria. In gastric pH environment, the electrostatic interaction between fucose-CS and HP was weaker resulting in larger NPs (particle size of 232.8±9.8 nm), leading to the burst release of the encapsulated low molecular weight and hydrophilic amoxicillin. This problem was addressed by using genipin, a naturally occurring cross-linking agent (as already mentioned), to cross-link the amine groups on the fucose-CS conjugate, in order to improve the stability of the NPS and reduce the amoxicillin release ratio.

The amoxicillin release profiles with or without genipin cross-link were analyzed at different pH, in 120 min: at pH 1.2 the amount of amoxicillin released from the cross-linked NPs was 17.1±1.9%, compared to 43.9±3.6% from the non cross-linked NPs; at pH 6.0, a very small amount of antibiotic was released from the cross-linked NPs; and at pH 7.0 (*H. pylori* survival situation), the NPs became unstable or collapsed, allowing a rapid release. This suggested that the genipin cross-linked fucose-CS/HP NPs had the potential to completely release the drug locally. The clearance of *H. pylori in vivo* was accessed using 30 mg/Kg amoxicillin (free or loaded in genipin cross-linked fucose-CS/HP). The treatment with amoxicillin alone gave a mean bacterial count of 278.3±31.5 CFU/stomach while the treatment with the NPs gave a mean bacterial count of 75.8±18.5 CFU/stomach, with significantly increased inhibitory effects on *H. pylori*-infected mice compared to the antibiotic solution alone. Further, histological analysis proved that the treatment with the loaded antibiotic was effective in reducing the *H. pylori*-associated gastric inflammation and in keeping the *H. pylori* density at low levels.

2.3.2. Chitosan (CS)-Based Metallic Nanoparticles (NPs)

Iron oxide NPs can be used as a MRI agent for specific tumor targeting. However, and has described before, these NPs present short-half lives *in vivo* and also reduced biocompatibility. In 2011, Yuk and collaborators described the preparation of glycol CS/HP immobilized iron oxide NPs (composite NPs) as a MRI agent with tumor-targeting ability [76]. Passive targeting of the iron oxide NPs-polymer composites to solid tumors is predictable due to the EPR effect and enhanced cellular permeability. However, the EPR effect is not a persistent characteristic of tumor vessels.

An alternative strategy exploited in this work is the targeting of specific molecular receptors in the tumor blood vessels. The preparation of the composite NPs included four steps as portrayed in Figure 8: 1) formulation of gold-deposited iron oxide NPs; 2) inclusion of gold-deposited iron oxide NPs into the glycol CS matrix; 3) stimulation of ionic interaction between glycol CS and HP to form the composite NPs; and 4) the freeze-drying of the composite NPS in a Pluronic F-68 aqueous solution, to improve the stability of the NPs in the aqueous media. The stability study – size distribution analysis over time – indicated that the disintegration of the composite NPs occurs after a certain period of time, which is ideal for MRI since the imaging agent should be cleared from the blood. Using a NIR fluorescence imaging system it was seen that composite NPs accumulate preferentially in tumor tissue and that it showed an enhancement in the T_2* negative images compared to Resovist (clinically approved superparamagnetic iron oxide for contrast-enhanced MRI).

Additionally, the composite NPs exhibited magnetic properties superior to Resovist, probably due to the accumulation of the magnetic NPs (Au-deposited iron oxide NPs) in the glycol CS/HP composite matrix. It was also demonstrated that, 6 h after injection of the composite NPs, a significant enhancement of the MR intensity at the tumor site occurred as compared to composite NPs without HP or Resovist. This occurrence indicated that the composite NPs could be used as MRI agent with improved targeting properties.

Many studies show that solid tumors have considerable amounts of fibrinogen-derived products, which suggests that fibrin is important in tumor formation. HP is not effective in inhibiting the fibrin formation in solid tumor but the localization of the HP-immobilized NPs at the tumor can be attributed to the interactions between the fibrinogen-derived products in the solid tumor and the HP immobilized in the composite NPs.

Plus, 12 h after injection, it was clearly shown by MRI that the composite NPs extravasated out of the tumor blood vessels and were predominantly distributed in the perivascular area. The targeting specificity was confirmed by histological studies of the tumors 12 h after injection (Figure 9). All the results suggest that the glycol CS/HP immobilized iron oxide NPs could be used in the development of MRI agents with tumor-targeting feature.

2.4. Chondroitin Sulfate

Chondroitin sulfate is a glycosaminoglycan (GAG) composed of interchanging units of β-1,3 linked glucuronic acid and (β-1,4)N-acetyl-D-galactosamine (GalNac) with sulfation occurring at either 4- or 6- position of GalNac. It is the most abundant GAG molecule in the human body and it presents itself mainly on the surface of the cells and in the extracellular matrix.

2.4.1. Chondroitin Sulfate-Based Nanoparticles (NPs)

A new type of hydrophilic NPs made of chondroitin sulfate and chitosan (CS) was reported and their potential for the delivery of proteins investigated [96]. This study showed that spherical and mono dispersed chondroitin sulfate-CS NPs were obtained by an ionic gelation due to the intermolecular linkages formed between the negatively charged sulfate and carboxylate groups of chondroitin sulfate and the positively charged amino group of CS.

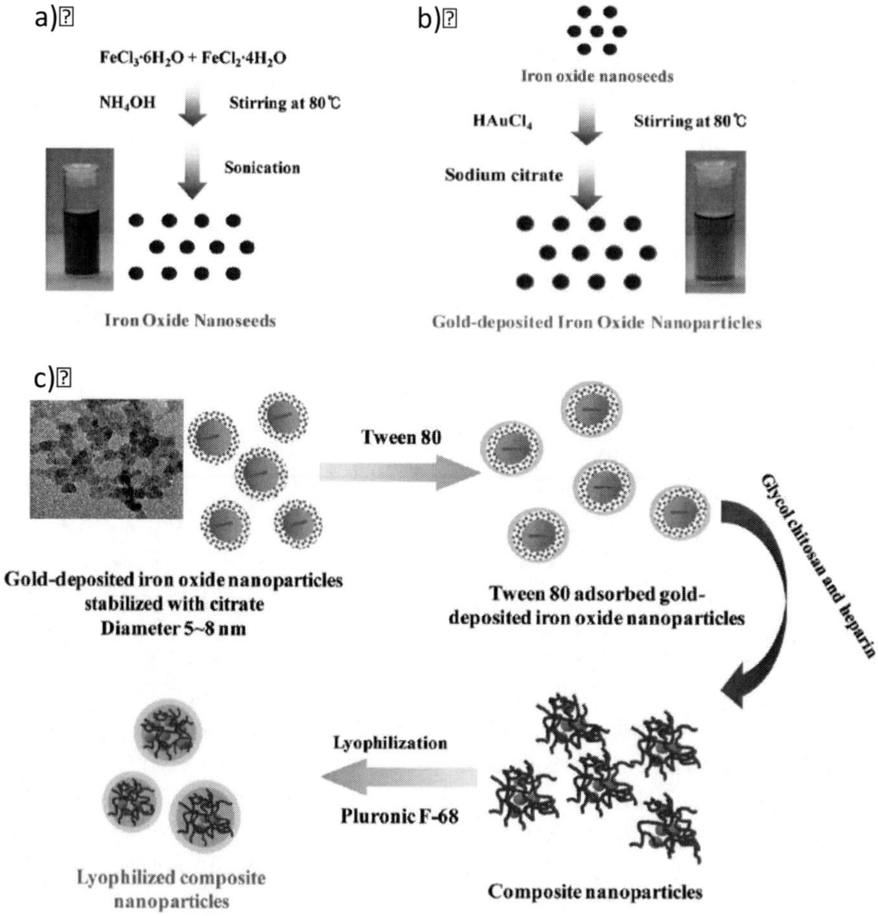

Reprinted from [76], with permission.

Figure 8. Illustration of the preparation of (a) iron oxide nanoseeds, (b) gold-deposited iron oxide NPs, and (c) gold-deposited iron oxide/glycol CS NPs with HP (composite NPs).

Protein encapsulation and release profiles were evaluated using fluorescein isothiocyanate conjugate bovine serum albumin (FITC-BSA) as a model protein. It was noticed that, the negative charge of FITC-BSA interacted with the positive charged CS *via* ionic cross-linking. Despite this, the concentration of the protein slightly influenced the zeta potential, which was in the range of +25 to 32 mV. The maximum protein load (~60 % of 2.0 mg/mL FITC-BSA) was obtained by incorporating the protein during the formation of the NPs, and the encapsulation efficiency increased with FITC-BSA concentration because the protein was involved in the ionic cross-linking.

In water, only about 6.9% FITC-BSA was detected after 1 h, a slow and sustained release of approximately 14.7% being observed by day 7. This behavior could be attributed to the strong ionic interaction between the counter ion of FITC-BSA and CS.

On the other hand, in PBS (pH 7.4), it was observed an initial burst release of 14.0%, that could be due to the decreased concentration of positively charged amino groups, which weakened the ionic interaction between FITC-BSA and CS, followed by a continuous and controlled release phase up to 4 days. Under all the conditions tested, the developed NPs presented no cytotoxicity towards Caco-2 cells.

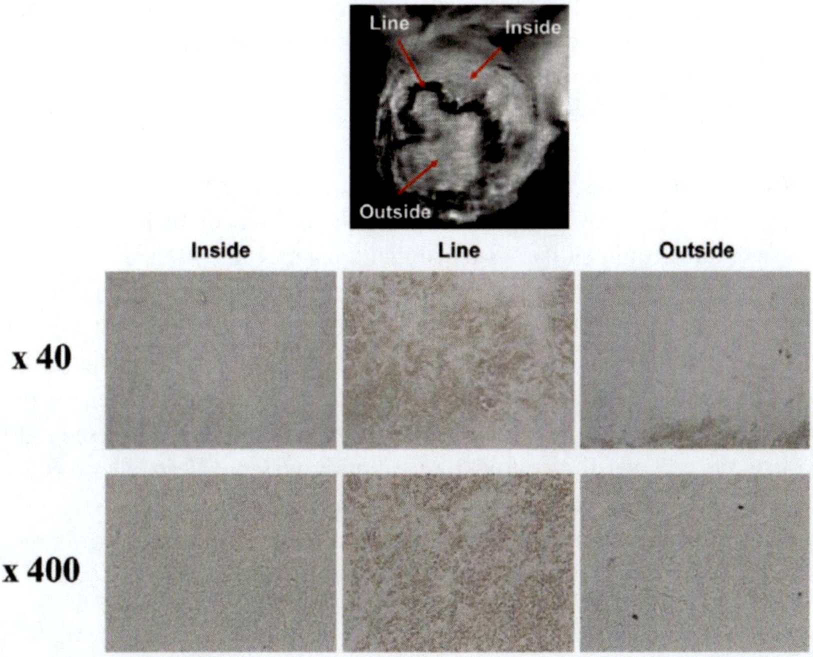

Adapted from [76], with permission.

Figure 9. Histological analysis (silver staining) of tumor tissue 12 h after i.v.injection of composite NPs. The tumors were sectioned into three parts (inside, line and outside) and stained with hematoxylin-eosin (H and E), Prussian blue and silver staining. The brown color observed at one part of the tumor tissue classified as line indicated the presence of the composite NPs in the tumor tissue.

Additional *ex vivo* cellular uptake studies with Caco 2 and HEK-293 cells revealed that the FITC-BSA loaded chondroitin sulfate-CS NPs were successfully endocytosed into the cells. The authors concluded that the described NPs could be potentially used as a carrier for hydrophilic compounds such as vaccines.

Polymeric micelles of block copolymers in an aqueous phase have attracted much attention not only due to their unique morphological characteristics but also for their potential applications in drug delivery. In analogy to surfactants of low molecular weights, several polymeric micelles have a core-corona structure, with a hydrophobic internal core and a hydrophilic corona. This amphiphilicity permits the formation of micelles in aqueous solutions and the inner core serving as a microcontainer for several substances. Lee et al. reported the synthesis and characterization of self-assembled amphiphilic micelle composed by chondroitin sulfate as a hydrophilic segment and poly(L-lactide) (PLLA) as a hydrophobic moiety [97]. Micelles of the copolymers were formed in an aqueous phase and characterized by ^{1}H NMR spectra, fluorescence techniques, DLS, atomic force microscopy (AFM) and confocal microscopy. The self-assembled amphiphilic micelles were spherical with a mean diameter of less than 200 nm with narrow size distribution. The critical aggregation concentrations (CAC) values depended in the content of hydrophobic segments and were in the range of 0.0043-0.0091 mg/mL. The very low CAC values indicated a strong tendency of the graft copolymer toward formation of micelles in aqueous solution. To obtain information about the internalization of the chondroitin sulfate/PLLA micelles into cells, the intracellular distribution of the micelles in HS68 human fibroblast cells was analyzed by laser confocal microscopy. It was verified that the cells readily uptake the micelles and that the fluorescence-labeled micelles were distributed over the intracellular space of the cytoplasm. After 4 h, the fluorescence images showed a high concentration of the micelles in the cytoplasm and in the perinuclear region.

2.4.2. Chondroitin Sulfate-Based Metallic Nanoparticles (NPs)

In 2011, Li et al. presented a method to synthetize biocompatible Au NPs at room temperature using chondroitin sulfate as a stabilizing agent [98]. As already mentioned, due to their unique optical, electrical and catalytic properties, chemical production of Au NPs have attracted great interest. The Au NPs were prepared by chemical reduction of Au salts resulting in the corresponding Au particles by using $NaBH_4$ as the reducing agent. The chondroitin sulfate stabilized Au NPs revealed no sign of aggregation after several months. The obtained Au NPs were spherical (or close-to-spherical) and well dispersed. Furthermore, the stability tests proved that the colloidal Au solution was not destabilized at various salt physiological conditions, even in high concentrations of NaCl (1M).

This stability was attributed to the existence of negatively charged environment around the Au NPs provided by the chondroitin sulfate molecules.

2.5. Fucoidan

Fucoidan is also entitled fucan or sulfated fucan. It is extracted from cell wall of marine brown algae such as Laminariales, Fucales, Chordariales, Desmarestiales and Dictyotales, and it has several biological activities (Table 1). The fucoidan is composed of α-1,3-linked fucopyranose with additional sulfate group and can present a branched structure [14, 99].

2.5.1. Fucoidan-Based Nanoparticles (NPs)

Lira et al. conceived new core-corona poly(isobutylcyanoacrylate) (PIBCA) nanoparticles (NPs) with a corona of fucoidan and investigated if the incorporation of fucoidan had relevant effect in the interactions of with cells [100]. The NPs were obtained using two methods: anionic emulsion polymerization (AEP) and redox radical emulsion polymerization (RREP). Mixtures of fucoidan and dextran were used to decorate the NPs. All the NPs obtained with the AEP method were stable. On the contrary, by the RREP method only formulations with low percentage of fucoidan (0 to 25%) resulted in stable suspensions. The NPs size prepared by AEP increased from 193±4 to 399±0.7 nm when the fucoidan:dextran ratio increased from 25 to 100%, respectively. On the other hand, in RREP

NPs, the size diminished from 307±4 nm to 223±1 nm with increasing fucoidan percentage (0 and 15%, correspondingly). In all formulations prepared, the zeta potential of the NPs became more negative when the amount of fucoidan increased, as expected, from -5 to -44 mV. It was observed, by TEM, that the NPs had spherical shape independently of the method of preparation and fucoidan content, and presented long-term stability: no significant changes in size were detected over 6 months at 4ºC. The cytotoxicity and the penetration capacity of the developed NPs were accessed in two different cell lines, macrophages J774 and fibroblasts NIH-3T3. It was observed that the synthesis method had an important effect in the cytotoxicity of the NPs on the macrophages J774: for example, the AEP-Fuc-25 NPs were sevenfold more cytotoxic (IC_{50} = 1.3±0.2 μg/mL) than the analogous NPs obtained by RREP method (IC_{50} = 9.6±0.5 μg/mL). However, no significant difference was noticed with the fibroblasts. The higher internalization of the NPs by the macrophages may justify the more acute sensitivity to toxic effects. Internalization of the NPs by both types of cells was examined by confocal fluorescence microscopy using fluorescent-labeled NPs. It has been showed that all types of NPs were incorporated by J774 macrophages in less than 1 h incubation and, as expected, the intensity of the fluorescence in the cells increased with the incubation time.

The results also showed that the presence of fucoidan at the surface of the NPs influenced the interaction with J774 cells and the fate of the NPs within the cells: as an example, the AEP NPs without fucoidan were homogenously distributed in the cells while those with fucoidan showed as bright and concentrated punctuations disseminated in the cells.

Although fibroblasts are not professional phagocytes they were able to internalize all the polymeric NPs tested as well.

However no apparent differences were noticeable in the internalization by fibroblasts NIH-3T3 of fucoidan-coated and non fucoidan-coated NPs as it was verified with macrophages J774. This study was the first to give evidence that the introduction of fucoidan could modulate the interaction of PIBCA NPs with macrophages.

2.5.2. Fucoidan-Based Metallic Nanoparticles (NPs)

Recently, a research study reported the preparation of fucan-coated magnetite (Fe_3O_4) NPs by the coprecipitation method [101]. The main objective was to improve the magnetite NPs stability and biocompatibility, making these fucan-coated NPs attractive for medical applications. The developed NPs were characterized by SEM, TEM, XRD, FTIR, Mössbauer spectroscopy and magnetic measurements. Most of the uncoated magnetite NPs were quasi-spherical with mean size of about 10 nm, the fucan coating preventing aggregation. It also proved that fucan did not degrade the magnetic core and that the magnitude of magnetization was higher comparing to uncoated NPs, an effect assigned to the fucan coating.

2.6. Ulvan

Ulvan is a complex acidic SP obtained from the cell walls of the green seaweed Ulvales (Chlorophyta). It has a repeating disaccharide unit of β-D-glucuronosly-(1,4)-α-L-rhamnose 3-sulfate (called aldobiuronic acid) [3, 102]. Usually, the aldobiuronic acid blocks of ulvan are constituted by a Type A_{3s}glucuronorhamnose and a Type B_{3s}iduronorhamnose, disposed in regular sequences within the heteropolymer chain.

2.6.1. Ulvan-Based Nanofibers

Toskas et al. reported the first nanofiber obtained by electrospinning of an ulvan rich extract [102]. Nanofiber matrices obtained with natural polymers have attracted great attention due to their potential application in the biomedical field as tissue engineering scaffolds, wound dressing and drug delivery devices. Poly(vinyl alcohol) (PVA) was added in order to reduce the viscosity of the initial ulvan gel, allowing it to be spinnable at higher total polymer concentrations.

As it can be seen in Figure 10 the mixed ulvan/PVA solutions, namely ratios 50:50 and 70:30, resulted in defect-free electrospun nanofibers with very good uniformity. When the 50:50 ulvan/PVA solution was used, perfectly uniform nanofibers with a diameter of 105±4 nm were obtained (Figure 10a). With the 70:30 solution smaller nanofibers with about 84±4 nm were observed (Figure 10b). With a higher ulvan ratio 85:15 ulvan/PVA, larger beads of roughly 1-2 μm were interconnected by thinner fibers of 60±5 nm with smaller beads of approximately 220±8 nm (Figure 10c).

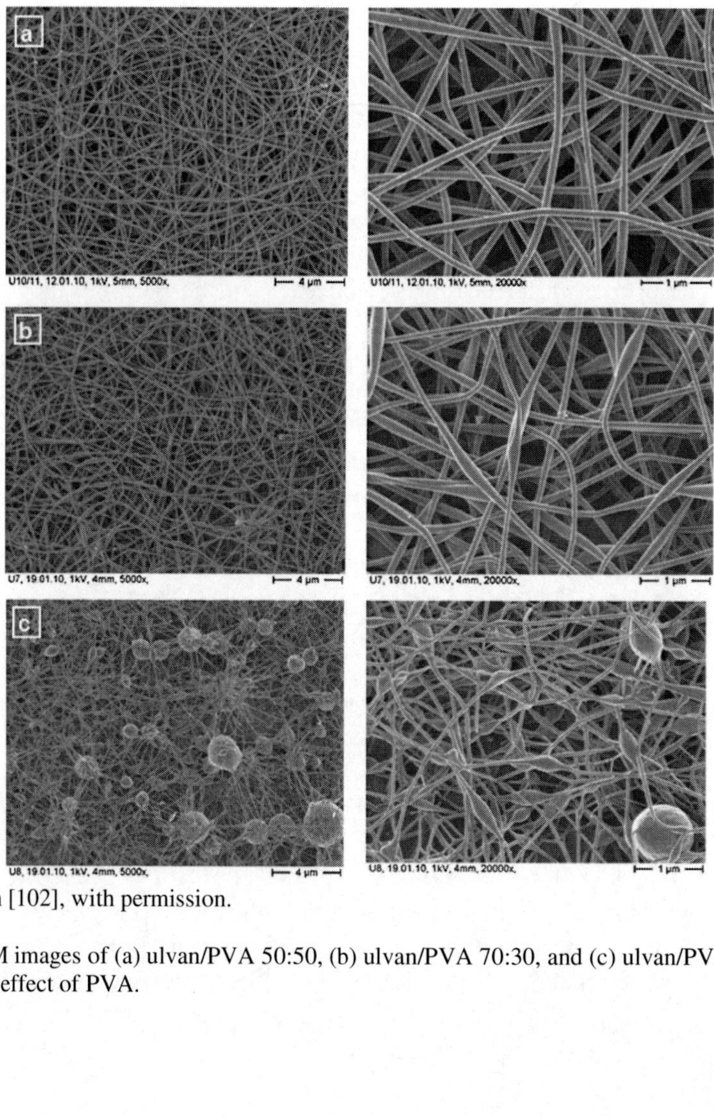

Reprinted from [102], with permission.

Figure 10. SEM images of (a) ulvan/PVA 50:50, (b) ulvan/PVA 70:30, and (c) ulvan/PVA 85:15 illustrating the effect of PVA.

Generally, fiber diameters decreased with increasing ulvan concentration and additionally bead formations diminished as the total polymer concentration (ulvan + PVA) increased.

The electrospun nanofibers were soluble in Dulbecco's Modified Eagle's Medium (DMEM) after immersion for 48 h, which in an important feature (degradability) for applications such as controlled drug release. This fibrous material presented characteristics that combined with ulvan biological activities (Table 1) could lead to their application in several biomedical fields including drug delivery, wound dressing and tissue engineering. In addition, this anionic polymer fiber can be combined with other biomaterials, which can broaden its potential applications.

2.7. Dermatan Sulfate

Dermatan sulfate (DmS) is one of the principal glycosaminoglycan (GAG) found in the mammalian skin. It is linear GAG composed of alternating disaccharide units of hexosamine, GalNac, and iduronic acid linked *via* β (1,4) or (1,3) glycosidic bonds [3, 103].

2.7.1. Dermatan Sulfate (DmS)-Based Nanoparticles (NPs)

One of the main challenges in tumor targeting and imaging is the deep penetration of the interstitial matrix. The penetration of hydrophobic, amphiphilic and positively charged drugs and drug carriers in the interstitial matrix is considerably hampered due to their binding to proximal membrane lipids and negatively charged phospholipids.

So, it is imperative to develop drug carriers that are able to homogeneously target and actively transport across tumor neovasculature and optimally reach the tumor matrix. Ranney et al. described improved *in vivo* probes to noninvasively evaluate drug penetration in individual animal and patient lesions [104], namely a new approach to functional tumor imaging and deep interstitial penetration of therapeutics agents was developed with agents formulated with 435-type dermatan sulfate (DmS 435) [105].

The DmS 435 combines four relevant biochemical and pharmaceutical properties: i) elevated HP cofactor binding (which is associated with high tumor uptake); ii) reversible binding of its low-affinity multivalent oligosaccharide determinants (which allows selective neovascular targeting, transport and release into the tumor matrix); iii) a molecular weight of 22.2 kDa, that restricts early biodistribution to the vascular compartment, maximizes tumor uptake and enables renal clearance of the non-metabolized form; and iv) a hydrophilic, polyolshielded carbohydrate surface, which facilitates deep site penetration, offers protection against non-target uptake and nonimmunogenicity. The MRI agent was obtained by self-assembly of Fe^{3+} with deferoxamine (Fe:Df) bound by strong ion pairing to DmS 435. The Fe:Df-DmS 435 carrier was physiochemical stable in solution up to 5 weeks at 4ºC and its lyophilized form was stable for longer than 1 year at 22ºC. In order to evaluate the carrier distribution at high spatial resolution, MRI was performed at high field with a small coil positioned directly around large (2 cm) Dunning prostate R3327 AT1 rat tumors (Figure 11).

It was shown that Fe:Df-DmS 435 NPs selectively targeted and crossed the neovascular endothelium in less than 7 min. In this work, a therapeutic formulation of DmS 435 and DOX (DmS 435-DOX) was also tested *in vivo*. The DmS 435-DOX NPs were prepared by aqueous high-pressure homogenization of DOX and DmS 435 producing NPs of about 11 nm of DOX cores coated with DmS 435, which were stable to lyophilization. The DOX encapsulation efficiency was of 99.5% and the physical stabilization partly resulted from paired-ion binding of DOX positively charged sugar amines to DmS 435 sulfate groups.

Drug release in water was less than 2% in 8 h, and the particle size did not change. Under mimicked body-fluid conditions, the drug was released with a $t_{1/2}$ of 1.9 h.

Microscopic analysis of AT1 tumor sections 3 h after i.v. injection demonstrated greater overall tumor fluorescence and deeper matrix penetration when using DmS 435-DOX NPs comparing to conventional DOX (Figure 12). The improved tumor-cell internalization of DmS-DOX over conventional DOX indicated that DOX remained coated with DmS as it migrates through the interstitium.

In fast growing subregions of the tumor, the nuclear fluorescence that resulted from DmS 435-DOX was even more intense than its already high cytoplasmic levels indicating that DmS 435-DOX enhanced the levels of DOX in tumor-cell nuclei.

Furthermore, in nude mice bearing large MX-1 human breast-tumor xenografts, conventional DOX administration prolonged the median time for 7-fold tumor growth from 8.3 days to 25.6, however no long-term tumor regression were observed. In turn, the DmS 435-DOX extended the tumor growth time to 43.2 days and produced 4 out of 10 complete tumor regressions at 90 days. These results correlated to the enhanced ability of DmS 435-DOX to reach deep tumor subregions inaccessible to conventional DOX. Together, these results demonstrated the potential of DmS 435-based NPs for targeted neovascular transport agents for sensitive, high-resolution MRI and enhanced treatment of large size human tumor metastases.

2.8. Dextran Sulfate

Dextran sulfate (DS) is a biodegradable and biocompatible polymer widely used for pharmaceutical formulations. DS has a branched carbohydrate backbone and negative charged sulfate groups with α-1,6 linked glucose residues with variant chain length and branching at α-1,3 and α-1,4 positions.

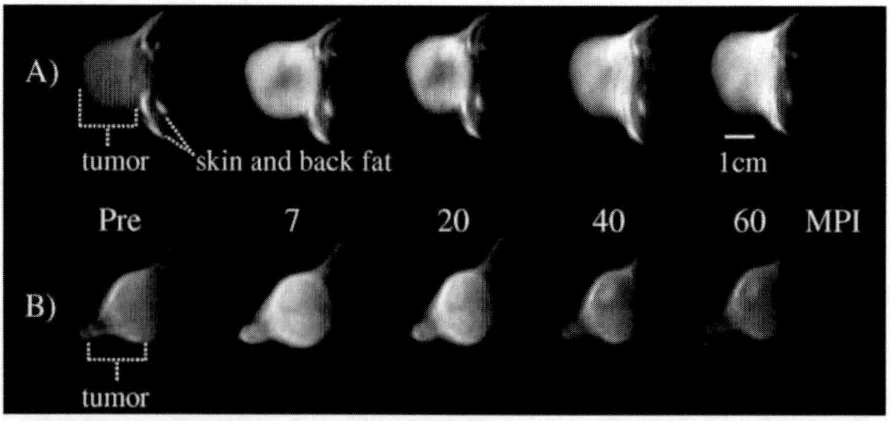

Reprinted from [105], with permission.

Figure 11. MRI images of AT1 tumor enhancement by (a) Fe:Df-DmS 435 and (b) Fe:Df. T1-weighted images, SE (TR/TE = 250/8 ms) (Pre – pre-injection; MPI – minutes post-injection).

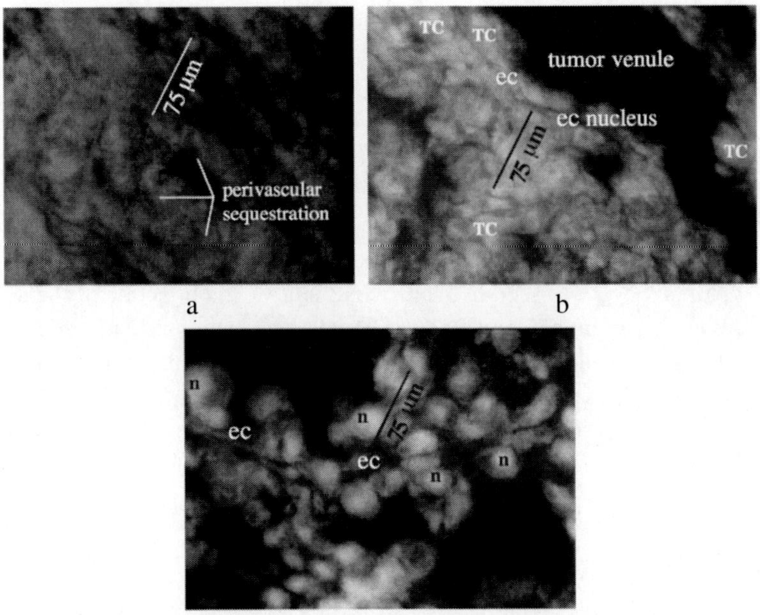

a b c

(n – nucleus of tumor cells; ec – endothelila cell; TC – tumor cell). Reprinted from [105], with permission.

Figure 12. Photomicrographs (100X mag) of DOX fluorescence intensity and distribution in AT1 tumors 3 h after i.v. injection (5 mg/kg, drug basis): dense cellular subregions for convential DOX (a) and DmS-DOX (b); rapidly proliferating subregion for DOX (c), showing a sungle endothelial stalk and nuclear fluorescence.

2.8.1. Dextran Sulfate (DS)-Based Nanoparticles (NPs)

In 2001, Tiyaboonchai et al. described the preparation of NPs derived from the oppositely charged polymers polyethylenimine (PEI) and DS, with zinc sulfate as a stabilizing agent [106]. PEI is a linear or branched chain polymer with primary, secondary and/or tertiary amines and it has been used as a versatile agent for *in vivo* gene delivery [107]. Amphotericin B (AmB) was used as a model drug to study this system. The PEI-DS NPs were produced from phase separation induced by electrostatic interactions between PEI and DS. The ideal conditions to obtain PEI-DS AmB-loaded NPs were: PEI solution at pH 8, polymer ratio PEI:DS of 2:1 and 25 μM zinc sulfate. Using these parameters, NPs with mean particle size of approximately 260 nm and size distribution in the range of 100-600 nm were produced.

The NPs showed high entrapment efficiency (85%). *In vitro* studies demonstrated that the AmB released rapidly from the developed NPs but despite this rapid release, the AmB-loaded NPs displayed significantly reduced toxicity in tissue culture in contrast to free AmB and were almost as efficient as the free drug in killing *Candida albicans*.

Later, the same research group used insulin to study the possibility of incorporating a protein into the PEI-DS NPs described earlier [108]. The encapsulation of insulin was an efficient process as most of the formulations showed an encapsulation efficiency of 80-90%. Circular dichroism studies demonstrated that under optimized formulation conditions, there were no conformational changes of the protein.

In general, *in vitro* release of insulin was very fast (about 5 min) for most of the formulations tested. However, the encapsulated protein displayed a prolonged hypoglycemic effect in streptozotocin-induced diabetic rats (Figure 13).

After subcutaneous injection of free insulin (0.35 IU/kg) into fasted induced diabetic rats, about 20% of the initial blood glucose levels (BGL) were reached in 1-2 h (hypoglycemic effect).

However, the BGL increased 2 h after the administration of free insulin indicating the expected rapid onset and short duration of action of the free protein. In the absence of zinc sulfate (the stabilizing agent as seen in the previous study), the injection of the developed NPs resulted in the same trend in temporal BGL, suggesting an analogous short duration of action.

In contrast, when insulin-loaded PEI-DS NPs in the presence of zinc were injected, a gradual decrease in BGL over 4 h was observed demonstrating a more sustained effect.

According to the authors, the system developed [106, 108] presented five great advantages: i) straightforward preparation under mild and aqueous conditions; ii) use of biocompatible polymers; iii) controllability of particle size; iv) high drug incorporation levels; and v) ability to stabilize encapsulated proteins, maintaining its secondary structure and biological activity.

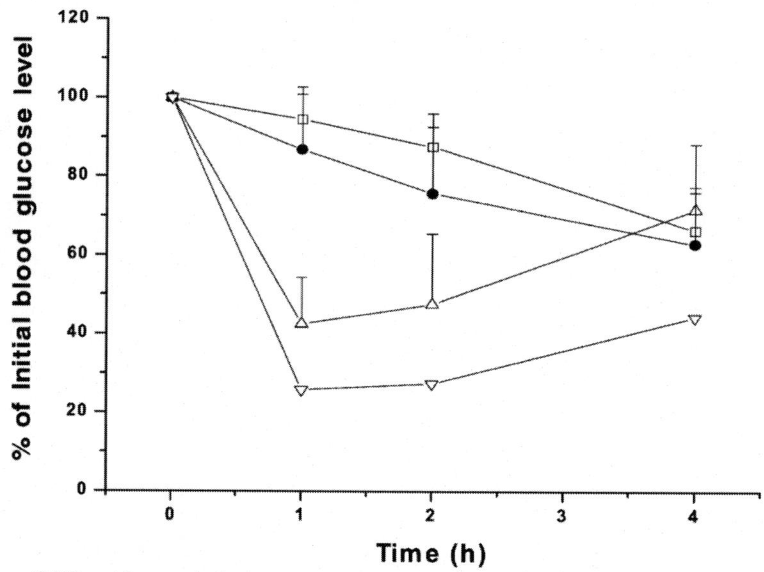

Reprinted from [108], with permission.

Figure 13. Effect of free insulin and insulin-loaded PEI-DS NPs injected subcutaneously to SZT-induced diabetic rats on the reduction of BGL. Mean values ± SEM, n = 3-4 rats [108]. (Inverted triangle – free insulin; full circle – PEI-DS ratio of 2.5:1; square – PEI-DS ration of 1.5:1; triangle – insulin-loaded NPs formulated without zinc sulfate with PEI-DS ration of 1.5:1).

Conclusion

This chapter gives a broad review of nanotechnological applications of important natural and chemically modified SPs. The structural diversity of SPs depends largely on their botanical or biological sources and their unique structural diversity and physiochemical properties, and potentiates their application in various biotechnological fields.

The main biomedical applications of SPs rely on their use for controlled drug delivery systems. NPs made of, coated by or functionalized with SPs were widely studied in medicine and pharmacotherapy: as drug delivery systems, replacing several chemical drugs that induce serious side effects; for enhanced tumor targeting and MRI; and for tissue engineering. Nanocomposites, nanogels and nanolayers of SPs have also been described.

Due to their outstanding properties, biological activities and availability, SPs will certainly contribute to the (nano)biotechnological advances in several fields potentiating the development of highly efficient biocompatible devices, such as NPs, with multifunctional applications.

AKNOWLEDGMENTS

Vera Carvalho acknowledge support from Fundação para a Ciência e a Tecnologia (FCT), Portugal (grant SFRH/BPD/73850/2010).

REFERENCES

[1] Wijesinghe, W. A. J. P., Jeon, Y. J., 2012. Biological activities and potential industrial applications of fucose rich sulfated polysaccharides and fucoidans isolated from brown seaweeds: A review. *Carbohyd. Polym.* 88, 13-20.

[2] Wijesekara, I., Pangestuti, R., Kim, S. K., 2011. Biological activities and potential health benefits of sulfated polysaccharides derived from marine algae. *Carbohyd. Polym.* 84, 14-21.

[3] Raveendran, S., Yoshida, Y., Maekawa, T., Kumar, S., 2013. Pharmaceutically versatile sulfated polysaccharide based bionano platforms. *Nanomed-Nanotechnol.* 9, 605-626.

[4] Haag, R., Kratz, F., 2006. Polymer therapeutics: Concepts and applications. *Angew. Chem. Int. Edit.* 45, 1198-1215.

[5] Devalapally, H., Chakilam, A., Amiji, M. M., 2007. Role of nanotechnology in pharmaceutical product development. *J. Pharm. Sci.-US.* 96, 2547-2565.

[6] Nayak, S., Lyon, L. A., 2005. Soft nanotechnology with soft nanoparticles. *Angew. Chem. Int. Edit.* 44, 7686-7708.

[7] Sundar, S., Kundu, J., Kundu, S. C., 2010. Biopolymeric nanoparticles. *Sci. Technol. Adv. Mat.* 11.

[8] Zong, A. Z., Cao, H. Z., Wang, F. S., 2012. Anticancer polysaccharides from natural resources: A review of recent research. *Carbohyd. Polym.* 90, 1395-1410.

[9] Jin, M. L., Huang, Q. S., Zhao, K., Shang, P., 2013. Biological activities and potential health benefit effects of polysaccharides isolated from Lycium barbarum L. *Int. J. Biol. Macromol.* 54, 16-23.

[10] Toida, T., Chaidedgumjorn, A., Linhardt, R. J., 2003. Structure and bioactivity of sulfated polysaccharides. *Trends Glycosci. Glyc.* 15, 29-46.

[11] Jiao, G. L., Yu, G. L., Zhang, J. Z., Ewart, H. S., 2011. Chemical Structures and Bioactivities of Sulfated Polysaccharides from Marine Algae. *Mar. Drugs*. 9, 196-223.

[12] Poli, A., Anzelmo, G., Nicolaus, B., 2010. Bacterial Exopolysaccharides from Extreme Marine Habitats: Production, Characterization and Biological Activities. *Mar. Drugs*. 8, 1779-1802.

[13] Berteau, O., Mulloy, B., 2003. Sulfated fucans, fresh perspectives: structures, functions, and biological properties of sulfated fucans and an overview of enzymes active toward this class of polysaccharide. *Glycobiology*. 13, 29R-40R.

[14] Pomin, V. H., Mourao, P. A. S., 2008. Structure, biology, evolution, and medical importance of sulfated fucans and galactans. *Glycobiology*. 18, 1016-1027.

[15] Cumashi, A., Ushakova, N. A., Preobrazhenskaya, M. E., D'Incecco, A., Piccoli, A., Totani, L., Tinari, N., Morozevich, G. E., Berman, A. E., Bilan, M. I., Usov, A. I., Ustyuzhanina, N. E., Grachev, A. A., Sanderson, C. J., Kelly, M., Rabinovich, G. A., Iacobelli, S., Nifantiev, N. E., 2007. A comparative study of the anti-inflammatory, anticoagulant, antiangiogenic, and antiadhesive activities of nine different fucoidans from brown seaweeds. *Glycobiology*. 17, 541-552.

[16] Hayashi, K., Nakano, T., Hashimoto, M., Kanekiyo, K., Hayashi, T., 2008. Defensive effects of a fucoidan from brown alga Undaria pinnatifida against herpes simplex virus infection. *Int. Immunopharmacol*. 8, 109-116.

[17] de Souza, M. C. R., Marques, C. T., Dore, C. M. G., da Silva, F. R. F., Rocha, H. A. O., Leite, E. L., 2007. Antioxidant activities of sulfated polysaccharides from brown and red seaweeds. *J. Appl. Phycol*. 19, 153-160.

[18] O'donnell, V., LaRocco, M., Baxt, B., 2008. Heparan sulfate-binding foot-and-mouth disease virus enters cells via caveola-mediated endocytosis. *J. Virol*. 82, 9075-9085.

[19] Bender, F. C., Whitbeck, J. C., Lou, H., Cohen, G. H., Eisenberg, R. J., 2005. Herpes simplex virus glycoprotein B binds to cell surfaces independently of heparan sulfate and blocks virus entry. *J. Virol*. 79, 11588-11597.

[20] Ghosh, T., Chattopadhyay, K., Marschall, M., Karmakar, P., Mandal, P., Ray, B., 2009. Focus on antivirally active sulfated polysaccharides: From structure-activity analysis to clinical evaluation. *Glycobiology*. 19, 2-15.

[21] Witvrouw, M., DeClercq, E., 1997. Sulfated polysaccharides extracted from sea algae as potential antiviral drugs. *Gen. Pharmacol*. 29, 497-511.

[22] Carre, V., Mbemba, E., Letourneur, D., Jozefonvicz, J., Gattegno, L., 1995. Interactions of Hiv-1 Envelope Glycoproteins with Derivatized Dextrans. *Bba-Gen Subjects*. 1243, 175-180.

[23] Baba, M., Snoeck, R., Pauwels, R., Declercq, E., 1988. Sulfated Polysaccharides Are Potent and Selective Inhibitors of Various Enveloped Viruses, Including Herpes-Simplex Virus, Cytomegalo-Virus, Vesicular Stomatitis-Virus, and Human Immunodeficiency Virus. *Antimicrob. Agents Ch*. 32, 1742-1745.

[24] Alban, S., Schauerte, A., Franz, G., 2002. Anticoagulant sulfated polysaccharides: Part I. Synthesis and structure-activity relationships of new pullulan sulfates. *Carbohyd. Polym*. 47, 267-276.

[25] Becker, C. F., Guimaraes, J. A., Mourao, P. A. S., Verli, H., 2007. Conformation of sulfated galactan and sulfated fucan in aqueous solutions: Implications to their anticoagulant activities. *J. Mol. Graph. Model*. 26, 391-399.

[26] Linhardt, R. J., Hileman, R. E., 1995. Dermatan Sulfate as a Potential Therapeutic Agent. *Gen. Pharmacol.* 26, 443-451.

[27] Jayakumar, R., Nwe, N., Tokura, S., Tamura, H., 2007. Sulfated chitin and chitosan as novel biomaterials. *Int. J. Biol. Macromol.* 40, 175-181.

[28] Mourao, P. A. S., Pereira, M. S., Pavao, M. S. G., Mulloy, B., Tollefsen, D. M., Mowinckel, M. C., Abildgaard, U., 1996. Structure and anticoagulant activity of a fucosylated chondroitin sulfate from echinoderm - Sulfated fucose branches on the polysaccharide account for its high anticoagulant action. *J. Biol. Chem.* 271, 23973-23984.

[29] Wang, J., Zhang, Q. B., Zhang, Z. S., Li, Z., 2008. Antioxidant activity of sulfated polysaccharide fractions extracted from Laminaria japonica. *Int. J. Biol. Macromol.* 42, 127-132.

[30] Gustafson, S., 1997. The influence of sulfated polysaccharides on the circulating levels of hyaluronan. *Glycobiology.* 7, 1209-1214.

[31] Leung, M. Y. K., Liu, C., Koon, J. C. M., Fung, K. P., 2006. Polysaccharide biological response modifiers. *Immunol. Lett.* 105, 101-114.

[32] Kemp, M. M., Linhardt, R. J., 2010. Heparin-based nanoparticles. *Wires Nanomed. Nanobi.* 2, 77-87.

[33] McGowan, I., 2006. Microbicides: A new frontier in HIV prevention. *Biologicals.* 34, 241-255.

[34] Paulert, R., Talamini, V., Cassolato, J. E. F., Duarte, M. E. R., Noseda, M. D., Smania, A., Stadnik, M. J., 2009. Effects of sulfated polysaccharide and alcoholic extracts from green seaweed Ulva fasciata on anthracnose severity and growth of common bean (Phaseolus vulgaris L.). *J. Plant Dis. Protect.* 116, 263-270.

[35] Roberts, M. A., Quemener, B., 1999. Measurement of carrageenans in food: challenges, progress, and trends in analysis. *Trends Food Sci. Tech.* 10, 169-181.

[36] Prajapati, V. D., Maheriya, P. M., Jani, G. K., Solanki, H. K., 2014. Carrageenan: A natural seaweed polysaccharide and its applications. *Carbohydr. Polym.* 105C, 97-112.

[37] Daniel-da-Silva, A. L., Ferreira, L., Gil, A. M., Trindade, T., 2011. Synthesis and swelling behavior of temperature responsive kappa-carrageenan nanogels. *J. Colloid Interf. Sci.* 355, 512-517.

[38] Salgueiro, A. M., Daniel-da-Silva, A. L., Girao, A. V., Pinheiro, P. C., Trindade, T., 2013. Unusual dye adsorption behavior of kappa-carrageenan coated superparamagnetic nanoparticles. *Chem. Eng. J.* 229, 276-284.

[39] Gupta, A. K., Gupta, M., 2005. Synthesis and surface engineering of iron oxide nanoparticles for biomedical applications. *Biomaterials.* 26, 3995-4021.

[40] Pardoe, H., Chua-anusorn, W., St Pierre, T. G., Dobson, J., 2001. Structural and magnetic properties of nanoscale iron oxide particles synthesized in the presence of dextran or polyvinyl alcohol. *J. Magn. Magn. Mater.* 225, 41-46.

[41] Bonacchi, D., Caneschi, A., Gatteschi, D., Sangregorio, C., Sessoli, R., Falqui, A., 2004. Synthesis and characterisation of metal oxides nanoparticles entrapped in cyclodextrin. *J. Phys. Chem. Solids.* 65, 719-722.

[42] Kim, D. K., Mikhaylova, M., Wang, F. H., Kehr, J., Bjelke, B., Zhang, Y., Tsakalakos, T., Muhammed, M., 2003. Starch-coated superparamagnetic nanoparticles as MR contrast agents. *Chem. Mater.* 15, 4343-4351.

[43] Daniel-da-Silva, A. L., Trindade, T., Goodfellow, B. J., Costa, B. F. O., Correia, R. N., Gil, A. M., 2007. In situ synthesis of magnetite nanoparticles in carrageenan gels. *Biomacromolecules*. 8, 2350-2357.

[44] Daniel-Da-Silva, A. L., Loio, R., Lopes-Da-Silva, J. A., Trindade, T., Goodfellow, B. J., Gil, A. M., 2008. Effects of magnetite nanoparticles on the thermorheological properties of carrageenan hydrogels. *J. Colloid Interf. Sci*. 324, 205-211.

[45] Tang, J., Myers, M., Bosnick, K. A., Brus, L. E., 2003. Magnetite Fe3O4 nanocrystals: Spectroscopic observation of aqueous oxidation kinetics. *Journal of Physical Chemistry B*. 107, 7501-7506.

[46] Daniel-da-Silva, A. L., Fateixa, S., Guiomar, A. J., Costa, B. F. O., Silva, N. J. O., Trindade, T., Goodfellow, B. J., Gil, A. M., 2009. Biofunctionalized magnetic hydrogel nanospheres of magnetite and kappa-carrageenan. *Nanotechnology*. 20.

[47] Daniel-da-Silva, A. L., Moreira, J., Neto, R., Estrada, A. C., Gil, A. M., Trindade, T., 2012. Impact of magnetic nanofillers in the swelling and release properties of kappa-carrageenan hydrogel nanocomposites. *Carbohyd. Polym*. 87, 328-335.

[48] Lim, S. Y., Lee, J. S., Park, C. B., 2010. In Situ Growth of Gold Nanoparticles by Enzymatic Glucose Oxidation Within Alginate Gel Matrix. *Biotechnol. Bioeng*. 105, 210-214.

[49] Matteini, P., Ratto, F., Rossi, F., Centi, S., Dei, L., Pini, R., 2010. Chitosan Films Doped with Gold Nanorods as Laser-Activatable Hybrid Bioadhesives. *Adv. Mater*. 22, 4313-+.

[50] Choi, W. I., Kim, J. Y., Kang, C., Byeon, C. C., Kim, Y. H., Tee, G., 2011. Tumor Regression In Vivo by Photothermal Therapy Based on Gold-Nanorod-Loaded, Functional Nanocarriers. *Acs Nano*. 5, 1995-2003.

[51] Salgueiro, A. M., Daniel-da-Silva, A. L., Fateixa, S., Trindade, T., 2013. kappa-Carrageenan hydrogel nanocomposites with release behavior mediated by morphological distinct Au nanofillers. *Carbohyd. Polym*. 91, 100-109.

[52] Hezaveh, H., Muhamad, II, 2012. The effect of nanoparticles on gastrointestinal release from modified kappa-carrageenan nanocomposite hydrogels. *Carbohydr. Polym*. 89, 138-45.

[53] Jayaramudu, T., Raghavendra, G. M., Varaprasad, K., Sadiku, R., Ramam, K., Raju, K. M., 2013. Iota-Carrageenan-based biodegradable Ag-0 nanocomposite hydrogels for the inactivation of bacteria. *Carbohyd. Polym*. 95, 188-194.

[54] Dai, W. G., Dong, L. C., Song, Y. Q., 2007. Nanosizing of a drug/carrageenan complex to increase solubility and dissolution rate. *Int. J. Pharm*. 342, 201-207.

[55] Jinno, J., Kamada, N., Miyake, M., Yamada, K., Mukai, T., Odomi, M., Toguchi, H., Liversidge, G. G., Higaki, K., Kimura, T., 2006. Effect of particle size reduction on dissolution and oral absorption of a poorly water-soluble drug, cilostazol, in beagle dogs. *J. Control Release*. 111, 56-64.

[56] Muller, R. H., Jacobs, C., Kayser, O., 2001. Nanosuspensions as particulate drug formulations in therapy Rationale for development and what we can expect for the future. *Adv. Drug Deliver. Rev*. 47, 3-19.

[57] Hancock, B. C., Parks, M., 2000. What is the true solubility advantage for amorphous pharmaceuticals? *Pharmaceutical research*. 17, 397-404.

[58] Liang, Y., Kiick, K. L., 2014. Heparin-functionalized polymeric biomaterials in tissue engineering and drug delivery applications. *Acta biomaterialia*. 10, 1588-600.

[59] Nader, H. B., Chavante, S. F., dos-Santos, E. A., Oliveira, T. W., de-Paiva, J. F., Jeronimo, S. M., Medeiros, G. F., de-Abreu, L. R., Leite, E. L., de-Sousa-Filho, J. F., Castro, R. A., Toma, L., Tersariol, I. L., Porcionatto, M. A., Dietrich, C. P., 1999. Heparan sulfates and heparins: similar compounds performing the same functions in vertebrates and invertebrates? *Brazilian journal of medical and biological research = Revista brasileira de pesquisas medicas e biologicas / Sociedade Brasileira de Biofisica* ... [et al.]. 32, 529-38.

[60] Harrison, L., McGinnis, J., Crowther, M., Ginsberg, J., Hirsh, J., 1998. Assessment of outpatient treatment of deep-vein thrombosis with low-molecular-weight heparin. *Arch. Intern. Med.* 158, 2001-2003.

[61] Hyers, T. M., Agnelli, G., Hull, R. D., Weg, J. G., Morris, T. A., Samama, M., Tapson, V., 1998. Antithrombotic therapy for venous thromboembolic disease. *Chest.* 114, 561S-578S.

[62] Jiao, Y., Ubrich, N., Marchand-Arvier, M., Vigneron, C., Hoffman, M., Lecompte, T., Maincent, P., 2002. In vitro and in vivo evaluation of oral heparin-loaded polymeric nanoparticles in rabbits. *Circulation.* 105, 230-5.

[63] Hoffart, V., Lamprecht, A., Maincent, P., Lecompte, T., Vigneron, C., Ubrich, N., 2006. Oral bioavailability of a low molecular weight heparin using a polymeric delivery system. *J. Control Release.* 113, 38-42.

[64] Lamprecht, A., Koenig, P., Ubrich, N., Maincent, P., Neumann, D., 2006. Low molecular weight heparin nanoparticles: mucoadhesion and behaviour in Caco-2 cells. *Nanotechnology.* 17, 3673-3680.

[65] Reyes-Ortega, F., Rodriguez, G., Aguilar, M. R., Lord, M., Whitelock, J., Stenzel, M. H., San Roman, J., 2013. Encapsulation of low molecular weight heparin (bemiparin) into polymeric nanoparticles obtained from cationic block copolymers: properties and cell activity. *J. Mater. Chem. B.* 1, 850-860.

[66] Wang, Y., Xin, D. C., Liu, K. J., Zhu, M. Q., Xiang, J. N., 2009. Heparin-Paclitaxel Conjugates as Drug Delivery System: Synthesis, Self-Assembly Property, Drug Release, and Antitumor Activity. *Bioconjugate Chem.* 20, 2214-2221.

[67] She, W. C., Li, N., Luo, K., Guo, C. H., Wang, G., Geng, Y. Y., Gu, Z. W., 2013. Dendronized heparin-doxorubicin conjugate based nanoparticle as pH-responsive drug delivery system for cancer therapy. *Biomaterials.* 34, 2252-2264.

[68] Li, L., Huh, K. M., Lee, Y. K., Kim, S. Y., 2011. Biofunctional self-assembled nanoparticles of folate-PEG-heparin/PBLA copolymers for targeted delivery of doxorubicin. *J. Mater. Chem.* 21, 15288-15297.

[69] Sakiyama-Elbert, S. E., Hubbell, J. A., 2000. Development of fibrin derivatives for controlled release of heparin-binding growth factors. *J. Control Release.* 65, 389-402.

[70] Choi, J. H., Jang, J. Y., Joung, Y. K., Kwon, M. H., Park, K. D., 2010. Intracellular delivery and anti-cancer effect of self-assembled heparin-Pluronic nanogels with RNase A. *J. Control Release.* 147, 420-427.

[71] Joung, Y. K., Jang, J. Y., Choi, J. H., Han, D. K., Park, K. D., 2013. Heparin-Conjugated Pluronic Nanogels as Muiti-Drug Nanocarriers for Combination Chemotherapy. *Mol. Pharmaceut.* 10, 685-693.

[72] Chung, Y. I., Tae, G., Yuk, S. H., 2006. A facile method to prepare heparin-functionalized nanoparticles for controlled release of growth factors. *Biomaterials.* 27, 2621-2626.

[73] Nillesen, S. T. M., Geutjes, P. J., Wismans, R., Schalkwijk, J., Daamen, W. F., van Kuppevelt, T. H., 2007. Increased angiogenesis and blood vessel maturation in acellular collagen-heparin scaffolds containing both FGF2 and VEGF. *Biomaterials*. 28, 1123-1131.

[74] Tang, D. W., Yu, S. H., Ho, Y. C., Mi, F. L., Kuo, P. L., Sung, H. W., 2010. Heparinized chitosan/poly(gamma-glutamic acid) nanoparticles for multi-functional delivery of fibroblast growth factor and heparin. *Biomaterials*. 31, 9320-9332.

[75] Huang, H. Z., Yang, X. R., 2004. Synthesis of polysaccharide-stabilized gold and silver nanoparticles: a green method. *Carbohydrate research*. 339, 2627-2631.

[76] Yuk, S. H., Oh, K. S., Cho, S. H., Lee, B. S., Kim, S. Y., Kwak, B. K., Kim, K., Kwon, I. C., 2011. Glycol Chitosan/Heparin Immobilized Iron Oxide Nanoparticles with a Tumor-Targeting Characteristic for Magnetic Resonance Imaging. *Biomacromolecules*. 12, 2335-2343.

[77] Lee, J. H., Jung, M. J., Hwang, Y. H., Lee, Y. J., Lee, S., Lee, D. Y., Shin, H., 2012. Heparin-coated superparamagnetic iron oxide for in vivo MR imaging of human MSCs. *Biomaterials*. 33, 4861-4871.

[78] Lee, K., Lee, H., Bae, K. H., Park, T. G., 2010. Heparin immobilized gold nanoparticles for targeted detection and apoptotic death of metastatic cancer cells. *Biomaterials*. 31, 6530-6536.

[79] Kemp, M. M., Kumar, A., Mousa, S., Park, T. J., Ajayan, P., Kubotera, N., Mousa, S. A., Linhardt, R. J., 2009. Synthesis of Gold and Silver Nanoparticles Stabilized with Glycosaminoglycans Having Distinctive Biological Activities. *Biomacromolecules*. 10, 589-595.

[80] Kemp, M. M., Kumar, A., Mousa, S., Dyskin, E., Yalcin, M., Ajayan, P., Linhardt, R. J., Mousa, S. A., 2009. Gold and silver nanoparticles conjugated with heparin derivative possess anti-angiogenesis properties. *Nanotechnology*. 20.

[81] Tsigos, I., Martinou, A., Kafetzopoulos, D., Bouriotis, V., 2000. Chitin deacetylases: new, versatile tools in biotechnology. *Trends Biotechnol*. 18, 305-12.

[82] Pereira, P., Carvalho, V., Ramos, R., Gama, F. M., *Chitosan NanoParticles for biomedical applications.*, In: *Chitosan: Manufacture, Properties and Usage*. 2010, Nova Science Publishers: New York.

[83] Grenha, A., Gomes, M. E., Rodrigues, M., Santo, V. E., Mano, J. F., Neves, N. M., Reis, R. L., 2010. Development of new chitosan/carrageenan nanoparticles for drug delivery applications. *J. Biomed. Mater. Res. A*. 92A, 1265-1272.

[84] Rodrigues, S., da Costa, A. M. R., Grenha, A., 2012. Chitosan/ carrageenan nanoparticles: Effect of cross-linking with tripolyphosphate and charge ratios. *Carbohyd. Polym*. 89, 282-289.

[85] Chen, Y., Mohanraj, V. J., Parkin, J. E., 2003. Chitosan-dextran sulfate nanoparticles for delivery of an anti-angiogenesis peptide. *Lett. Pept. Sci*. 10, 621-629.

[86] Huang, Y. C., Liu, T. J., 2012. Mobilization of mesenchymal stem cells by stromal cell-derived factor-1 released from chitosan/tripolyphosphate/fucoidan nanoparticles. *Acta biomaterialia*. 8, 1048-1056.

[87] Fermas, S., Gonnet, F., Sutton, A., Charnaux, N., Mulloy, B., Du, Y., Baleux, F., Daniel, R., 2008. Sulfated oligosaccharides (heparin and fucoidan) binding and dimerization of stromal cell-derived factor-1 (SDF-1/CXCL 12) are coupled as evidenced by affinity CE-MS analysis. *Glycobiology*. 18, 1054-64.

[88] Huang, Y. C., Yang, Y. T., 2013. Effect of basic fibroblast growth factor released from chitosan-fucoidan nanoparticles on neurite extension. *Journal of tissue engineering and regenerative medicine*.

[89] Matou, S., Helley, D., Chabut, D., Bros, A., Fischer, A. M., 2002. Effect of fucoidan on fibroblast growth factor-2-induced angiogenesis in vitro. *Thrombosis research*. 106, 213-21.

[90] Huang, Y. C., Li, R. Y., 2014. Preparation and characterization of antioxidant nanoparticles composed of chitosan and fucoidan for antibiotics delivery. *Mar Drugs*. 12, 4379-98.

[91] Tan, Q., Tang, H., Hu, J. G., Hu, Y. R., Zhou, X. M., Tao, Y. M., Wu, Z. S., 2011. Controlled release of chitosan/heparin nanoparticle-delivered VEGF enhances regeneration of decellularized tissue-engineered scaffolds. *Int. J. Nanomed*. 6, 929-942.

[92] Santo, V. E., Duarte, A. R. C., Gomes, M. E., Mano, J. F., Reis, R. L., 2010. Hybrid 3D structure of poly(D,L-lactic acid) loaded with chitosan/chondroitin sulfate nanoparticles to be used as carriers for biomacromolecules in tissue engineering. *J. Supercrit. Fluid*. 54, 320-327.

[93] Chung, Y. I., Kim, J. C., Kim, Y. H., Tae, G., Lee, S. Y., Kim, K., Kwon, I. C., 2010. The effect of surface functionalization of PLGA nanoparticles by heparin- or chitosan-conjugated Pluronic on tumor targeting. *J. Control Release*. 143, 374-382.

[94] Lin, Y. H., Chang, C. H., Wu, Y. S., Hsu, Y. M., Chiou, S. F., Chen, Y. J., 2009. Development of pH-responsive chitosan/heparin nanoparticles for stomach-specific anti-Helicobacter pylori therapy. *Biomaterials*. 30, 3332-3342.

[95] Lin, Y. H., Tsai, S. C., Lai, C. H., Lee, C. H., He, Z. S., Tseng, G. C., 2013. Genipin-cross-linked fucose-chitosan/heparin nanoparticles for the eradication of Helicobacter pylori. *Biomaterials*. 34, 4466-4479.

[96] Yeh, M. K., Cheng, K. M., Hu, C. S., Huang, Y. C., Young, J. J., 2011. Novel protein-loaded chondroitin sulfate-chitosan nanoparticles: preparation and characterization. *Acta biomaterialia*. 7, 3804-12.

[97] Lee, C. T., Huang, C. P., Lee, Y. D., 2006. Preparation of amphiphilic poly(L-lactide)-graft-chondroitin sulfate copolymer self-aggregates and its aggregation behavior. *Biomacromolecules*. 7, 1179-1186.

[98] Li, W., Li, X., Su, H., Zhao, S. J., Li, Y. Y., Hu, J. D., 2011. Facile synthesis of chondroitin sulfate-stabilized gold nanoparticles. *Mater. Chem. Phys*. 125, 518-521.

[99] Li, B., Lu, F., Wei, X. J., Zhao, R. X., 2008. Fucoidan: Structure and bioactivity. *Molecules*. 13, 1671-1695.

[100] Lira, M. C. B., Santos-Magalhaes, N. S., Nicolas, V., Marsaud, V., Silva, M. P. C., Ponchel, G., Vauthier, C., 2011. Cytotoxicity and cellular uptake of newly synthesized fucoidan-coated nanoparticles. *Eur. J. Pharm. Biopharm*. 79, 162-170.

[101] Silva, V. A. J., Andrade, P. L., Silva, M. P. C., Bustamante, A., Valladares, L. D., Aguiar, J. A., 2013. Synthesis and characterization of Fe3O4 nanoparticles coated with fucan polysaccharides. *J. Magn. Magn. Mater*. 343, 138-143.

[102] Toskas, G., Hund, R. D., Laourine, E., Cherif, C., Smyrniotopoulos, V., Roussis, V., 2011. Nanofibers based on polysaccharides from the green seaweed Ulva Rigida. *Carbohyd. Polym*. 84, 1093-1102.

[103] Dhahri, M., Abed, A., Lajimi, R. H., Ben Mansour, M., Gueguen, V., Ben Abdesselem, S., Chaubet, F., Letourneur, D., Meddahi-Pelle, A., Maaroufi, R. M., 2011. Grafting of

dermatan sulfate on polyethylene terephtalate to enhance biointegration. *J. Biomed. Mater. Res. A*. 98A, 114-121.

[104] Ranney, D. F., *Endothelial envelopment drug carriers*, 1992, Google Patents.

[105] Ranney, D., Antich, P., Dadey, E., Mason, R., Kulkarni, P., Singh, O., Chen, H. G., Constantinescu, A., Parkey, R., 2005. Dermatan carriers for neovascular transport targeting, deep tumor penetration and improved therapy. *J. Control Release*. 109, 222-235.

[106] Tiyaboonchai, W., Woiszwillo, J., Middaugh, C. R., 2001. Formulation and characterization of amphotericin B-polyethylenimine-dextran sulfate nanoparticles. *J. Pharm. Sci*. 90, 902-14.

[107] Lemkine, G. F., Demeneix, B. A., 2001. Polyethylenimines for in vivo gene delivery. *Current opinion in molecular therapeutics*. 3, 178-82.

[108] Tiyaboonchai, W., Woiszwillo, J., Sims, R. C., Middaugh, C. R., 2003. Insulin containing polyethylenimine-dextran sulfate nanoparticles. *Int. J. Pharm*. 255, 139-51.

ABOUT THE EDITORS

MIGUEL GAMA, PhD
Associated Professor with Habilitation
Minho University – Deparment of Biological Engineering, Portugal

Miguel Gama received his PhD in Biological Engineering at Minho University, in Portugal, in 1996. He is currently Associate Professor with Habilitation at Minho University, where he directs the FUNCARB Laboratory, a research group with about 20 researchers. He has authored/edited 4 books, over 100 papers in international journals, and 2 patents. Was the co-founder of a spinoff from Minho University, BNCTechnologies, dedicated to the development of applications for bacterial cellulose. His research interests include enzyme technology - enzymatic degradation of biomass - and the development of biomaterials, in particular bacterial cellulose, self-assembled amphiphilic nanogels and injectable hydrogels, aiming the development of applications for tissue regeneration and drug delivery.

HELENA B. NADER, PhD
Full Professor, Department of Biochemistry,
University of Sao Paulo, Brasil

Helena B. Nader completed her undergraduate studies in Biomedical Sciences (bachelor degree) at the Federal University of São Paulo (Unifesp), in 1970, a bachelor of education at the University of São Paulo (USP) in 1971, and a PhD in Molecular Biology also at the Unifesp, in 1974. In 1977 she concluded her post-doctoral studies at the University of South California, with a Fogarty grant (NIH). Helena is fellow of the Brazilian Academy of Sciences (ABC), of the São Paulo Academy of Sciences (ACIESP), and of the World Academy of Sciences (TWAS). Throughout her scientific career she was honored with various national and international prizes and academic titles. She has been dean of undergraduate programs and of research and graduate programs. Before her present term as president of the Brazilian Society for the Advancement of Science (SBPC), Nader was president of the Brazilian Society of Biochemistry and Molecular Biology (SBBq, 2009-2010).

She is professor and head of the Molecular Biology Program of UNIFESP and her academic career is dedicated to the understanding of structure, function and pharmacological activities of heparin and heparan sulfate. Her group developed new methods for the separation and analysis of the different glycosaminoglycans (GAGs) together with the characterization of different enzymes involved in the degradation of these compounds. The chemical and physicochemical analysis of the fragments and of the native polymers allowed them to detect, identify and determine their structural characteristics as well as propose their functions.

HUGO ROCHA, PhD

Associate Professor, Departament of Biochemistry,
at the Federal University of Rio Grande do Norte, Brasil

Hugo Rocha received his BS in Pharmacy in 1994 at Federal University of Rio Grande do Norte (UFRN) at Natal, Brazil. He received his MS and PhD in Molecular Biology and Science in 1998 and 2002, respectively from Federal University of São Paulo (UNIFESP). Hugo is now an associate professor of Biochemistry, in Departament of Biochemistry, at the Federal University of Rio Grande do Norte. His principal research interest is in the investigation of structural features and biological activities of sulfated poysacharides such as fucans, fucoidan, and sulfated galactans extracted from seaweed and other sources. He was advisor of twenty five MSc and seven PhD thesis. He is a Fellow of Conselho Nacional de Pesquisa (CNPq), Brazil. Professor Rocha has published over 60 peer-reviewed publications.

INDEX

D

E

F

G

H

I

J

K

L

M

N

O

P